ORCCA

Open Resources for Community College Algebra

ORCCA

Open Resources for Community College Algebra

Portland Community College Faculty

August 20, 2018

Project Leads: Ann Cary and Alex Jordan

Technology Engineer: Alex Jordan

Contributing Authors: Ann Cary, Alex Jordan, Ross Kouzes, Scot Leavitt, Cara Lee, Carl Yao, and Ralf Youtz

WeBWorK Problem Coding: Chris Hughes, Alex Jordan, Carl Yao

Other Contributors: Kara Colley

Cover Image: Ralf Youtz

Edition: 1.0

Website: pcc.edu/ORCCA

Acknowledgements

This book has been made possible through Portland Community College's Strategic Investment Funding, approved by PCC's Budget Planning Advisory Council and the Board of Directors. Without significant funding to provide the authors with the adequate time, an ambitious project such as this one would not be possible.

The technology that makes it possible to create synced print, eBook, and WeBWorK content is PreTeXt, created by Rob Beezer. David Farmer and the American Institute of Mathematics have worked to make the PreTeXt eBook layout functional, yet simple. A grant from OpenOregon funded the original bridge between WeBWorK and PreTeXt.

This book uses WeBWorK to provide most of its exercises, which may be used for online homework. WeBWorK was created by Mike Gage and Arnie Pizer, and has benefited from over 25 years of contributions from open source developers. In 2013, Chris Hughes, Alex Jordan, and Carl Yao programmed most of the WeBWorK questions in this book with a PCC curriculum development grant.

The javascript library MathJax, created and maintained by David Cervone, Volker Sorge, Christian Lawson-Perfect, and Peter Krautzberger allows math content to render nicely on screen in the eBook. Additionally, MathJax makes web accessible mathematics possible.

The print edition (PDF) is built using the typesetting software LaTeX, created by Donald Knuth and enhanced by Leslie Lamport.

Each of these open technologies, along with many that we use but have not listed here, has been enhanced by many additional contributors spanning the past 40 years. We are grateful for all of these contributions.

To All

HTML, PDF, and print This book is available as an eBook, a free PDF, or printed and bound. All versions offer the same content and are synchronized such that cross-references match across versions. They can each be found at pcc.edu/orcca.

There are some differences between the eBook, PDF, and printed versions.

- The eBook is recommended, as it offers interactive elements and easier navigation than print. It requires no more than internet access and a modern web browser.

- A PDF version can be downloaded and then accessed without the internet. Some content is in color, but most of the colorized content from the eBook has been converted to black and white to ensure adequate contrast when printing in black and white. The exceptions are the graphs generated by WeBWorK.

- Printed and bound copies are available online. Up-to-date information about purchasing a copy should be available at pcc.edu/orcca. Contact the authors if you have trouble finding the latest version online. For each online sale, all royalties go to a PCC Foundation account, where roughly half will fund student scholarships, and half will fund continued maintenance of this book and other OER.

Copying Content The graphs and other images that appear in this manual may be copied in various file formats using the eBook version. Below each image are links to .png, .eps, .svg, .pdf, and .tex files that contain the image.

Mathematical content can be copied from the eBook. To copy math content into *MS Word*, right-click or control-click over the math content, and click to Show Math As MathML Code. Copy the resulting code, and Paste Special into *Word*. In the Paste Special menu, paste it as Unformatted Text. To copy math content into LATEX source, right-click or control-click over the math content, and click to Show Math As TeX Commands.

Tables can be copied from the eBook version and pasted into applications like *MS Word*. However, mathematical content within tables will not always paste correctly without a little extra effort as described above.

Accessibility The HTML version is intended to meet or exceed web accessibility standards. If you encounter an accessibility issue, please report it.

- All graphs and images should have meaningful alt text that communicates what a sighted person would see, without necessarily giving away anything that is intended to be deduced from the image.

- All math content is rendered using MathJax. MathJax has a contextual menu that can be accessed in several ways, depending on what operating system and browser you are using. The most common way is to right-click or control-click on some piece of math content.

- In the MathJax contextual menu, you may set options for triggering a zoom effect on math content, and also by what factor the zoom will be. Also in the MathJax contextual menu, you can enable the Explorer, which allows for sophisticated navigation of the math content.

- A screen reader will generally have success verbalizing the math content from MathJax. With certain screen reader and browser combinations, you may need to set some configuration settings in the MathJax contextual menu.

Tablets and Smartphones   PreTeXt documents like this book are "mobile-friendly." When you view the HTML version, the display adapts to whatever screen size or window size you are using. A math teacher will always recommend that you do not study from the small screen on a phone, but if it's necessary, the eBook gives you that option.

WeBWorK for Online Homework   Most exercises are available in a ready-to-use collection of WeBWorK problem sets. Visit webwork.pcc.edu/webwork2/orcca-demonstration to see a demonstration WeBWorK course where guest login is enabled. Anyone interested in using these problem sets should contact the project leads.

Odd Answers   The answers to the odd homework exercises at the end of each section are not contained in the PDF or print versions. As the eBook evolves, they may or may not be contained in an appendix there. In any case, odd answers are available *somewhere*. Check pcc.edu/orcca to see where.

Interactive and Static Examples   Traditionally, a math textbook has examples throughout each section. This textbook uses two types of "example":

Static These are labeled "Example." Static examples may or may not be subdivided into a "statement" followed by a walk-through solution. This is basically what traditional examples from math textbooks do.

Active These are labeled " Checkpoint," not to be confused with the exercises that come at the end of a section that might be assigned for homework, etc. In the HTML output, active examples have WeBWorK answer blanks where a reader could try submitting an answer. In the PDF output, active examples are almost indistinguishable from static examples, but there is a WeBWorK icon indicating that a reader could interact more actively using the eBook. Generally, a walk-through solution is provided immediately following the answer blank.

Some HTML readers will skip the opportunity to try an active example and go straight to its solution. Some readers will try an active example once and then move on to the solution. Some readers will tough it out for a period of time and resist reading the solution.

For readers of the PDF, it is expected that they would read the example and its solution just as they would read a static example.

A reader is *not* required to try submitting an answer to an active example before moving on. A reader *is* expected to read the solution to an active example, even if they succeed on their own at finding an answer.

Interspersed through a section there are usually several exercises that are intended as active reading exercises. A reader can work these examples and submit answers to WeBWorK to see if they are correct. The important thing is to keep the reader actively engaged instead of providing another static written example. In most cases, it is expected that a reader will read the solutions to these exercises just as they would be expected to read a more traditional static example.

Pedagogical Decisions

The authors and the greater PCC faculty have taken various stances on certain pedagogical and notational questions that arise in basic algebra instruction. We attempt to catalog these decisions here, although this list is certainly incomplete. If you find something in the book that runs contrary to these decisions, please let us know.

- Interleaving is our preferred approach, compared to a proficiency-based approach. To us, this means that once the book covers a topic, that topic will be appear in subsequent sections and chapters in indirect ways.

- Chapter 1 is mostly written as a *review*, and is not intended to teach all of these topics from first principles.

- We round decimal results to four significant digits, or possibly fewer leaving out trailing zeros. We do this to maintain consistency with the most common level of precision that WeBWorK uses to assess decimal answers. We *round*, not *truncate*. And we use the $\approx$ symbol. For example $\pi \approx 3.142$ and Portland's population is ≈ 609500.

- We offer *alternative* video lessons associated with each section, found in most sections in the eBook. We hope these videos provide readers with an alternative to whatever is in the reading, but there may be discrpancies here and there between the video content and reading content.

- We believe in always trying to open a topic with some level of application rather than abstract examples. From applications and practical questions, we move to motivate more abstract definitions and notation. This approach is perhaps absent in the first chapter, which is intended to be a review only. At first this may feel backwards to some instructors, with some "easier" examples (with no context) appearing after "more difficult" contextual examples.

- Linear inequalities are not strictly separated from linear equations. The section that teaches how to solve $2x + 3 = 8$ is immediately followed by the section teaching how to solve $2x + 3 < 8$.

  Our aim is to not treat inequalities as an add-on optional topic, but rather to show how intimately related they are to corresponding equations.

- When issues of "proper formatting" of student work arise, we value that the reader understand *why* such things help the reader to communicate outwardly. We believe that mathematics is about more than understanding a topic, but also about understanding it well enough to communicate results to others.

  For example we promote progression of equations like

  $$1 + 1 + 1 = 2 + 1$$
  $$= 3$$

  instead of

  $$1 + 1 + 1 = 2 + 1 = 3.$$

  And we want students to *understand* that the former method makes their work easier for a reader to read. It is not simply a matter of "this is the standard and this is how it's done."

- When solving equations (or systems of linear equations), most examples should come with a check, intended to communicate to students that checking is part of the process. In Chapters 1–4, these checks will be complete simplifications using order of operations one step at a time. The later sections will often have more summary checks where either order of operations steps are skipped in groups, or we promote entering expressions into a calculator. Occasionally in later sections the checks will still have finer details, especially when there are issues like with negative numbers squared.

- Within a section, any first example of solving some equation (or system) should summarize with some variant of both "the solution is…" and "the solution set is…." Later examples can mix it up, but always offer at least one of these.

- There is a section on very basic arithmetic (five operations on natural numbers) in an appendix, not in the first chapter. This appendix is only available in the eBook.

- With applications of linear equations (as opposed to linear systems), we limit applications to situations where the setup will be in the form $x + f(x) = C$ and also certain rate problems where the setup will be in the form $5t + 4t = C$. There are other classes of application problem (mixing problems, interest problems, …) which can be handled with a system of two equations, and we reserve these until linear systems are covered.

- With simplifications of rational expressions in one variable, we always include domain restrictions that are lost in the simplification. For example, we would write $\frac{x(x+1)}{x+1} = x$, for $x \neq -1$. With multivariable rational expressions, we are content to ignore domain restrictions lost during simplification.

Entering WeBWorK Answers

This preface offers some guidance with syntax for WeBWorK answers. WeBWorK answer blanks appear in the active reading examples (called "checkpoints") in the HTML version of the book. If you are using WeBWorK for online homework, then you will also enter answers into WeBWorK answer blanks there.

Basic Arithemtic   The five basic arithmetic operations are: addition, subtraction, multiplication, and raising to a power. The symbols for addition and subtraction are + and −, and both of these are directly avialable on most keyboards as + and -.

On paper, multiplication is sometimes written using × and sometimes written using · (a centered dot). Since these symbols are not available on most keyboards, WeBWorK uses * instead, which is often shift-8 on a full keyboard.

On paper, division is sometimes written using ÷, sometimes written using a fraction layout like $\frac{4}{2}$, and sometimes written just using a slash, /. The slash is available on most full keyboards, near the question mark. WeBWorK uses / to indicate division.

On paper, raising to a power is written using a two-dimensional layout like 4^2. Since we don't have a way to directly type that with a simple keyboard, calculators and computers use the caret character, ^, as in 4^2. The character is usually shift-6.

Roots and Radicals   On paper, a square root is represented with a radical symbol like $\sqrt{}$. Since a keyboard does not usually have this symbol, WeBWorK and many computer applications use sqrt( ) instead. For example, to enter $\sqrt{17}$, type sqrt(17).

Higher-index radicals are written on paper like $\sqrt[4]{12}$. Again we have no direct way to write this using most keyboards. In *some* WeBWorK problems it is possible to type something like root(4, 12) for the fourth root of twelve. However this is not enabled for all WeBWorK problems.

As an alternative that you may learn about in a later chapter, $\sqrt[4]{12}$ is mathematically equal to $12^{1/4}$, so it can be typed as 12^(1/4). Take note of the parentheses, which very much matter.

Common Hiccups with Grouping Symbols   Suppose you wanted to enter $\frac{x+1}{2}$. You might type x+1/2, but this is not right. The computer will use the order of operations (see Section 1.4) and do your division first, dividing 1 by 2. So the computer will see $x + \frac{1}{2}$. To address this, you would need to use grouping symbols like parentheses, and type something like (x+1)/2.

Suppose you wanted to enter $6^{1/4}$, and you typed 6^1/4. This is not right. The order of operations places a higher priority on exponentiation than division, so it calculates 6^1 first and then divides the result by 4. That is simply not the same as raising 6 to the $\frac{1}{4}$ power. Again the way to address this is to use grouping symbols, like 6^(1/4).

Entering Decimal Answers   Often you will find a decimal answer with decimal places that go on and on. You are allowed to round, but not by too much. WeBWorK generally looks at how many *significant digits* you use, and generally expects you to use *four or more* correct significant digits.

"Significant digits" and "places past the decimal" are not the same thing. To count significant digits, read the number left to right and look for the first nonzero digit. Then count all the digits to the right including that first one.

The number 102.3 has four significant digits, but only one place past the decimal. This number could be a correct answer to a WeBWorK question. The number 0.0003 has one significant digit and four places past the decimal. This number might cause you trouble if you enter it, because maybe the "real" answer was 0.0003091, and rounding to 0.0003 was too much rounding.

Special Symbols   There are a handful of special symbols that are easy to write on paper, but it's not clear how to type them. Here are WeBWorK's expectations.

Symbol	Name	How to Type
∞	infinity	infinity or inf
π	pi	pi
$\cup$	union	U
$\mathbb{R}$	the real numbers	R
$\mid$	such that	| (shift-\, where \ is above the enter key)
$\leq$	less than or equal to	<=
$\geq$	greater than or equal to	>=
$\neq$	not equal to	!=

Contents

Contents

CHAPTER 10

Functions and Their Representations

10.1 Function Basics

In Section 9.1 there is a light introduction to functions. This chapter introduces functions more thoroughly, and is independent from Section 9.1.

10.1.1 Informal Definition of a Function

We are familiar with the $\sqrt{\ }$ symbol. This symbol is used to turn numbers into their square roots. Sometimes it's simple to do this on paper or in our heads, and sometimes it helps to have a calculator. We can see some calculations in Table 10.1.2.

$$\sqrt{9} \quad = 3$$
$$\sqrt{1/4} \quad = 1/2$$
$$\sqrt{2} \quad \approx 1.41$$

Table 10.1.2: Values of $\sqrt{x}$

The $\sqrt{\ }$ symbol represents a *process*; it's a way for us to turn numbers into other numbers. This idea of having a process for turning numbers into other numbers is the fundamental topic of this chapter.

Definition 10.1.3 Function (Informal Definition). A **function** is a process for turning numbers into (potentially) different numbers. The process must be *consistent*, in that whenever you apply it to some particular number, you always get the same result.

Section 10.5 covers a more technical definition for functions, and gets into some topics that are more appropriate when using that definition. Definition 10.1.3 is so broad that you probably use functions all the time.

Example 10.1.4 Think about each of these examples, where some process is used for turning one number into another.

- If you convert a person's birth year into their age, you are using a function.

- If you look up the Kelly Blue Book value of a Honda Odyssey based on how old it is, you are using a function.

- If you use the expected guest count for a party to determine how many pizzas you should order, you are using a function.

The process of using $\sqrt{}$ to change numbers might feel more "mathematical" than these examples. Let's continue thinking about $\sqrt{}$ for now, since it's a formula-like symbol that we are familiar with. One concern with $\sqrt{}$ is that although we live in the modern age of computers, this symbol is not found on most keyboards. And yet computers still tend to be capable of producing square roots. Computer technicians write sqrt( ) when they want to compute a square root, as we see in Table 10.1.5.

sqrt(9)	= 3
sqrt(1/4)	= 1/2
sqrt(2)	$\approx$ 1.41

Table 10.1.5: Values of sqrt(x)

The parentheses are very important. To see why, try to put yourself in the "mind" of a computer, and look closely at sqrt49. The computer will recognize sqrt and know that it needs to compute a square root. But computers have myopic vision and they might not see the entire number 49. A computer might think that it needs to compute sqrt4 and then append a 9 to the end, which would produce a final result of 29. This is probably not what was intended. And so the purpose of the parentheses in sqrt(49) is to denote exactly what number needs to be operated on.

This demonstrates the standard notation that is used worldwide to write down most functions. By having a standard notation for communicating about functions, people from all corners of the earth can all communicate mathematics with each other more easily, even when they don't speak the same language.

Functions have their own names. We've seen a function named sqrt, but any name you can imagine is allowable. In the sciences, it is common to name functions with whole words, like weight or health_index. In mathematics, we often abbreviate such function names to w or h. And of course, since the word "function" itself starts with "f," we will often name a function f.

It's crucial to continue reminding ourselves that functions are *processes* for changing numbers; they are not numbers themselves. And that means that we have a potential for confusion that we need to stay aware of. In some contexts, the symbol t might represent a variable—a number that is represented by a letter. But in other contexts, t might represent a function—a process for changing numbers into other numbers. By staying conscious of the context of an investigation, we avoid confusion.

Next we need to discuss how we go about using a function's name.

Definition 10.1.6 Function Notation. The standard notation for referring to functions involves giving the function itself a name, and then writing:

$$\begin{matrix} \text{name} \\ \text{of} \\ \text{function} \end{matrix} \left(\text{input} \right)$$

Example 10.1.7 $f(13)$ is pronounced "f of 13." The word "of" is very important, because it reminds us that f is a process and we are about to apply that process to the input value 13. So f is the function, 13 is the input, and $f(13)$ is the output we'd get from using 13 as input.

$f(x)$ is pronounced "f of x." This is just like the previous example, except that the input is not any specific number. The value of x could be 13 or any other number. Whatever x's value, $f(x)$ means the corresponding output from the function f.

BudgetDeficit(2017) is pronounced "BudgetDeficit of 2017." This is probably about a function that takes a year as input, and gives that year's federal budget deficit as output. The process here of changing a year into a dollar amount might not involve any mathematical formula, but rather looking up information from the Congressional Budget Office's website.

Celsius(*F*) is pronounced "Celsius of *F*." This is probably about a function that takes a Fahrenheit temperature as input and gives the corresponding Celsius temperature as output. Maybe a formula is used to do this; maybe a chart or some other tool is used to do this. Here, Celsius is the function, *F* is the input variable, and Celsius(*F*) is the output from the function.

Note 10.1.8. While a function has a name like f, and the input to that function often has a variable name like x, the expression $f(x)$ represents the output of the function. To be clear, $f(x)$ is *not* a function. Rather, f is a function, and $f(x)$ its output when the number x was used as input.

Checkpoint 10.1.9. Suppose you see the sentence, "If x is the number of software licenses you buy for your office staff, then $c(x)$ is the total cost of the licenses."

    a. In the function notation, what represents input? ⬚.

    b. What is the function here? ⬚.

    c. What represents output? ⬚.

Explanation. The input is x, the function is c, and $c(x)$ is the output from c when the input is x.

Warning 10.1.10 Notation Ambiguity. As mentioned earlier, we need to remain conscious of the context of any symbol we are using. It's possible for f to represent a function (a process), but it's also possible for f to represent a variable (a number). Similarly, parentheses might indicate the input of a function, or they might indicate that two numbers need to be multiplied. It's up to our judgment to interpret algebraic expressions in the right context. Consider the expression $a(b)$. This could easily mean the output of a function a with input b. It could also mean that two numbers a and b need to be multiplied. It all depends on the context in which these symbols are being used.

Sometimes it's helpful to think of a function as a machine, as in Figure 10.1.11. This illustrates how complicated functions can be. A number is just a number. But a *function* has the capacity to take in all kinds of different numbers into it's hopper (feeding tray) as inputs and transform them into their outputs.

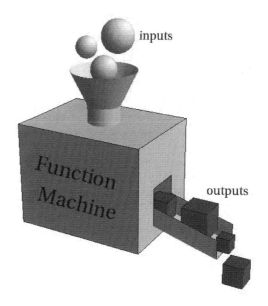

Figure 10.1.11: Imagining a function as a machine. (Image by Duane Nykamp using Mathematica.)

10.1.2 Tables and Graphs

Since functions are potentially complicated, we want ways to understand them more easily. Two basic tools for understanding a function better are tables and graphs.

Example 10.1.12 A Table for the Budget Deficit Function. Consider the function BudgetDeficit, that takes a year as its input and outputs the US federal budget deficit for that year. For example, the Congressional Budget Office's website tells us that BudgetDeficit(2009) is $1.41 trillion. If we'd like to understand this function better, we might make a table of all the inputs and outputs we can find. Using the CBO's website (www.cbo.gov/topics/budget), we can put together Table 10.1.13.

input x (year)	output BudgetDeficit(x) ($trillion)
2007	0.16
2008	0.46
2009	1.4
2010	1.3
2011	1.3
2012	1.1
2013	0.68
2014	0.48
2015	0.44
2016	0.59

Table 10.1.13

How is this table helpful? There are things about the function that we can see now by looking at the numbers in this table.

- We can see that the budget deficit had a spike between 2008 and 2009.
- And it fell again between 2012 and 2013.
- It appears to stay roughly steady for several years at a time, with occasional big jumps or drops.

These observations help us understand the function BudgetDeficit a little better.

Checkpoint 10.1.14. According to Table 10.1.13, what is the value of BudgetDeficit(2015)?

Explanation. Table 10.1.13 shows that when the input is 2015, the output is 0.44. So BudgetDeficit(2015) = 0.44. In context, that means that in 2015 the budget deficit was $0.44 trillion.

Example 10.1.15 A Table for the Square Root Function. Let's return to our example of the function sqrt. Tabulating some inputs and outputs reveals 10.1.16

input, x	output, sqrt(x)
0	0
1	1
2	≈ 1.41
3	≈ 1.73
4	2
5	≈ 2.24
6	≈ 2.45
7	≈ 2.65
8	≈ 2.83
9	3

Table 10.1.16

How is this table helpful? Here are some observations that we can make now.

- We can see that when input numbers increase, so do output numbers.
- We can see even though outputs are increasing, they increase by less and less with each step forward in x.

These observations help us understand sqrt a little better. For instance, based on these observations which do you think is larger: the difference between sqrt(23) and sqrt(24), or the difference between sqrt(85) and sqrt(86)?

Checkpoint 10.1.17. According to Table 10.1.16, what is the value of sqrt(6)?

Explanation. Table 10.1.16 shows that when the input is 6, the output is about 2.45. So sqrt(6) ≈ 2.45.

Another powerful tool for understanding functions better is a graph. Given a function f, one way to make its graph is to take a table of input and output values, and read each row as the coordinates of a point in the xy-plane.

> **Example 10.1.18 A Graph for the Budget Deficit Function.** Returning to the function BudgetDeficit that we studied in Example 10.1.12, in order to make a graph of this function we view Table 10.1.13 as a list of points with x and y coordinates, as in Table 10.1.19. We then plot these points on a set of coordinate axes, as in Figure 10.1.20. The points have been connected with a curve so that we can see the overall pattern given by the progression of points. Since there was not any actual data for inputs in between any two years, the curve is dashed. That is, this curve is dashed because it just represents someone's best guess as to how to connect the plotted points. Only the plotted points themselves are precise.

(input, output) $(x, \text{BudgetDeficit}(x))$
(2007, 0.16)
(2008, 0.46)
(2009, 1.4)
(2010, 1.3)
(2011, 1.3)
(2012, 1.1)
(2013, 0.68)
(2014, 0.48)
(2015, 0.44)
(2016, 0.59)

Table 10.1.19

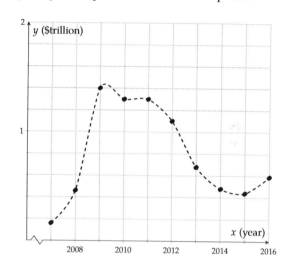

Figure 10.1.20: $y = \text{BudgetDeficit}(x)$

How has this graph helped us to understand the function better? All of the observations that we made in Example 10.1.12 are perhaps even more clear now. For instance, the spike in the deficit between 2008 and 2009 is now visually apparent. Seeking an explanation for this spike, we recall that there was a financial crisis in late 2008. Revenue from income taxes dropped at the same time that federal money was spent to prevent further losses.

Example 10.1.21 A Graph for the Square Root Function. Let's now construct a graph for sqrt. Tabulating inputs and outputs gives the points in Table 10.1.22, which in turn gives us the graph in Figure 10.1.23.

(input, output)
$(x, \text{sqrt}(x))$
$(0, 0)$
$(1, 1)$
$\approx (2, 1.41)$
$\approx (3, 1.73)$
$(4, 2)$
$\approx (5, 2.24)$
$\approx (6, 2.45)$
$\approx (7, 2.65)$
$\approx (8, 2.83)$
$(9, 3)$

Table 10.1.22

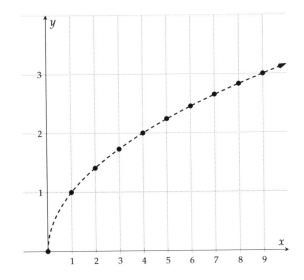

Figure 10.1.23: $y = \text{sqrt}(x)$

Just as in the previous example, we've plotted points where we have concrete coordinates, and then we have made our best attempt to connect those points with a curve. Unlike the previous example, here we believe that points will continue to follow the same pattern indefinitely to the right, and so we have added an arrowhead to the graph.

What has this graph done to improve our understanding of sqrt? As inputs (x-values) increase, the outputs (y-values) increase too, although not at the same rate. In fact we can see that our graph is steep on its left, and less steep as we move to the right. This confirms our earlier observation in Example 10.1.15 that outputs increase by smaller and smaller amounts as the input increases.

Note 10.1.24 Graph of a Function. Given a function f, when we refer to a **graph of** f we are *not* referring to an entire picture, like Figure 10.1.23. A graph of f is only *part* of that picture—the curve and the points that it connects. Everything else: axes, tick marks, the grid, labels, and the surrounding white space is just useful decoration, so that we can read the graph more easily.

It is also common to refer to the graph of f as the **graph of the equation** $y = f(x)$. However, we should avoid saying "the graph of $f(x)$." That would indicate a fundamental misunderstanding of our notation. We have decided that $f(x)$ is the output for a certain input x. That means that $f(x)$ is just a number; a relatively uninteresting thing compared to f the function, and not worthy of a two-dimensional picture.

While it is important to be able to make a graph of a function f, we also need to be capable of looking at a graph and reading it well. A graph of f provides us with helpful specific information about f; it tells us what f does to its input values. When we were making graphs, we plotted points of the form

$$(\text{input}, \text{output})$$

Now given a graph of f, we interpret coordinates in the same way.

Example 10.1.25 In Figure 10.1.26 we have a graph of a function f. If we wish to find $f(1)$, we recognize that 1 is being used as an input. So we would want to find a point of the form $(1, \ )$. Seeking out x-coordinate 1 in Figure 10.1.26, we find that the only such point is $(1, 2)$. Therefore the output for 1 is 2; in other words $f(1) = 2$.

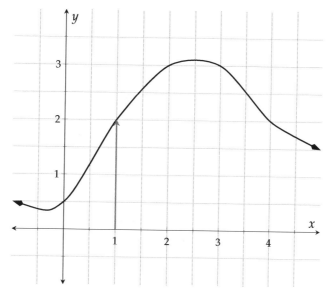

Figure 10.1.26: $y = f(x)$

Checkpoint 10.1.27. Use the graph of f in Figure 10.1.26 to find $f(0)$, $f(3)$, and $f(4)$.

a. $f(0) =$ [ ]

b. $f(3) =$ [ ]

c. $f(4) =$ [ ]

Explanation.

a. $f(0) = 0.5$, since $(0, 0.5)$ is on the graph.

b. $f(3) = 3$, since $(3, 3)$ is on the graph.

c. $f(4) = 2$, since $(4, 2)$ is on the graph.

Suppose that u is the unemployment function of time. That is, $u(t)$ is the unemployment rate in the United States in year t. The graph of the equation $y = u(t)$ is given in Figure 10.1.29 (data.bls.gov/timeseries/LNS14000000).

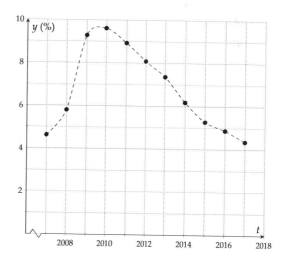

Figure 10.1.29: Unemployment in the United States

Example 10.1.28 Unemployment Rates. What was the unemployment in 2008? It is a straightforward matter to use Figure 10.1.29 to find that unemployment was about 6% in 2008. Asking this question is exactly the same thing as asking to find $u(2008)$. That is, we have one question that can either be asked in an everyday-English way or which can be asked in a terse, mathematical notation-heavy way:

"What was unemployment in 2008?"   "Find $u(2008)$."

If we use the table to establish that $u(2009) \approx 9.25$, then we should be prepared to translate that into everyday-English using the context of the function: In 2009, unemployment in the u.s. was about 9.25%.

If we ask the question "when was unemployment at 5%," we can read the graph and see that there were two such times: mid-2007 and about 2016. But there is again a more mathematical notation-heavy way to ask this question. Namely, since we are being told that the output of u is 5, we are being asked to solve the equation $u(t) = 5$. So the following communicate the same thing:

"When was unemployment at 5%?"   "Solve the equation $u(t) = 5$."

And our answer to this question is:

"Unemployment was at 5% in mid-2007 and about 2016."   "$t \approx 2007.5$ or $t \approx 2016$."

Checkpoint 10.1.30. Use the graph of u in Figure 10.1.29 to answer the following.

a. Find $u(2011)$ and interpret it.

$u(2011) \approx$ ▭

Interpretation:

b. Solve the equation $u(t) = 6$ and interpret your solution(s).

$t \approx$ ▭ or $t \approx$ ▭

Interpretation:

Explanation.

a. $u(2011) \approx 9$; In 2011 the US unemployment rate was about 9%.

b. $t \approx 2008$ or $t \approx 2014$; The points at which unemployment was 6% were in early 2008 and early 2014.

10.1.3 Translating Between Four Descriptions of the Same Function

We have noted that functions are complicated, and we want ways to make them easier to understand. It's common to find a problem involving a function and not know how to find a solution to that problem. Most functions have at least four standard ways to think about them, and if we learn how to translate between these four perspectives, we often find that one of them makes a given problem easier to solve.

The four modes for working with a given function are

- a verbal description

- a table of inputs and outputs

- a graph of the function

- a formula for the function

This has been visualized in Figure 10.1.31.

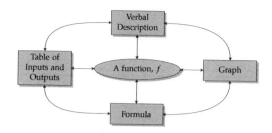

Figure 10.1.31: Function Perspectives

Example 10.1.32 Consider a function f that squares its input and then adds 1. Translate this verbal description of f into a table, a graph, and a formula.

Explanation.

To make a table for f, we'll have to select some input x-values. These choices are left entirely up to us, so we might as well choose small, easy-to-work-with values. However we shouldn't shy away from negative input values. Given the verbal description, we should be able to compute a column of output values. Table 10.1.33 is one possible table that we might end up with.

x	$f(x)$
-2	$(-2)^2 + 1 = 5$
-1	$(-1)^2 + 1 = 2$
0	$0^2 + 1 = 1$
1	$1^2 + 1 = 2$
2	5
3	10
4	17

Table 10.1.33

Once we have a table for f, we can make a graph for f as in Figure 10.1.34, using the table to plot points.

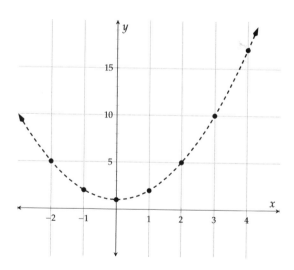

Figure 10.1.34: $y = f(x)$

Lastly, we must find a formula for f. This means we need to write an algebraic expression that says the same thing about f as the verbal description, the table, and the graph. For this example, we can focus on the verbal description. Since f takes its input, squares it, and adds 1, we have that

$$f(x) = x^2 + 1.$$

Example 10.1.35 Let F be the function that takes a Celsius temperature as input and outputs the corresponding Fahrenheit temperature. Translate this verbal description of F into a table, a graph, and a formula.

Explanation. To make a table for F, we will need to rely on what we know about Celsius and Fahrenheit temperatures. It is a fact that the freezing temperature of water at sea level is $0\,°C$, which equals $32\,°F$. Also, the boiling temperature of water at sea level is $100\,°C$, which is the same as $212\,°F$. One more piece of information we might have is that standard human body temperature is $37\,°C$, or $98.6\,°F$. All of this is compiled in Table 10.1.36. Note that we tabulated inputs and outputs by working with the context of the function, not with any computations.

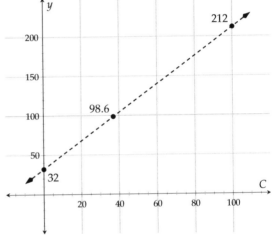

Once a table is established, making a graph by plotting points is a simple matter, as in Figure 10.1.37. The three plotted points seem to be in a straight line, so we think it is reasonable to connect them in that way.

C	$F(C)$
0	32
37	98.6
100	212

Table 10.1.36

Figure 10.1.37: $y = F(C)$

To find a formula for F, the verbal definition is not of much direct help. But F's graph does seem to be a straight line. And linear equations are familiar to us. This line has a y-intercept at $(0, 32)$ and a slope we can calculate: $\frac{212-32}{100-0} = \frac{180}{100} = \frac{9}{5}$. So the equation of this line is $y = \frac{9}{5}C + 32$. On the other hand, the equation of this graph is $y = F(C)$, since it is a graph of the function F. So evidently,

$$F(C) = \frac{9}{5}C + 32.$$

Exercises

Review and Warmup

1. Evaluate $\dfrac{9r - 1}{r}$ for $r = -6$.

2. Evaluate $\dfrac{3r - 4}{r}$ for $r = 8$.

3. Evaluate the following expressions.

    a. Evaluate $2t^2$ when $t = 3$. $2t^2 = \boxed{}$

    b. Evaluate $(2t)^2$ when $t = 3$. $(2t)^2 = \boxed{}$

4. Evaluate the following expressions.

    a. Evaluate $2t^2$ when $t = 5$. $2t^2 = \boxed{}$

    b. Evaluate $(2t)^2$ when $t = 5$. $(2t)^2 = \boxed{}$

5. Locate each point in the graph:

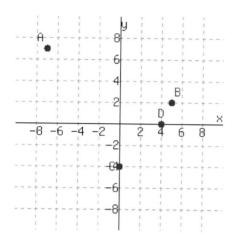

Write each point's position as an ordered pair, like $(1, 2)$.

    $A = \underline{}$ $B = \underline{}$

    $C = \underline{}$ $D = \underline{}$

6. Locate each point in the graph:

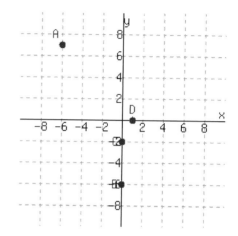

Write each point's position as an ordered pair, like $(1, 2)$.

    $A = \underline{}$ $B = \underline{}$

    $C = \underline{}$ $D = \underline{}$

Function Formulas and Evaluation   Evaluate the function at the given values.

7. $g(x) = x - 4$

    a. $g(1) = \boxed{}$

    b. $g(-4) = \boxed{}$

    c. $g(0) = \boxed{}$

8. $K(x) = x - 2$

    a. $K(4) = \boxed{}$

    b. $K(-2) = \boxed{}$

    c. $K(0) = \boxed{}$

9. $F(x) = 6x$

    a. $F(1) = \boxed{}$

    b. $F(-2) = \boxed{}$

    c. $F(0) = \boxed{}$

10. $G(x) = 2x$

    a. $G(3) = \boxed{}$

    b. $G(-3) = \boxed{}$

    c. $G(0) = \boxed{}$

11. $H(x) = -2x + 10$

    a. $H(1) = \boxed{}$

    b. $H(-1) = \boxed{}$

    c. $H(0) = \boxed{}$

12. $H(x) = -4x + 4$

    a. $H(5) = \boxed{}$

    b. $H(-1) = \boxed{}$

    c. $H(0) = \boxed{}$

13. $K(x) = -x + 1$

    a. $K(5) = \boxed{}$

    b. $K(-2) = \boxed{}$

    c. $K(0) = \boxed{}$

14. $f(x) = -x + 8$

    a. $f(2) = \boxed{}$

    b. $f(-4) = \boxed{}$

    c. $f(0) = \boxed{}$

15. $g(y) = y^2 + 3$

    a. $g(1) = \boxed{}$

    b. $g(-1) = \boxed{}$

    c. $g(0) = \boxed{}$

16. $h(x) = x^2 - 9$

    a. $h(4) = \boxed{}$

    b. $h(-1) = \boxed{}$

    c. $h(0) = \boxed{}$

17. $h(r) = -r^2 - 1$

    a. $h(3) = \boxed{}$

    b. $h(-1) = \boxed{}$

    c. $h(0) = \boxed{}$

18. $F(y) = -y^2 + 8$

    a. $F(2) = \boxed{}$

    b. $F(-2) = \boxed{}$

    c. $F(0) = \boxed{}$

19. $G(t) = -4$

    a. $G(5) = \boxed{}$

    b. $G(-4) = \boxed{}$

    c. $G(0) = \boxed{}$

20. $H(r) = 5$

    a. $H(4) = \boxed{}$

    b. $H(5) = \boxed{}$

    c. $H(0) = \boxed{}$

21. $H(x) = \dfrac{3x}{-8x - 2}$

    a. $H(7) = \boxed{}$.

    b. $H(-3) = \boxed{}$.

22. $K(x) = \dfrac{8x}{x - 8}$

    a. $K(6) = \boxed{}$.

    b. $K(-7) = \boxed{}$.

23. $f(x) = \dfrac{20}{x + 7}$.

    a. $f(3) = \boxed{}$.

    b. $f(-7) = \boxed{}$.

24. $g(x) = \dfrac{12}{x + 1}$.

    a. $g(-3) = \boxed{}$.

    b. $g(-1) = \boxed{}$.

25. $g(x) = -x + 5$

    a. $g(3) = \boxed{}$

    b. $g(-3) = \boxed{}$

26. $h(x) = -4x - 5$

    a. $h(8) = \boxed{}$

    b. $h(-3) = \boxed{}$

27. $F(x) = x^2 - 3x + 2$

    a. $F(4) = \boxed{}$

    b. $F(-2) = \boxed{}$

28. $G(x) = x^2 + 4x - 6$

    a. $G(2) = \boxed{}$

    b. $G(-2) = \boxed{}$

29. $H(x) = -3x^2 - 4x + 2$

    a. $H(3) = \boxed{}$

    b. $H(-3) = \boxed{}$

30. $H(x) = -3x^2 + 4x + 5$

    a. $H(3) = \boxed{}$

    b. $H(-5) = \boxed{}$

31. $K(x) = \sqrt{x}$.

    a. $K(81) = \boxed{}$

    b. $K\left(\frac{16}{49}\right) = \boxed{}$

    c. $K(-4) = \boxed{}$

32. $f(x) = \sqrt{x}$.

    a. $f(36) = \boxed{}$

    b. $f\left(\frac{64}{25}\right) = \boxed{}$

    c. $f(-4) = \boxed{}$

33. $g(x) = \sqrt[3]{x}$

    a. $g(-125) = \boxed{}$

    b. $g\left(\frac{1}{8}\right) = \boxed{}$

34. $g(x) = \sqrt[3]{x}$

    a. $g(-8) = \boxed{}$

    b. $g\left(\frac{27}{125}\right) = \boxed{}$

35. $h(x) = -4$

    a. $h(8) = \boxed{}$

    b. $h(-4) = \boxed{}$

36. $F(x) = -18$

    a. $F(4) = \boxed{}$

    b. $F(-7) = \boxed{}$

Function Formulas and Solving Equations

37. Solve for x, where $G(x) = 20x + 2$.

    a. If $G(x) = -78$, then $x =$ ⬚.

    b. If $G(x) = 7$, then $x =$ ⬚.

38. Solve for x, where $H(x) = 6x - 9$.

    a. If $H(x) = -27$, then $x =$ ⬚.

    b. If $H(x) = -7$, then $x =$ ⬚.

39. Solve for x, where $H(x) = x^2 - 10$.

    a. If $H(x) = 6$, then $x =$ ⬚.

    b. If $H(x) = -17$, then $x =$ ⬚.

40. Solve for x, where $K(x) = x^2 + 4$.

    a. If $K(x) = 5$, then $x =$ ⬚.

    b. If $K(x) = -5$, then $x =$ ⬚.

41. Solve for x, where $f(x) = x^2 + 3x - 26$.

    If $f(x) = 2$, then $x =$ ⬚.

42. Solve for x, where $g(x) = x^2 + 18x + 81$.

    If $g(x) = 1$, then $x =$ ⬚.

43. If G is a function defined by $G(y) = -3y - 5$,

    Find $G(0)$. _____

    Solve $G(y) = 0$. _____

44. If g is a function defined by $g(y) = 4y - 11$,

    Find $g(0)$. _____

    Solve $g(y) = 0$. _____

45. If K is a function defined by $K(y) = y^2 - 9$,

    Find $K(0)$. _____

    Solve $K(y) = 0$. _____

46. If G is a function defined by $G(r) = 3r^2 - 6$,

    Find $G(0)$. _____

    Solve $G(r) = 0$. _____

47. If g is a function defined by $g(r) = r^2 + 2r - 35$,

    Find $g(0)$. _____

    Solve $g(r) = 0$. _____

48. If K is a function defined by $K(t) = t^2 - 11t + 18$,

    Find $K(0)$. _____

    Solve $K(t) = 0$. _____

Functions and Points on a Graph

49. If $K(6) = 9$, then the point $\boxed{}$ is on the graph of K.

If $(7, 6)$ is on the graph of K, then $K(7) = \boxed{}$.

50. If $f(2) = -1$, then the point $\boxed{}$ is on the graph of f.

If $(4, 3)$ is on the graph of f, then $f(4) = \boxed{}$.

51. If $g(t) = r$, then the point $\boxed{}$ is on the graph of g.

The answer is not a specific numerical point, but one with variables for coordinates.

52. If $g(y) = x$, then the point $\boxed{}$ is on the graph of g.

The answer is not a specific numerical point, but one with variables for coordinates.

53. If (x, y) is on the graph of h, then $h(x) = \boxed{}$.

54. If (t, x) is on the graph of F, then $F(t) = \boxed{}$.

55. For the function $G(x)$, when $x = -2$, its y-value is 9.

Choose all true statements.

□ The function's value is 9 at -2. □ The point $(9, -2)$ is on the graph of the function. □ $G(9) = -2$ □ The point $(-2, 9)$ is on the graph of the function. □ $G(-2) = 9$ □ The function's value is -2 at 9.

56. For the function $H(x)$, when $x = 2$, its y-value is -5.

Choose all true statements.

□ $H(2) = -5$ □ $H(-5) = 2$ □ The function's value is 2 at -5. □ The function's value is -5 at 2. □ The point $(-5, 2)$ is on the graph of the function. □ The point $(2, -5)$ is on the graph of the function.

Function Graphs

57. Use the graph of H below to evaluate the given expressions. (Estimates are OK.)

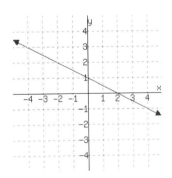

$H(-2) = \boxed{}$

$H(2) = \boxed{}$

58. Use the graph of K below to evaluate the given expressions. (Estimates are OK.)

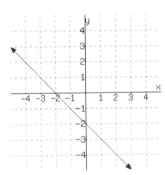

$K(-3) = \boxed{}$

$K(2) = \boxed{}$

59. Use the graph of f below to evaluate the given expressions. (Estimates are OK.)

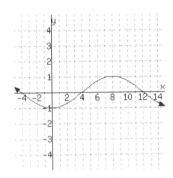

$f(0) = \boxed{}$

$f(12) = \boxed{}$

60. Use the graph of *g* below to evaluate the given expressions. (Estimates are OK.)

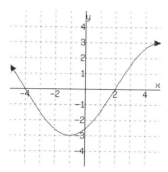

$g(-1) =$ [_____]

$g(2) =$ [_____]

61. Use the graph of *g* below to evaluate the given expressions. (Estimates are OK.)

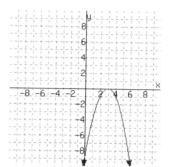

$g(3) =$ [_____]

$g(5) =$ [_____]

62. Use the graph of *h* below to evaluate the given expressions. (Estimates are OK.)

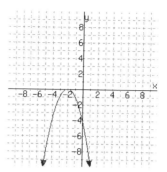

$h(-2) =$ [_____]

$h(-1) =$ [_____]

63. Function *f* is graphed.

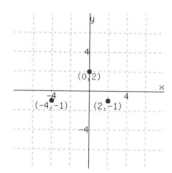

a. $f(0) =$ [_____]

b. Solve $f(x) = -1$.

64. Function *f* is graphed.

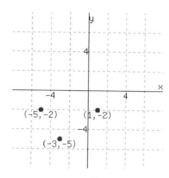

a. $f(-3) =$ [_____]

b. Solve $f(x) = -2$.

65. Function *f* is graphed.

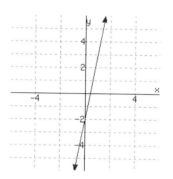

a. $f(1) =$ [_____]

b. Solve $f(x) = -2$.

66. Function *f* is graphed.

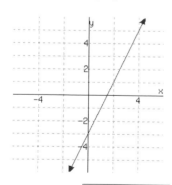

a. $f(2) =$ [_____]

b. Solve $f(x) = 3$.

67. Function *f* is graphed.

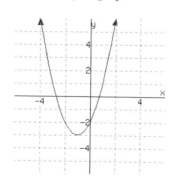

a. $f(-3) =$ [_____]

b. Solve $f(x) = -2$.

68. Function *f* is graphed.

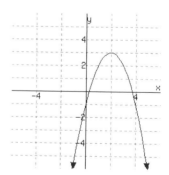

a. $f(3) =$ [_____]

b. Solve $f(x) = -1$.

Function Tables

69. Use the table of values for g below to evaluate the given expressions.

x	-4	-1	2	5	8
$g(x)$	-1	-0.5	4.1	9.9	-0.4

$g(-4) = $ ⬚

$g(5) = $ ⬚

70. Use the table of values for g below to evaluate the given expressions.

x	-1	1	3	5	7
$g(x)$	7.7	-0.6	-2	-1.1	4.5

$g(1) = $ ⬚

$g(7) = $ ⬚

71. Make a table of values for the function g, defined by $g(x) = -2x^2$. Based on values in the table, sketch a graph of g.

x	$g(x)$
___	___
___	___
___	___
___	___
___	___

72. Make a table of values for the function K, defined by $K(x) = \dfrac{2^x - 2}{x^2 + 3}$. Based on values in the table, sketch a graph of K.

x	$K(x)$
___	___
___	___
___	___
___	___
___	___

Translating Between Different Representations of a Function

73. Here is a verbal representation of a function G.

Square the input x to obtain the output y. Give a numeric representation of G:

x	0	1	2	3	4
$G(x)$	___	___	___	___	___

Give a symbolic representation of G:

$G(x) = $ ⬚

74. Here is a verbal representation of a function G.

Cube the input x to obtain the output y. Give a numeric representation of G:

x	0	1	2	3	4
$G(x)$	___	___	___	___	___

Give a symbolic representation of G:

$G(x) = $ ⬚

75. Here is a verbal representation of a function H.

Triple the input x and then subtract nine to obtain the output y. Give a numeric representation of H:

x	0	1	2	3	4
$H(x)$	——	——	——	——	——

Give a symbolic representation of H:

$H(x) =$ [＿＿＿＿＿]

76. Here is a verbal representation of a function K.

Double the input x and then subtract two to obtain the output y. Give a numeric representation of K:

x	0	1	2	3	4
$K(x)$	——	——	——	——	——

Give a symbolic representation of K:

$K(x) =$ [＿＿＿＿＿]

77. Express the function f numerically with the table.

$$f(x) = 2x^2 - \frac{1}{3}x$$

x	−3	−2	−1	0	1	2	3
$f(x)$	—	—	—	—	—	—	—

On graphing paper, you should be able to give a graphical representation of f too.

78. Express the function g numerically with the table.

$$g(x) = x^3 - \frac{1}{3}x^2$$

x	−3	−2	−1	0	1	2	3
$g(x)$	—	—	—	—	—	—	—

On graphing paper, you should be able to give a graphical representation of g too.

79. Express the function g numerically with the table.

$$g(x) = \frac{1 - x}{5 + x}$$

x	−3	−2	−1	0	1	2	3
$g(x)$	—	—	—	—	—	—	—

On graphing paper, you should be able to give a graphical representation of g too.

80. Express the function h numerically with the table.

$$h(x) = \frac{7 - x}{7 + x}$$

x	−3	−2	−1	0	1	2	3
$h(x)$	—	—	—	—	—	—	—

On graphing paper, you should be able to give a graphical representation of h too.

Functions in Context

81. Virginia started saving in a piggy bank on her birthday. The function $f(x) = 3x + 1$ models the amount of money, in dollars, in Virginia's piggy bank. The independent variable represents the number of days passed since her birthday.

    Interpret the meaning of $f(1) = 4$.

    ⊙ A. The piggy bank started with $4 in it, and Virginia saves $1 each day.

    ⊙ B. Four days after Virginia started her piggy bank, there were $1 in it.

    ⊙ C. The piggy bank started with $1 in it, and Virginia saves $4 each day.

    ⊙ D. One days after Virginia started her piggy bank, there were $4 in it.

82. Timothy started saving in a piggy bank on his birthday. The function $f(x) = 2x + 2$ models the amount of money, in dollars, in Timothy's piggy bank. The independent variable represents the number of days passed since his birthday.

    Interpret the meaning of $f(3) = 8$.

    ⊙ A. Eight days after Timothy started his piggy bank, there were $3 in it.

    ⊙ B. Three days after Timothy started his piggy bank, there were $8 in it.

    ⊙ C. The piggy bank started with $8 in it, and Timothy saves $3 each day.

    ⊙ D. The piggy bank started with $3 in it, and Timothy saves $8 each day.

83. An arcade sells multi-day passes. The function $g(x) = \frac{1}{4}x$ models the number of days a pass will work, where x is the amount of money paid, in dollars.

    Interpret the meaning of $g(16) = 4$.

    ⊙ A. Each pass costs $16, and it works for 4 days.

    ⊙ B. If a pass costs $4, it will work for 16 days.

    ⊙ C. If a pass costs $16, it will work for 4 days.

    ⊙ D. Each pass costs $4, and it works for 16 days.

84. An arcade sells multi-day passes. The function $g(x) = \frac{1}{4}x$ models the number of days a pass will work, where x is the amount of money paid, in dollars.

    Interpret the meaning of $g(12) = 3$.

    ⊙ A. If a pass costs $12, it will work for 3 days.

    ⊙ B. Each pass costs $3, and it works for 12 days.

    ⊙ C. Each pass costs $12, and it works for 3 days.

    ⊙ D. If a pass costs $3, it will work for 12 days.

85. Kayla will spend \$270 to purchase some bowls and some plates. Each bowl costs \$1, and each plate costs \$6. The function $p(b) = -\frac{1}{6}b + 45$ models the number of plates Kayla will purchase, where b represents the number of bowls Kayla will purchase.

Interpret the meaning of $p(180) = 15$.

⊙ A. If 180 bowls are purchased, then 15 plates will be purchased.

⊙ B. \$180 will be used to purchase bowls, and \$15 will be used to purchase plates.

⊙ C. If 15 bowls are purchased, then 180 plates will be purchased.

⊙ D. \$15 will be used to purchase bowls, and \$180 will be used to purchase plates.

86. Aleric will spend \$210 to purchase some bowls and some plates. Each bowl costs \$5, and each plate costs \$6. The function $p(b) = -\frac{5}{6}b + 35$ models the number of plates Aleric will purchase, where b represents the number of bowls Aleric will purchase.

Interpret the meaning of $p(6) = 30$.

⊙ A. If 30 bowls are purchased, then 6 plates will be purchased.

⊙ B. If 6 bowls are purchased, then 30 plates will be purchased.

⊙ C. \$6 will be used to purchase bowls, and \$30 will be used to purchase plates.

⊙ D. \$30 will be used to purchase bowls, and \$6 will be used to purchase plates.

87. Kara will spend \$400 to purchase some bowls and some plates. Each plate costs \$1, and each bowl costs \$8. The function $q(x) = -\frac{1}{8}x + 50$ models the number of bowls Kara will purchase, where x represents the number of plates to be purchased.

Interpret the meaning of $q(24) = 47$.

⊙ A. \$47 will be used to purchase bowls, and \$24 will be used to purchase plates.

⊙ B. 47 plates and 24 bowls can be purchased.

⊙ C. \$24 will be used to purchase bowls, and \$47 will be used to purchase plates.

⊙ D. 24 plates and 47 bowls can be purchased.

88. Cheryl will spend \$160 to purchase some bowls and some plates. Each plate costs \$9, and each bowl costs \$8. The function $q(x) = -\frac{9}{8}x + 20$ models the number of bowls Cheryl will purchase, where x represents the number of plates to be purchased.

Interpret the meaning of $q(16) = 2$.

⊙ A. 2 plates and 16 bowls can be purchased.

⊙ B. 16 plates and 2 bowls can be purchased.

⊙ C. \$2 will be used to purchase bowls, and \$16 will be used to purchase plates.

⊙ D. \$16 will be used to purchase bowls, and \$2 will be used to purchase plates.

89. Find the rule of the linear function f that gives the number of minutes in x weeks.

$f(x) =$ ⬚

90. Find the rule of the linear function f that gives the number of seconds in x weeks.

$f(x) =$ ⬚

91. Suppose that M is the function that computes how many miles are in x feet. Find the algebraic rule for M. (If you do not know how many feet are in one mile, you can look it up on Google.)

$M(x) =$ ⬚

Evaluate $M(22000)$ and interpret the result:

There are about ⬚ miles in ⬚ feet.

92. Suppose that K is the function that computes how many kilograms are in x pounds. Find the algebraic rule for K. (If you do not know how many pounds are in one kilogram, you can look it up on Google.)

$K(x) =$ ⬚

Evaluate $K(241)$ and interpret the result.

Something that weighs ⬚ pounds would weigh about ⬚ kilograms.

93. Suppose that f is the function that the phone company uses to determine what your bill will be (in dollars) for a long-distance phone call that lasts t minutes. Each call costs a fixed price of $4.95 plus 10 cents per minute. Write a formula for this linear function f.

94. Suppose that f is the function that gives the total cost (in dollars) of downhill skiing x times during a season with a $500 season pass. Write a formula for f.

95. Suppose that f is the function that tells you how many dimes are in x dollars. Write a formula for f.

The following figure has the graph $y = d(t)$, which models a particle's distance from the starting line in feet, where t stands for time in seconds since timing started.

96.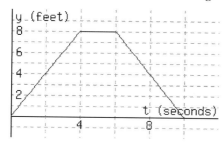

97.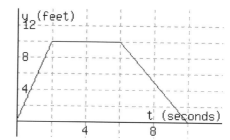

a. $d(3) =$ [_____]

b. Interpret the meaning of $d(3)$:

⊙ A. The particle was 3 feet away from the starting line 6 seconds since timing started.

⊙ B. The particle was 6 feet away from the starting line 3 seconds since timing started.

⊙ C. In the first 6 seconds, the particle moved a total of 3 feet.

⊙ D. In the first 3 seconds, the particle moved a total of 6 feet.

c. Solve $d(t) = 2$ for t. $t =$ [_____]

d. Interpret the meaning of part c's solution(s):

⊙ A. The article was 2 feet from the starting line 9 seconds since timing started.

⊙ B. The article was 2 feet from the starting line 1 seconds since timing started, and again 9 seconds since timing started.

⊙ C. The article was 2 feet from the starting line 1 seconds since timing started, or 9 seconds since timing started.

⊙ D. The article was 2 feet from the starting line 1 seconds since timing started.

a. $d(4) =$ [_____]

b. Interpret the meaning of $d(4)$:

⊙ A. The particle was 4 feet away from the starting line 10 seconds since timing started.

⊙ B. In the first 10 seconds, the particle moved a total of 4 feet.

⊙ C. The particle was 10 feet away from the starting line 4 seconds since timing started.

⊙ D. In the first 4 seconds, the particle moved a total of 10 feet.

c. Solve $d(t) = 5$ for t. $t =$ [_____]

d. Interpret the meaning of part c's solution(s):

⊙ A. The article was 5 feet from the starting line 1 seconds since timing started.

⊙ B. The article was 5 feet from the starting line 1 seconds since timing started, and again 8 seconds since timing started.

⊙ C. The article was 5 feet from the starting line 1 seconds since timing started, or 8 seconds since timing started.

⊙ D. The article was 5 feet from the starting line 8 seconds since timing started.

98. The function C models the the number of customers in a store *t* hours since the store opened.

t	0	1	2	3	4	5	6	7
C(t)	0	43	81	102	98	83	43	0

a. $C(2) =$ [   ]

b. Interpret the meaning of C(2):

 ⊙ A. There were 81 customers in the store 2 hours after the store opened.

 ⊙ B. There were 2 customers in the store 81 hours after the store opened.

 ⊙ C. In 2 hours since the store opened, the store had an average of 81 customers per hour.

 ⊙ D. In 2 hours since the store opened, there were a total of 81 customers.

c. Solve $C(t) = 43$ for t. $t =$ [   ]

d. Interpret the meaning of Part c's solution(s):

 ⊙ A. There were 43 customers in the store 1 hours after the store opened.

 ⊙ B. There were 43 customers in the store either 1 hours after the store opened, or 6 hours after the store opened.

 ⊙ C. There were 43 customers in the store 6 hours after the store opened.

 ⊙ D. There were 43 customers in the store 1 hours after the store opened, and again 6 hours after the store opened.

99. Chicago's average monthly rainfall, $R = f(t)$ inches, is given as a function of the month, *t*, where January is $t = 1$, in the table below.

t, month	1	2	3	4	5	6	7	8
R, inches	1.8	1.8	2.7	3.1	3.5	3.7	3.5	3.4

(a) Solve $f(t) = 3.1$.

$t =$ [   ]

The solution(s) to $f(t) = 3.1$ can be interpreted as saying

 ⊙ Chicago's average rainfall in the month of April is 3.1 inches.

 ⊙ Chicago's average rainfall is greatest in the month of February.

 ⊙ Chicago's average rainfall increases by 3.1 inches in the month of February.

 ⊙ Chicago's average rainfall is least in the month of April.

 ⊙ None of the above

(b) Solve $f(t) = f(5)$.

$t =$ [   ]

The solution(s) to $f(t) = f(5)$ can be interpreted as saying

 ⊙ Chicago's average rainfall is 3.5 inches in the months of May and July.

 ⊙ Chicago's average rainfall is 3.5 inches in the month of July.

 ⊙ Chicago's average rainfall is greatest in the month of May.

 ⊙ Chicago's average rainfall is 3.5 inches in the month of May.

 ⊙ None of the above

100. Let $f(t)$ denote the number of people eating in a restaurant t minutes after 5 PM. Answer the following questions:

a) Which of the following statements best describes the significance of the expression $f(4) = 19$?

⊙ There are 19 people eating at 9:00 PM

⊙ Every 4 minutes, 19 more people are eating

⊙ There are 19 people eating at 5:04 PM

⊙ There are 4 people eating at 5:19 PM

⊙ None of the above

b) Which of the following statements best describes the significance of the expression $f(a) = 30$?

⊙ At 5:30 PM there are a people eating

⊙ Every 30 minutes, the number of people eating has increased by a people

⊙ a hours after 5 PM there are 30 people eating

⊙ a minutes after 5 PM there are 30 people eating

⊙ None of the above

c) Which of the following statements best describes the significance of the expression $f(30) = b$?

⊙ b hours after 5 PM there are 30 people eating

⊙ Every 30 minutes, the number of people eating has increased by b people

⊙ At 5:30 PM there are b people eating

⊙ b minutes after 5 PM there are 30 people eating

⊙ None of the above

d) Which of the following statements best describes the significance of the expression $n = f(t)$?

⊙ n hours after 5 PM there are t people eating

⊙ t hours after 5 PM there are n people eating

⊙ Every t minutes, n more people have begun eating

⊙ n minutes after 5 PM there are t people eating

⊙ None of the above

101. Let $s(t) = 12t^2 - 2t + 300$, where s is the position (in mi) of a car driving on a straight road at time t (in hr). The car's velocity (in mi/hr) at time t is given by $v(t) = 24t - 2$.

  a. *Using function notation,* express the car's position after 2.5 hours. The answer here is not a formula, it's just something using function notation like $f(8)$.

  b. Where is the car then? The answer here is a number with units.

  c. *Use function notation* to express the question, "When is the car going 65 $\frac{mi}{hr}$?" The answer is an equation that uses function notation; something like $f(x)=23$. You are not being asked to actually solve the equation, just to write down the equation.

  d. Where is the car when it is going 22 $\frac{mi}{hr}$? The answer here is a number with units. You are being asked a question about its position, but have been given information about its speed.

102. Let $s(t) = 13t^2 - 3t + 300$, where s is the position (in mi) of a car driving on a straight road at time t (in hr). The car's velocity (in mi/hr) at time t is given by $v(t) = 26t - 3$.

  a. *Using function notation,* express the car's position after 3.9 hours. The answer here is not a formula, it's just something using function notation like $f(8)$.

  b. Where is the car then? The answer here is a number with units.

  c. *Use function notation* to express the question, "When is the car going 61 $\frac{mi}{hr}$?" The answer is an equation that uses function notation; something like $f(x)=23$. You are not being asked to actually solve the equation, just to write down the equation.

  d. Where is the car when it is going 75 $\frac{mi}{hr}$? The answer here is a number with units. You are being asked a question about its position, but have been given information about its speed.

103. Describe your own example of a function that has real context to it. You will need some kind of input variable, like "number of years since 2000" or "weight of the passengers in my car." You will need a process for using that number to bring about a different kind of number. The process does not need to involve a formula; a verbal description would be great, as would a formula.

Give your function a name. Write the symbol(s) that you would use to represent input. Write the symbol(s) that you would use to represent output.

104. Use the graph of h in the figure to fill in the table.

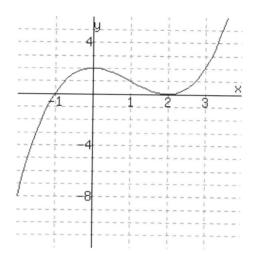

x	-2	-1	0	1	2
$h(x)$	_____	_____	_____	_____	_____

a. Evaluate $h(3) - h(0)$. [ ]

b. Evaluate $h(2) - h(-1)$. [ ]

c. Evaluate $2h(-1)$. [ ]

d. Evaluate $h(0) + 3$. [ ]

105. Use the given graph of a function f, along with $a, b, c, d, e,$ and h to answer the following questions. Some answers are points, and should be entered as ordered pairs. Some answers ask you to solve for x, so the answer should be in the form $x=...$

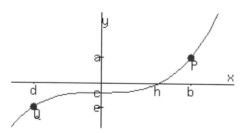

a. What are the coordinates of the point P? [ ]

b. What are the coordinates of the point Q? [ ]

c. Evaluate $f(b)$. (The answer is symbolic, not a specific number.) [ ]

d. Solve $f(x) = e$ for x. (The answer is symbolic, not a specific number.) [ ]

e. Suppose $c = f(z)$. Solve the equation $z = f(x)$ for x. [ ]

10.2 Domain and Range

A function is a process for turning input values into output values. Occasionally a function f will have input values for which the process breaks down.

10.2.1 Domain

Example 10.2.2 Let P be the population of Portland as a function of the year. According to Google[a] we can say that:

$$P(2016) = 639863 \qquad P(1990) = 487849$$

But what if we asked to find $P(1600)$? The question doesn't really make sense anymore. The Multnomah tribe lived in villages in the area, but the city of Portland was not incorporated until 1851. We say that $P(1600)$ is *undefined*.

---

[a]https://www.google.com/publicdata/explore?ds=kf7tgg1uo9ude_&met_y=population&hl=en&dl=en#!ctype=l&strail=false
&bcs=d&nselm=h&met_y=population&scale_y=lin&ind_y=false&rdim=country&idim=place:4159000&ifdim=country
&hl=en_US&dl=en&ind=false

Example 10.2.3 If m is a person's mass in kg, let $w(m)$ be their weight in lb. There is an approximate formula for w:

$$w(m) \approx 2.2m$$

From this formula we can find:

$$w(50) \approx 110 \qquad w(80) \approx 176$$

which tells us that a 50-kg person weighs 110 lb, and an 80-kg person weighs 176 lb.

What if we asked for $w(-100)$? In the context of this example, we would be asking for the weight of a person whose mass is -100 kg. This is clearly nonsense. That means that $w(-100)$ is *undefined*. Note that the *context* of the example is telling us that $w(-100)$ is undefined even though the formula alone might suggest that $w(-100) = -220$.

Example 10.2.4 Let g have the formula

$$g(x) = \frac{x}{x - 7}.$$

For most x-values, $g(x)$ is perfectly computable:

$$g(2) = -\frac{2}{5} \qquad g(14) = 2.$$

But if we try to compute $g(7)$, we run into an issue of arithmetic.

$$g(7) = \frac{7}{7 - 7}$$
$$= \frac{7}{0}$$

The expression $\frac{7}{0}$ is *undefined*. There is no number that this could equal.

Checkpoint 10.2.5. If $f(x) = \dfrac{x+2}{x+8}$, find an input for f that would cause an undefined output.

The number $\boxed{}$ would cause an undefined output.

Explanation. Trying -8 as an input value would not work out; it would lead to division by 0.

These examples should motivate the following definition.

Definition 10.2.6 Domain. The **domain** of a function f is the collection of all of its valid input values.

Example 10.2.7 Referring to the functions from Examples 10.2.2–10.2.4

- The domain of P is all years starting from 1851 and later. It would also be reasonable to say that the domain is actually all years from 1851 up to the current year, since we cannot guarantee that Portland will exist forever.

- The domain of w is all positive real numbers. It is nonsensical to have a person with negative mass or even one with zero mass. While there is some lower bound for the smallest mass a person could have, and also an upper bound for the largest mass a person could have, these boundaries are gray. We can say for sure that non-positive numbers should never be used as inputs for w.

- The domain of g is all real numbers except 7. This is the only number that causes a breakdown in g's formula.

10.2.2 Interval, Set, and Set-Builder Notation

Communicating the domain of a function can be wordy. In mathematics, we can communicate the same information using concise notation that is accepted for use almost everywhere. Table 10.2.8 contains example functions from this section and their domains, and demonstrates *interval notation* for these domains. Basic interval notation is covered in Section 1.6, but some of our examples here go beyond what that section covers.

Function	Verbal Domain	Number Line Illustration	Interval Notation
P from Example 10.2.2	all years 1851 and greater		$[1851, \infty)$
w from Example 10.2.3	all real numbers greater than 0		$(0, \infty)$
g from Example 10.2.4	all real numbers except 7		$(-\infty, 7) \cup (7, \infty)$

Table 10.2.8: Domains from Earlier Examples

Again, basic interval notation is covered in Section 1.6, but one thing appears in Table 10.2.8 that is not explained in that earlier section: the $\cup$ symbol, which we see in the domain of g.

Occasionally there is a need to consider number line pictures such as Figure 10.2.9, where two or more intervals appear.

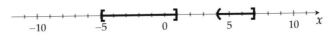

Figure 10.2.9: A number line with a union of two intervals

This picture is trying to tell you to consider numbers that are between −5 and 1, together with numbers that are between 4 and 7. That word "together" is related to the word "union," and in math the **union symbol**, ∪, captures this idea. So we can write the numbers in this picture as

$$[-5, 1] \cup (4, 7]$$

(which uses interval notation).

With the domain of g in Table 10.2.8, the number line picture shows us another "union" of two intervals. They are very close together, but there are still two separated intervals in that picture: $(-\infty, 7)$ and $(7, \infty)$. Their union is represented by $(-\infty, 7) \cup (7, \infty)$.

Checkpoint 10.2.10. What is the domain of the function sqrt, where $\text{sqrt}(x) = \sqrt{x}$, using interval notation?

Explanation. The function sqrt cannot take a negative number as an input. It can however take any positive number as input, or the number 0 as input. Representing this on a number line, we find the domain is $[0, \infty)$ in interval notation.

Checkpoint 10.2.11. What is the domain of the function ℓ where $\ell(x) = \frac{2}{x-3}$, using interval notation?

Explanation. The function ℓ cannot take a 3 as an input. It can however take any other number as input. Representing this on a number line, we have an interval $(-\infty, 3)$ to the left of 3, and $(3, \infty)$ to the right of 3. So we find the domain is $(-\infty, 3) \cup (3, \infty)$.

Sometimes we will consider collections of only a short list of numbers. In those cases, we use **set notation** (first introduced in Section 1.5). With set notation, we have a list of numbers in mind, and we simply list all of those numbers. Curly braces are standard for surrounding the list. Table 10.2.12 illustrates set notation in use.

Picture of Set	Set Notation
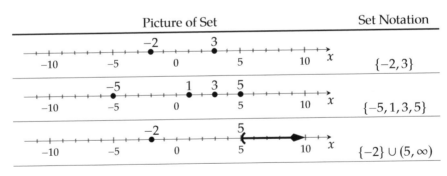	$\{-2, 3\}$
	$\{-5, 1, 3, 5\}$
	$\{-2\} \cup (5, \infty)$

Table 10.2.12: Set Notation

Checkpoint 10.2.13. A change machine lets you put in an x-dollar bill, and gives you $f(x)$ nickels in return equal in value to x dollars. Any current, legal denomination of US paper money can be fed to the change machine. What is the domain of f?

Explanation. The current, legal denominations of US paper money are \$1, \$2, \$5, \$10, \$20, \$50, and \$100. So the domain of f is the set $\{1, 2, 5, 10, 20, 50, 100\}$.

While most collections of numbers that we will encounter can be described using a combination of interval notation and set notation, there is another commonly used notation that is very useful in algebra: **set-builder notation**, which was introduced in Section 1.6. Set-builder notation also uses curly braces. Set-builder notation provides a template for what a number that is under consideration might look like, and then it gives you restrictions on how to use that template. A very basic example of set-builder notation is

$$\{x \mid x \geq 3\}.$$

Verbally, this is "the set of all x such that x is greater than or equal to 3." Table 10.2.14 gives more examples of set-builder notation in use.

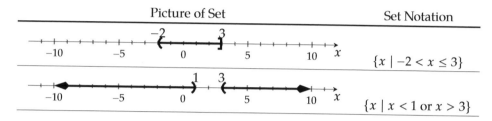

Picture of Set	Set Notation
	$\{x \mid -2 < x \leq 3\}$
	$\{x \mid x < 1 \text{ or } x > 3\}$

Table 10.2.14: Set-Builder Notation

Checkpoint 10.2.15. What is the domain of the function sqrt, where sqrt$(x) = \sqrt{x}$, using set-builder notation?

Explanation. The function sqrt cannot take a negative number as an input. It can however take any positive number as input, or the number 0 as input. Representing this on a number line, we find the domain is $\{x \mid x \geq 0\}$ in set-builder notation.

Example 10.2.16 What is the domain of the function A, where $A(x) = \frac{2x+1}{x^2-2x-8}$?

Note that if you plugged in some value for x, the only thing that might go wrong is if the denominator equals 0. So a *bad* value for x would be when

$$x^2 - 2x - 8 = 0$$
$$(x + 2)(x - 4) = 0$$

Here, we used a basic factoring technique from Section 7.3. To continue, either

$$x + 2 = 0 \qquad \text{or} \qquad x - 4 = 0$$
$$x = -2 \qquad \text{or} \qquad x = 4.$$

These are the *bad* x-values because they lead to division by 0 in the formula for A. So on a number line, if we wanted to picture the domain of A, we would make a sketch like:

So the domain is the union of three intervals: $(-\infty, -2) \cup (-2, 4) \cup (4, \infty)$.

Example 10.2.17 What is the domain of the function B, where $B(x) = \sqrt{7 - x} + 3$?

Note that if you plugged in some value for x, the only thing that might go wrong is if the value in the radical is negative. So the *good* values for x would be when

$$7 - x \geq 0$$
$$7 \geq x$$
$$x \leq 7$$

So on a number line, if we wanted to picture the domain of B, we would make a sketch like:

So the domain is the interval $(-\infty, 7]$.

There are three main properties of algebraic functions that cause numbers to be excluded from a domain, which are summarized here.

Denominators Division by zero is undefined. So if a function contains an expression in a denominator, it will only be defined where that expression is not equal to zero.

Example 10.2.16 demonstrates this.

Square Roots The square root of a negative number is undefined. So if a function contains a square root, it will only be defined when the expression inside that radical is greater than or equal to zero. (This is actually true for any even nth radical.)

Example 10.2.17 demonstrates this.

Context Some numbers are nonsensical in context. If a function has real-world context, then this may add additional restrictions on the input values.

Example 10.2.3 demonstrates this.

List 10.2.18: Summary of Algebraic Domain Restrictions

10.2.3 Range

The domain of a function is the collection of its valid inputs; there is a similar notion for *output*.

Definition 10.2.19 Range. The **range** of a function f is the collection of all of its possible output values.

Example 10.2.20 Let f be the function defined by the formula $f(x) = x^2$. Finding f's *domain* is straightforward. Any number anywhere can be squared to produce an output, so f has domain $(-\infty, \infty)$. What is the *range* of f?

Explanation. We would like to describe the collection of possible numbers that f can give as output. We will use a graphical approach. Figure 10.2.21 displays a graph of f, and the visualization that reveals f's range.

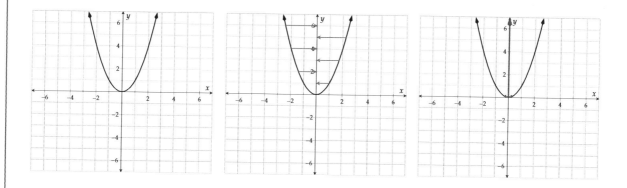

Figure 10.2.21: $y = f(x)$ where $f(x) = x^2$. The second graph illustrates how to visualize the range. In the third graph, the range is marked as an interval on the y-axis.

Output values are the y-coordinates in a graph. If we "slide the ink" left and right over to the y-axis to emphasize what the y-values in the graph are, we have y-values that start from 0 and continue upward forever. Therefore the range is $[0, \infty)$.

Warning 10.2.22 Finding range from a formula. Sometimes it is possible to compute a range without the aid of a graph. However, doing so can often require techniques covered in calculus. Therefore when you are asked to find the range of a function based on its formula, your approach will most often need to be a graphical one.

Example 10.2.23 Given the function g graphed in Figure 10.2.24, find the domain and range of g.

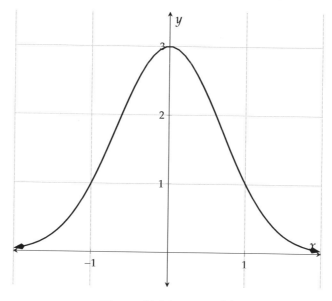

Figure 10.2.24: $y = g(x)$

Explanation. To find the domain, we can visualize all of the *x*-values that are valid inputs for this function by "sliding the ink" down onto the *x*-axis. The arrows at the far left and far right of the curve indicate that whatever pattern we see in the graph continues off to the left and right. Here, we see that the arms of the graph appear to be tapering down to the *x*-axis and extending left and right forever. Every *x*-value can be used to get an output for the function, so the domain is $(-\infty, \infty)$.

If we visualize the possible *outputs* by "sliding the ink" sideways onto the *y*-axis, we find that outputs as high as 3 are possible (including 3 itself). The outputs appear to get very close to 0 when *x* is large, but they aren't quite equal to 0. So the range is $(0, 3]$.

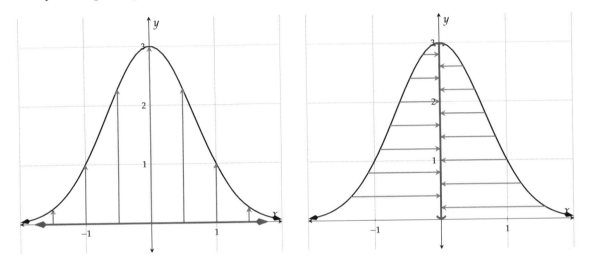

Figure 10.2.25: Domain of *g*   **Figure 10.2.26:** Range of *g*

Checkpoint 10.2.27. Given the function *h* graphed below, find the domain and range of *h*. Note there is an invisible vertical line at $x = 2$, and the two arms of the graph are extending downward (and upward) forever, getting arbitrarily close to that vertical line, but never touching it. Also note that the two arms extend forever to the left and right, getting arbitrarily close to the *y*-axis, but never touching it.

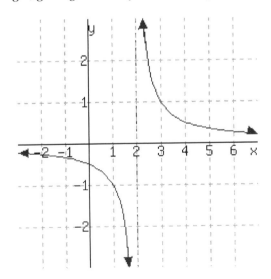

The domain of *h* is [ ] and the range of *h* is [ ].

Explanation.   To find the domain, we try to visualize all of the *x*-values that are valid inputs for this function. The arrows pointing left and right on the curve indicate that whatever pattern we see in the graph continues off to the left and right. So for *x*-values far to the right or left, we will be able to get an output for *h*.

The arrows pointing up and down are supposed to indicate that the curve will get closer and closer to the vertical line $x = 2$ after the curve leaves the viewing window we are using. So even when *x* is some number very close to 2, we will be able to get an output for *h*.

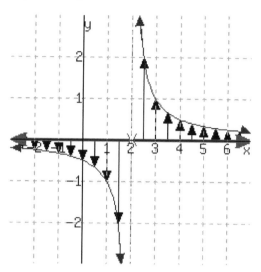

The one *x*-value that doesn't behave is $x = 2$. If we tried to use that as an input, there is no point on the graph directly above or below that on the *x*-axis. So the domain is $(-\infty, 2) \cup (2, \infty)$.

To find the range, we try to visualize all of the *y*-values that are possible outputs for this function. Sliding the ink of the curve left/right onto the *y*-axis reveals that $y = 0$ is the only *y*-value that we could never obtain as an output. So the range is $(-\infty, 0) \cup (0, \infty)$.

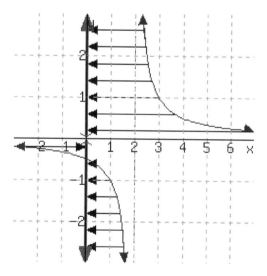

The examples of finding domain and range so far have all involved either a verbal description of a function, a formula for that function, or a graph of that function. Recall that there is a fourth perspective on functions:

a table. In the case of a table, we have very limited information about the function's inputs and outputs. If the table is all that we have, then there are a handful of input values listed in the table for which we know outputs. For any other input, the output is undefined.

Example 10.2.28 Consider the function k given in Table 10.2.29. What is the domain and range of k?

x	$k(x)$
3	4
8	5
10	5

Table 10.2.29

Explanation.   All that we know about k is that $k(3) = 4$, $k(8) = 5$, and $k(10) = 5$. Without any other information such as a formula for k or a context for k that tells us its verbal description, we must assume that its domain is $\{3, 8, 10\}$; these are the only valid input for k. Similarly, k's range is $\{4, 5\}$.

Note that we have used set notation, not interval notation, since the answers here were *lists* of x-values (for the domain) and y-values (for the range). Also note that we could graph the information that we have about k in Figure 10.2.30, and the visualization of "sliding ink" to determine domain and range still works.

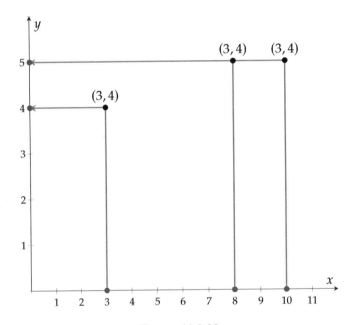

Figure 10.2.30

Exercises

Review and Warmup

1. Here is an interval:

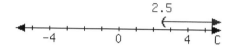

Write the interval using set-builder notation.

Write the interval using interval notation.

2. Here is an interval:

Write the interval using set-builder notation.

Write the interval using interval notation.

3. Here is an interval:

Write the interval using set-builder notation.

Write the interval using interval notation.

4. Here is an interval:

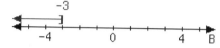

Write the interval using set-builder notation.

Write the interval using interval notation.

5. Solve this compound inequality, and write your answer in *interval notation*.

$x \geq 0$ and $x \leq 2$

6. Solve this compound inequality, and write your answer in *interval notation*.

$x \geq -3$ and $x < -2$

7. Solve this compound inequality, and write your answer in *interval notation*.

$x \geq 6$ or $x \leq 3$

8. Solve this compound inequality, and write your answer in *interval notation*.

$x > 2$ or $x < -2$

Domain and Range From a Graph A function is graphed.

9.

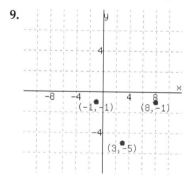

This function has domain ☐ and range ☐ .

10.

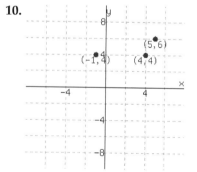

This function has domain ☐ and range ☐ .

11.

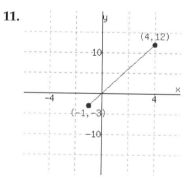

This function has domain ☐ and range ☐ .

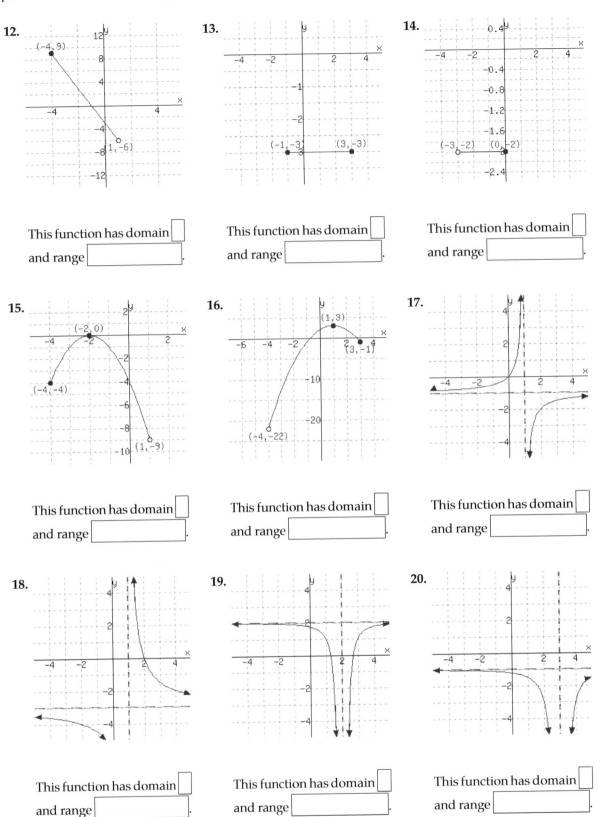

12.

This function has domain ☐
and range ☐.

13.

This function has domain ☐
and range ☐.

14.

This function has domain ☐
and range ☐.

15.

This function has domain ☐
and range ☐.

16.

This function has domain ☐
and range ☐.

17.

This function has domain ☐
and range ☐.

18.

This function has domain ☐
and range ☐.

19.

This function has domain ☐
and range ☐.

20.

This function has domain ☐
and range ☐.

21.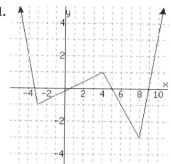

This function has domain ☐ and range ☐.

22.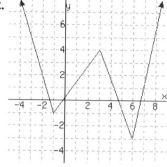

This function has domain ☐ and range ☐.

23.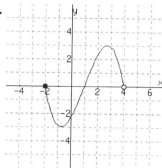

The function has domain ☐ and range ☐.

24.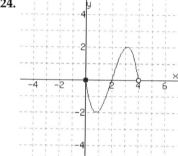

The function has domain ☐ and range ☐.

25.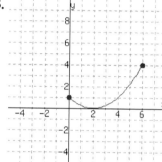

The function has domain ☐ and range ☐.

26.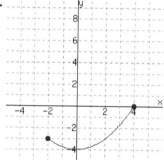

The function has domain ☐ and range ☐.

27.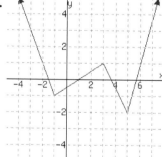

The function has domain ☐ and range ☐.

28.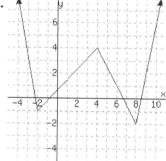

The function has domain ☐ and range ☐.

29.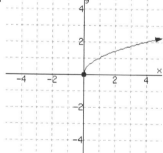

The function has domain ☐ and range ☐.

30.

The function has domain ☐
and range ☐.

31.

The function has domain ☐
and range ☐.

32.

The function has domain ☐
and range ☐.

33.

The function has domain ☐
and range ☐.

34.

The function has domain ☐
and range ☐.

35.

The function has domain ☐
and range ☐.

36.

The function has domain ☐
and range ☐.

Domain From a Formula

37. Find the domain of H where $H(x) = -x + 4$.

38. Find the domain of K where $K(x) = -8x - 9$.

39. Find the domain of f where $f(x) = \dfrac{8}{7}x^4$.

40. Find the domain of f where $f(x) = \dfrac{5}{7}x^2$.

41. Find the domain of g where $g(x) = |-9x - 3|$.

42. Find the domain of h where $h(x) = |5x + 5|$.

43. Find the domain of F where $F(x) = \dfrac{2x}{x + 2}$.

44. Find the domain of F where $F(x) = \dfrac{5x}{x + 10}$.

45. Find the domain of G where $G(x) = \dfrac{x}{7x - 4}$.

46. Find the domain of H where $H(x) = \dfrac{4x}{7x + 3}$.

47. Find the domain of K where $K(x) = \dfrac{10x + 7}{x^2 + 14x + 40}$.

48. Find the domain of f where $f(x) = \dfrac{3x - 5}{x^2 + 8x - 20}$.

49. Find the domain of f where $f(x) = \dfrac{4 - 4x}{x^2 - 7x}$.

50. Find the domain of g where $g(x) = \dfrac{9x - 9}{x^2 + 8x}$.

51. Find the domain of h where $h(x) = \dfrac{2x + 2}{x^2 - 1}$.

52. Find the domain of F where $F(x) = \dfrac{9 - 5x}{x^2 - 49}$.

53. Find the domain of F where $F(x) = \dfrac{9x - 3}{16x^2 - 25}$.

54. Find the domain of G where $G(x) = \dfrac{x + 5}{64x^2 - 25}$.

55. Find the domain of H where $H(x) = -\dfrac{6x + 7}{x^2 + 5}$.

56. Find the domain of K where $K(x) = \dfrac{8x + 2}{x^2 + 3}$.

57. Find the domain of the function. $f(x) = -\dfrac{10}{\sqrt{x+9}}$

58. Find the domain of the function. $f(x) = -\dfrac{7}{\sqrt{x-2}}$

59. Find the domain of the function. $g(x) = \sqrt{9 - x}$

60. Find the domain of the function. $h(x) = \sqrt{6 - x}$

61. Find the domain of the function. $F(x) = \sqrt{3 + 14x}$

62. Find the domain of the function. $F(x) = \sqrt{9 + 16x}$

63. Find the domain of A where $A(x) = \dfrac{x + 14}{x^2 - 81}$.

64. Find the domain of p where $p(x) = \dfrac{x + 16}{x^2 - 4}$.

65. Find the domain of a where $a(x) = \dfrac{16x + 6}{x^2 + 7x - 98}$.

66. Find the domain of m where $m(x) = \dfrac{16x - 3}{x^2 + 2x - 8}$.

67. Find the domain of r where $r(x) = \dfrac{\sqrt{3 + x}}{1 - x}$.

68. Find the domain of B where $B(x) = \dfrac{\sqrt{5 + x}}{7 - x}$.

Domain and Range Using Context

69. Ross bought a used car for $9,000. The car's value decreases at a constant rate each year. After 7 years, the value decreased to $6,900.

Use a function to model the car's value as the number of years increases. Find this function's domain and range in this context.

The function's domain in this context is ⬚.

The function's range in this context is ⬚.

70. Michael bought a used car for $8,400. The car's value decreases at a constant rate each year. After 7 years, the value decreased to $5,600.

Use a function to model the car's value as the number of years increases. Find this function's domain and range in this context.

The function's domain in this context is ⬚.

The function's range in this context is ⬚.

71. Assume a car uses gas at a constant rate. After driving 25 miles since a full tank of gas was purchased, there was 15.75 gallons of gas left; after driving 50 miles since a full tank of gas was purchased, there was 13.5 gallons of gas left.

Use a function to model the amount of gas in the tank (in gallons). Let the independent variable be the number of miles driven since a full tank of gas was purchased. Find this function's domain and range in this context.

The function's domain in this context is ⬚.

The function's range in this context is ⬚.

72. Assume a car uses gas at a constant rate. After driving 20 miles since a full tank of gas was purchased, there was 7.2 gallons of gas left; after driving 60 miles since a full tank of gas was purchased, there was 5.6 gallons of gas left.

Use a function to model the amount of gas in the tank (in gallons). Let the independent variable be the number of miles driven since a full tank of gas was purchased. Find this function's domain and range in this context.

The function's domain in this context is ⬚.

The function's range in this context is ⬚.

73. Henry inherited a collection of coins when he was 14 years old. Ever since, he has been adding into the collection the same number of coins each year. When he was 21 years old, there were 580 coins in the collection. When he was 29 years old, there were 900 coins in the collection. At the age of 51, Henry donated all his coins to a museum.

Use a function to model the number of coins in Henry's collection, starting in the year he inherited the collection, and ending in the year the collection was donated. Find this function's domain and range in this context.

The function's domain in this context is ⬚.

The function's range in this context is ⬚.

74. Scot inherited a collection of coins when he was 15 years old. Ever since, he has been adding into the collection the same number of coins each year. When he was 21 years old, there were 380 coins in the collection. When he was 29 years old, there were 540 coins in the collection. At the age of 57, Scot donated all his coins to a museum.

Use a function to model the number of coins in Scot's collection, starting in the year he inherited the collection, and ending in the year the collection was donated. Find this function's domain and range in this context.

The function's domain in this context is ⬚.

The function's range in this context is ⬚.

75. Assume a tree grows at a constant rate. When the tree was planted, it was 4 feet tall. After 8 years, the tree grew to 6.4 feet tall.

Use a function to model the tree's height as years go by. Assume the tree can live 200 years, find this function's domain and range in this context.

The function's domain in this context is ☐.

The function's range in this context is ☐.

76. Assume a tree grows at a constant rate. When the tree was planted, it was 2.2 feet tall. After 10 years, the tree grew to 10.2 feet tall.

Use a function to model the tree's height as years go by. Assume the tree can live 180 years, find this function's domain and range in this context.

The function's domain in this context is ☐.

The function's range in this context is ☐.

77. An object was shot up into the air at an initial vertical speed of 384 feet per second. Its height as time passes can be modeled by the quadratic function f, where $f(t) = -16t^2 + 384t$. Here t represents the number of seconds since the object's release, and $f(t)$ represents the object's height in feet.

Find the function's domain and range in this context.

The function's domain in this context is ☐.

The function's range in this context is ☐.

78. An object was shot up into the air at an initial vertical speed of 416 feet per second. Its height as time passes can be modeled by the quadratic function f, where $f(t) = -16t^2 + 416t$. Here t represents the number of seconds since the object's release, and $f(t)$ represents the object's height in feet.

Find the function's domain and range in this context.

The function's domain in this context is ☐.

The function's range in this context is ☐.

79. From a clifftop over the ocean 421.89 m above sea level, an object was shot straight up into the air with an initial vertical speed of 125.93 $\frac{m}{s}$. On its way down it missed the cliff and fell into the ocean. Its height (above sea level) as time passes can be modeled by the quadratic function f, where $f(t) = -4.9t^2 + 125.93t + 421.89$. Here t represents the number of seconds since the object's release, and $f(t)$ represents the object's height (above sea level) in meters.

Find the function's domain and range in this context.

The function's domain in this context is ☐.

The function's range in this context is ☐.

80. From a clifftop over the ocean 370.44 m above sea level, an object was shot straight up into the air with an initial vertical speed of 108.78 $\frac{m}{s}$. On its way down it missed the cliff and fell into the ocean. Its height (above sea level) as time passes can be modeled by the quadratic function f, where $f(t) = -4.9t^2 + 108.78t + 370.44$. Here t represents the number of seconds since the object's release, and $f(t)$ represents the object's height (above sea level) in meters.

Find the function's domain and range in this context.

The function's domain in this context is ☐.

The function's range in this context is ☐.

81. You will build a rectangular sheep pen next to a river. There is no need to build a fence along the river, so you only need to build three sides. You have a total of 470 feet of fence to use. Find the dimensions of the pen such that you can enclose the maximum area.

    Use a function to model the area of the rectangular pen, with respect to the length of the width (the two sides perpendicular to the river). Find the function's domain and range in this context.

    The function's domain is [＿＿＿＿＿].

    The function's range is [＿＿＿＿＿].

82. You will build a rectangular sheep pen next to a river. There is no need to build a fence along the river, so you only need to build three sides. You have a total of 480 feet of fence to use. Find the dimensions of the pen such that you can enclose the maximum area.

    Use a function to model the area of the rectangular pen, with respect to the length of the width (the two sides perpendicular to the river). Find the function's domain and range in this context.

    The function's domain is [＿＿＿＿＿].

    The function's range is [＿＿＿＿＿].

83. A student's first name is a function of their student identification number.

    (a) Describe the domain for this function in a sentence. Specifics are not needed.

    (b) Describe the range for this function in a sentence. Specifics are not needed.

84. The year a car was made is a function of its VIN (Vehicle Identification Number).

    (a) Describe the domain for this function in a sentence. Specifics are not needed.

    (b) Describe the range for this function in a sentence. Specifics are not needed.

Challenge

85. For each part, sketch the graph of a function with the given domain and range.

    a. The domain is $(0, \infty)$ and the range is $(-\infty, 0)$.

    b. The domain is $(1, 2)$ and the range is $(3, 4)$.

    c. The domain is $(0, \infty)$ and the range is $[2, 3]$.

    d. The domain is $(1, 2)$ and the range is $(-\infty, \infty)$.

    e. The domain is $(-\infty, \infty)$ and the range is $(-1, 1)$.

    f. The domain is $(0, \infty)$ and the range is $[0, \infty)$.

10.3 Using Technology to Explore Functions

Graphing technology allows us to explore the properties of functions more deeply than we can with only pencil and paper. It can quickly create a table of values, and quickly plot the graph of a function. Such technology can also evaluate functions, solve equations with functions, find maximum and minimum values, and explore other key features.

There are many graphing technologies currently available, including (but not limited to) physical (handheld) graphing calculators, *Desmos*, *GeoGebra*, *Sage*, and *WolframAlpha*.

This section will focus on *how* technology can be used to explore functions and their key features. Although the choice of particular graphing technology varies by each school and curriculum, the main ways in which technology is used to explore functions is the same and can be done with each of the technologies above.

10.3.1 Finding an Appropriate Window

With a simple linear equation like $y = 2x + 5$, most graphing technologies will show this graph in a good window by default. A common default window goes from $x = -10$ to $x = 10$ and $y = -10$ to $y = 10$.

What if we wanted to graph something with a much larger magnitude though, such as $y = 2000x + 5000$? If we tried to view this for $x = -10$ to $x = 10$ and $y = -10$ to $y = 10$, the function would appear as an almost vertical line since it has such a steep slope.

Using technology, we will create a table of values for this function as shown in Figure 10.1a. Then we will set the x-values for which we view the function to go from $x = -5$ to $x = 5$ and the y-values from $y = -20,000$ to $y = 20,000$. The graph is shown in Figure 10.1b.

x	$y = 2000x + 5000$
−5	−5000
−4	−3000
−3	−1000
−2	1000
−1	3000
0	5000
1	7000
2	9000
3	11000
4	13000
5	15000

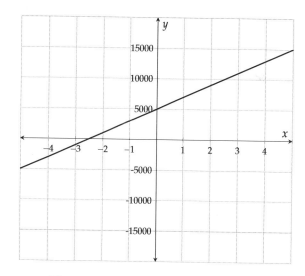

(a) A table of values

(b) Graphed with an appropriate window

Figure 10.3.1: Creating a table of values to determine an appropriate graphing window

Now let's practice finding an appropriate viewing window with a less familiar function.

Example 10.3.2 Find an appropriate window for $q(x) = \frac{x^3}{100} - 2x + 1$.

Entering this function into graphing technology, we input q(x)=(x^3)/100-2x+1. A default window will generally give us something like this:

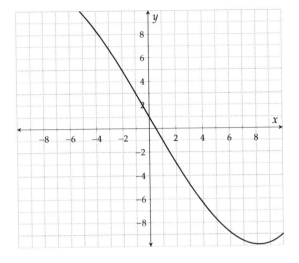

Figure 10.3.3: Function q graphed in the default window.

We can tell from the lower right corner of Figure 10.3.3 that we're not quite viewing all of the important details of this function. To determine a better window, we could use technology to make a table of values. Another more rudimentary option is to double the viewing constraints for x and y, as shown in Figure 10.3.4. Many graphing technologies have the ability to zoom in and out quickly.

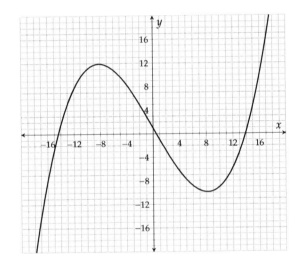

Figure 10.3.4: Function q graphed in an expanded window.

10.3.2 Using Technology to Determine Key Features of a Graph

The key features of a graph can be determined using graphing technology. Here, we'll show how to determine the x-intercepts, y-intercepts, maximum/minimum values, and the domain and range using technology.

Example 10.3.5 Graph the function given by $p(x) = -1000x^2 - 100x + 40$. Determine an appropriate viewing window, and then use graphing technology to determine the following:

  a. Determine the x-intercepts of the function.

  b. Determine the y-intercept of the function.

  c. Determine the maximum function value and where it occurs.

  d. State the domain and range of this function.

Explanation.

To start, we'll take a quick view of this function in a default window. We can see that we need to zoom in on the x-values, but we need to zoom out on the y-values.

From the graph we see that the x-values might as well run from about -0.5 to 0.5, so we will look at x-values in that window in increments of 0.1, as shown in Table 10.2a. This table allows us to determine an appropriate viewing window for $y = p(x)$ which is shown in Figure 10.2b. The table suggests we should go a little higher than 40 on the y-axis, and it would be OK to go the same distance in the negative direction to keep the x-axis centered.

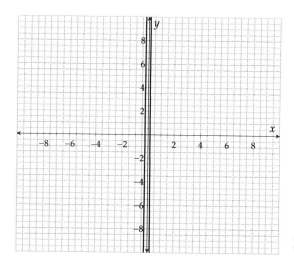

Figure 10.3.6: Graph of $y = p(x)$ in an inappropriate window

x	$p(x)$
-0.5	-160
-0.4	-80
-0.3	-20
-0.2	20
-0.1	40
0	40
0.1	20
0.2	-20
0.3	-80
0.4	-160
0.5	-260

(a) Function values for $y = p(x)$

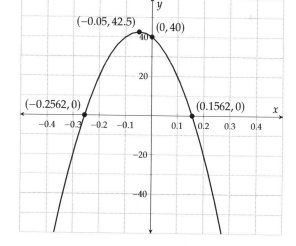

(b) Graph of $y = p(x)$ in an appropriate window showing key features

Figure 10.3.6: Creating a table of values to determine an appropriate graphing window

We can now use Figure 10.2b to determine the x-intercepts, the y-intercept, the maximum function value, and the domain and range.

    a. To determine the x-intercepts, we will find the points where y is zero. These are about $(-0.2562, 0)$ and $(0.1562, 0)$.

    b. To determine the y-intercept, we need the point where x is zero. This point is $(0, 40)$.

    c. The highest point on the graph is the vertex, which is about $(-0.05, 42.5)$. So the maximum function value is 42.5 and occurs at -0.05.

    d. We can see that the function is defined for all x-values, so the domain is $(-\infty, \infty)$. The maximum function value is 42.5, and there is no minimum function value. Thus the range is $(-\infty, 42.5]$.

If we use graphing technology to graph the function g where $g(x) = 0.0002x^2 + 0.00146x + 0.00266$, we may be mislead by the way values are rounded. Without technology, we know that this function is a quadratic function and therefore has at most two x-intercepts and has a vertex that will determine the minimum function value. However, using technology we could obtain a graph with the following key points:

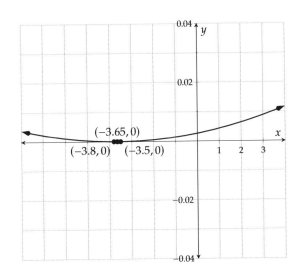

Figure 10.3.8: Misleading graph

Example 10.3.7 This *looks* like there are three x-intercepts, which we know is not possible for a quadratic function. We can evaluate g at $x = -3.65$ and determine that $g(-3.65) = -0.0000045$, which is *approximately* zero when rounded. So the true vertex of this function is $(-3.65, -0.0000045)$, and the minimum value of this function is -0.0000045 (not zero).

Every graphing tool generally has some type of limitation like this one, and it's good to be aware that these limitations exist.

10.3.3 Solving Equations and Inequalities Graphically Using Technology

To *algebraically* solve an equation like $h(x) = v(x)$ for

$$h(x) = -0.01(x - 90)(x + 20) \qquad \text{and} \qquad v(x) = -0.04(x - 10)(x - 80),$$

we'd start by setting up

$$-0.01(x - 90)(x + 20) = -0.04(x - 10)(x - 80)$$

To solve this, we'd then simplify each side of the equation, set it equal to zero, and finally use the quadratic formula.

An alternative is to *graphically* solve this equation, which is done by graphing

$$y = -0.01(x - 90)(x + 20)$$ and $$y = -0.04(x - 10)(x - 80).$$

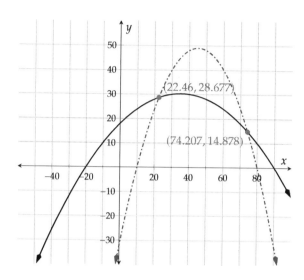

The points of intersection, $(22.46, 28.677)$ and $(74.207, 14.878)$, show where these functions are equal. This means that the x-values give the solutions to the equation $-0.01(x - 90)(x + 20) = -0.04(x - 10)(x - 80)$. So the solutions are approximately 22.46 and 74.207, and the solution set is approximately $\{22.46, 74.207\}$.

Figure 10.3.9: Points of intersection for $h(x) = v(x)$

Similarly, to *graphically* solve an equation like $h(x) = 25$ for

$$h(x) = -0.01(x - 90)(x + 20),$$

we can graph

$$y = -0.01(x - 90)(x + 20)$$ and $$y = 25$$

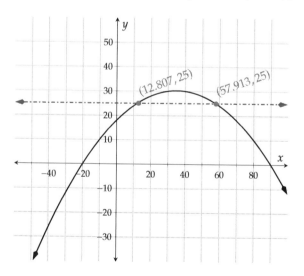

The points of intersection are $(12.807, 25)$ and $(57.913, 25)$, which tells us that the solutions to $h(x) = 25$ are approximately 12.807 and 57.913. The solution set is approximately $\{12.807, 57.913\}$.

Figure 10.3.10: Points of intersection for $h(x) = 25$

Example 10.3.11 Use graphing technology to solve the following inequalities:

a. $-20t^2 - 70t + 300 \geq -5t + 300$

b. $-20t^2 - 70t + 300 < -5t + 300$

Explanation. To solve these inequalities graphically, we will start by graphing the equations $y = -20t^2 - 70t + 300$ and $y = -5t + 300$ and determining the points of intersection:

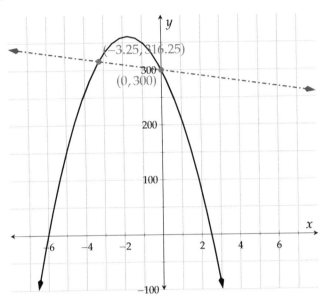

Figure 10.3.12: Points of intersection for $y = -20t^2 - 70t + 300$ and $y = -5t + 300$

To solve $-20t^2 - 70t + 300 \geq -5t + 300$, we need to determine where the y-values of the graph of $y = -20t^2 - 70t + 300$ are *greater* than the y-values of the graph of $y = -5t + 300$ in addition to the values where the y-values are equal. This region is highlighted in Figure 10.3.13.

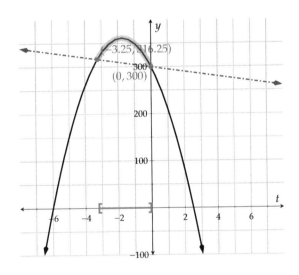

Figure 10.3.13

a. We can see that this region includes all values of t between, and including, $t = -3.25$ and $t = 0$. So the solutions to this inequality include all values of t for which $-3.25 \leq t \leq 0$. We can write this solution set in interval notation as $[-3.25, 0]$ or in set-builder notation as $\{t \mid -3.25 \leq t \leq 0\}$.

b. To now solve $-20t^2 - 70t + 300 < -5t + 300$, we will need to determine where the y-values of the graph of $y = -20t^2 - 70t + 300$ are *less than* the y-values of the graph of $y = -5t + 300$. This region is highlighted in Figure 10.3.14.

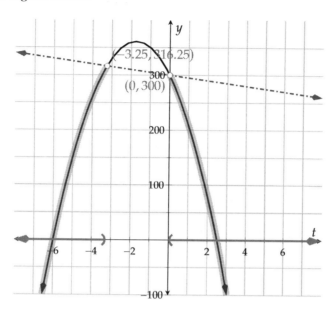

Figure 10.3.14

We can see that $-20t^2 - 70t + 300 < -5t + 300$ for all values of t where $t < -3.25$ or $t > 0$. We can write this solution set in interval notation as $(-\infty, -3.25) \cup (0, \infty)$ or in set-builder notation as $\{t \mid t < -3.25 \text{ or } t > 0\}$.

Exercises

Using Technology to Create a Table of Function Values Use technology to make a table of values for the function.

1. $K(x) = -4x^2 + 15x - 4$

x	$K(x)$

2. $K(x) = -3x^2 + 18x + 3$

x	$K(x)$

3. $f(x) = 2.25x^2 + 70x - 67$

x	$f(x)$
_____	_____
_____	_____
_____	_____
_____	_____
_____	_____
_____	_____
_____	_____

4. $g(x) = -0.5x^2 - 170x + 79$

x	$g(x)$
_____	_____
_____	_____
_____	_____
_____	_____
_____	_____
_____	_____
_____	_____

5. $h(x) = -10x^3 + 10x + 23$

x	$h(x)$
_____	_____
_____	_____
_____	_____
_____	_____
_____	_____
_____	_____
_____	_____

6. $F(x) = 6x^3 + 180x - 33$

x	$F(x)$
_____	_____
_____	_____
_____	_____
_____	_____
_____	_____
_____	_____
_____	_____

Determining Appropriate Windows

7. Let $f(x) = -5943x - 4132$. Choose an appropriate window for graphing f that shows its key features.

The x-interval could be [＿＿＿＿] and the y-interval could be [＿＿＿＿].

8. Let $f(x) = -663x + 767$. Choose an appropriate window for graphing f that shows its key features.

The x-interval could be [＿＿＿＿] and the y-interval could be [＿＿＿＿].

9. Let $f(x) = 772x^2 + 189x - 4162$. Choose an appropriate window for graphing f that shows its key features.

The x-interval could be [＿＿＿＿] and the y-interval could be [＿＿＿＿].

10. Let $f(x) = -882x^2 - 602x + 4033$. Choose an appropriate window for graphing f that shows its key features.

The x-interval could be [＿＿＿＿] and the y-interval could be [＿＿＿＿].

11. Let $f(x) = -0.0005x^2 + 0.001x - 0.41$. Choose an appropriate window for graphing f that shows its key features.

The x-interval could be [＿＿＿＿＿] and the y-interval could be [＿＿＿＿＿].

12. Let $f(x) = 0.00014x^2 + 0.0027x + 0.4$. Choose an appropriate window for graphing f that shows its key features.

The x-interval could be [＿＿＿＿＿] and the y-interval could be [＿＿＿＿＿].

Finding Points of Intersection

13. Use technology to determine how many times the equations $y = (350 - 12x)(-102 - 8x)$ and $y = -6000$ intersect. They intersect   ($\square$ zero times $\square$ one time $\square$ two times $\square$ three times) .

14. Use technology to determine how many times the equations $y = (-234 - 7x)(-380 + 19x)$ and $y = -6000$ intersect. They intersect   ($\square$ zero times $\square$ one time $\square$ two times $\square$ three times) .

15. Use technology to determine how many times the equations $y = -x^3 + x^2 + 3x$ and $y = 7x - 3$ intersect. They intersect   ($\square$ zero times $\square$ one time $\square$ two times $\square$ three times) .

16. Use technology to determine how many times the equations $y = x^3 - 2x^2 - 8x$ and $y = x - 3$ intersect. They intersect   ($\square$ zero times $\square$ one time $\square$ two times $\square$ three times) .

17. Use technology to determine how many times the equations $y = -0.7(6x^2 - 4)$ and $y = -0.11(9x - 4)$ intersect. They intersect   ($\square$ zero times $\square$ one time $\square$ two times $\square$ three times) .

18. Use technology to determine how many times the equations $y = -0.4(7x^2 + 8)$ and $y = 0.19(6x - 9)$ intersect. They intersect   ($\square$ zero times $\square$ one time $\square$ two times $\square$ three times) .

19. Use technology to determine how many times the equations $y = 1.55(x + 5)^2 - 2.7$ and $y = 0.95x - 1$ intersect. They intersect   ($\square$ zero times $\square$ one time $\square$ two times $\square$ three times) .

20. Use technology to determine how many times the equations $y = 2(x - 7)^2 + 4.85$ and $y = -0.15x - 1$ intersect. They intersect   ($\square$ zero times $\square$ one time $\square$ two times $\square$ three times) .

Using Technology to Find Key Features of a Graph

21. For the function j defined by

$$j(x) = -\frac{2}{5}(x-3)^2 + 6,$$

use technology to determine the following. Round answers as necessary.

    a. Any intercepts.

    b. The vertex.

    c. The domain.

    d. The range.

22. For the function k defined by

$$k(x) = 2(x+1)^2 + 10,$$

use technology to determine the following. Round answers as necessary.

    a. Any intercepts.

    b. The vertex.

    c. The domain.

    d. The range.

23. For the function L defined by

$$L(x) = 3000x^2 + 10x + 4,$$

use technology to determine the following. Round answers as necessary.

    a. Any intercepts.

    b. The vertex.

    c. The domain.

    d. The range.

24. For the function M defined by

$$M(x) = -(300x - 2950)^2,$$

use technology to determine the following. Round answers as necessary.

    a. Any intercepts.

    b. The vertex.

    c. The domain.

    d. The range.

25. For the function N defined by

$$N(x) = (300x - 1.05)^2,$$

use technology to determine the following. Round answers as necessary.

    a. Any intercepts.

    b. The vertex.

    c. The domain.

    d. The range.

26. For the function B defined by

$$B(x) = x^2 - 0.05x + 0.0006,$$

use technology to determine the following. Round answers as necessary.

    a. Any intercepts.

    b. The vertex.

    c. The domain.

    d. The range.

Solving Equations and Inequalities Graphically Using Technology

27. Let $s(x) = \frac{1}{5}x^2 - 2x + 10$ and $t(x) = -x + 40$. Use graphing technology to determine the following.

    a. What are the points of intersection for these two functions?

    b. Solve $s(x) = t(x)$.

    c. Solve $s(x) > t(x)$.

    d. Solve $s(x) \leq t(x)$.

28. Let $w(x) = \frac{1}{4}x^2 - 3x - 8$ and $m(x) = x + 12$. Use graphing technology to determine the following.

    a. What are the points of intersection for these two functions?

    b. Solve $w(x) = m(x)$.

    c. Solve $w(x) > m(x)$.

    d. Solve $w(x) \leq m(x)$.

29. Let $f(x) = 4x^2 + 5x - 1$ and $g(x) = 5$. Use graphing technology to determine the following.

    a. What are the points of intersection for these two functions?

    b. Solve $f(x) = g(x)$.

    c. Solve $f(x) < g(x)$.

    d. Solve $f(x) \geq g(x)$.

30. Let $p(x) = 6x^2 - 3x + 4$ and $k(x) = 7$. Use graphing technology to determine the following.

    a. What are the points of intersection for these two functions?

    b. Solve $p(x) = k(x)$.

    c. Solve $p(x) < k(x)$.

    d. Solve $p(x) \geq k(x)$.

31. Let $q(x) = -4x^2 - 24x + 10$ and $r(x) = 2x + 22$. Use graphing technology to determine the following.

    a. What are the points of intersection for these two functions?

    b. Solve $q(x) = r(x)$.

    c. Solve $q(x) > r(x)$.

    d. Solve $q(x) \leq r(x)$.

32. Let $h(x) = -10x^2 - 5x + 3$ and $j(x) = -3x - 9$. Use graphing technology to determine the following.

    a. What are the points of intersection for these two functions?

    b. Solve $h(x) = j(x)$.

    c. Solve $h(x) > j(x)$.

    d. Solve $h(x) \leq j(x)$.

33. Use graphing technology to solve the equation $0.4x^2 + 0.5x - 0.2 = 2.4$. Approximate the solution(s) if necessary.

34. Use graphing technology to solve the equation $-0.25x^2 - 2x + 1.75 = 4.75$. Approximate the solution(s) if necessary.

35. Use graphing technology to solve the equation $(200+5x)(100-2x) = 15000$. Approximate the solution(s) if necessary.

36. Use graphing technology to solve the equation $(200 - 5x)(100 + 10x) = 20000$. Approximate the solution(s) if necessary.

37. Use graphing technology to solve the equation $2x^3 - 5x + 1 = -\frac{1}{2}x + 1$. Approximate the solution(s) if necessary.

38. Use graphing technology to solve the equation $-x^3 + 8x = -4x + 16$. Approximate the solution(s) if necessary.

39. Use graphing technology to solve the equation $-0.05x^2 - 2.03x - 19.6 = 0.05x^2 + 1.97x + 19.4$. Approximate the solution(s) if necessary.

40. Use graphing technology to solve the equation $-0.02x^2 + 1.97x - 51.5 = 0.05(x - 50)^2 - 0.03(x - 50)$. Approximate the solution(s) if necessary.

41. Use graphing technology to solve the equation $-200x^2 + 60x - 55 = -20x - 40$. Approximate the solution(s) if necessary.

42. Use graphing technology to solve the equation $150x^2 - 20x + 50 = 100x + 40$. Approximate the solution(s) if necessary.

43. Use graphing technology to solve the inequality $2x^2 + 5x - 3 > -5$. State the solution set using interval notation, and approximate if necessary.

44. Use graphing technology to solve the inequality $-x^2 + 4x - 7 > -12$. State the solution set using interval notation, and approximate if necessary.

45. Use graphing technology to solve the inequality $10x^2 - 11x + 7 \leq 7$. State the solution set using interval notation, and approximate if necessary.

46. Use graphing technology to solve the inequality $-10x^2 - 15x + 4 \leq 9$. State the solution set using interval notation, and approximate if necessary.

47. Use graphing technology to solve the inequality $-x^2 - 6x + 1 > x + 5$. State the solution set using interval notation, and approximate if necessary.

48. Use graphing technology to solve the inequality $3x^2 + 5x - 4 > -2x + 1$. State the solution set using interval notation, and approximate if necessary.

49. Use graphing technology to solve the inequality $-10x + 4 \leq 20x^2 - 34x + 6$. State the solution set using interval notation, and approximate if necessary.

50. Use graphing technology to solve the inequality $-15x^2 - 6 \leq 10x - 4$. State the solution set using interval notation, and approximate if necessary.

51. Use graphing technology to solve the inequality $\frac{1}{2}x^2 + \frac{3}{2}x \geq \frac{1}{2}x - \frac{3}{2}$. State the solution set using interval notation, and approximate if necessary.

52. Use graphing technology to solve the inequality $\frac{3}{4}x \geq \frac{1}{4}x^2 - 3x$. State the solution set using interval notation, and approximate if necessary.

10.4 Simplifying Expressions with Function Notation

In this section, we will discuss algebra simplification that will appear in many facets of education. Simplification is a skill, like cooking noodles or painting a wall. It may not always be exciting, but it does serve a purpose. Also like cooking noodles or painting a wall, it isn't usually difficult, and yet there are common avoidable mistakes that people make. With practice from this section, you'll have experience to prevent yourself from overcooking the noodles or ruining your paintbrush.

10.4.1 Negative Signs in and out of Function Notation

Let's start by reminding ourselves about the meaning of function notation. When we write $f(x)$, we have a process f that is doing something to an input value x. Whatever is inside those parentheses is the input to the function. What if we use something for input that is not quite as simple as "x?"

Example 10.4.2 Find and simplify a formula for $f(-x)$, where $f(x) = x^2 + 3x - 4$.

Explanation. Those parentheses encase "$-x$," so we are meant to treat "$-x$" as the input. The rule that we have been given for f is

$$f(x) = x^2 + 3x - 4.$$

But the x's that are in this formula are just place-holders. What f does to a number can just as easily be communicated with

$$f(\ ) = (\ )^2 + 3(\ ) - 4.$$

So now that we are meant to treat "$-x$" as the input, we will insert "$-x$" into those slots, after which we can do more familiar algebraic simplification:

$$f(\ ) = (\ )^2 + 3(\ ) - 4$$
$$f(-x) = (-x)^2 + 3(-x) - 4$$
$$= x^2 - 3x - 4$$

The previous example contrasts nicely with this one:

Example 10.4.3 Find and simplify a formula for $-f(x)$, where $f(x) = x^2 + 3x - 4$.

Explanation. Here, the parentheses only encase "x." The negative sign is on the outside. So the way to see this expression is that first f will do what it does to x, and then that result will be negated:

$$-f(x) = -(x^2 + 3x - 4)$$
$$= -x^2 - 3x + 4$$

Note that the answer to this exercise, which was to simplify $-f(x)$, is different from the answer to Example 10.4.2, which was to simplify $f(-x)$. In general you cannot pass a negative sign in and out of function notation and still have the same quantity.

In Example 10.4.2 and Example 10.4.3, we are working with the expressions $f(-x)$ and $-f(x)$, and trying to find "simplified" formulas. If it seems strange to be doing these things, perhaps this applied example will help.

Checkpoint 10.4.4. The NASDAQ Composite Index measures how well a portion of the stock market is doing. Suppose $N(t)$ is the value of the index t days after January 1, 2018. A formula for N is $N(t) = 3.34t^2 + 26.2t + 6980$.

What if you wanted a new function, B, that gives the value of the NASDAQ index t days *before* January 1, 2018? Technically, t days *before* is the same as *negative t days after*. So $B(t)$ is the same as $N(-t)$, and now the expression $N(-t)$ means something. Find a simplified formula for $N(-t)$.

$N(-t) = $ [   ]

Explanation.

$$N(\ \ ) = 3.34(\ \ )^2 + 26.2(\ \ ) + 6980$$
$$N(-t) = 3.34(-t)^2 + 26.2(-t) + 6980$$
$$= 3.34t^2 - 26.2t + 6980$$

10.4.2 Other Nontrivial Simplifications

Example 10.4.5 Find and simplify a formula for $h(5x)$, where $h(x) = \frac{x}{x-2}$.

Explanation. The parentheses encase "$5x$," so we are meant to treat "$5x$" as the input.

$$h(\ \ ) = \frac{(\ \ )}{(\ \ ) - 2}$$
$$h(5x) = \frac{5x}{5x - 2}$$
$$= \frac{5x}{5x - 2}$$

Example 10.4.6 Find and simplify a formula for $\frac{1}{3}g(3x)$, where $g(x) = 2x^2 + 8$.

Explanation. Do the $\frac{1}{3}$ and the 3 cancel each other? No. The 3 is part of the input, affecting x right away. Then g does whatever it does to $3x$, and *then* we multiply the result by $\frac{1}{3}$. Since the function g acts "in between," we don't have the chance to cancel the 3 with the $\frac{1}{3}$. Let's see what actually happens:

Those parentheses encase "$3x$," so we are meant to treat "$3x$" as the input. We will keep the $\frac{1}{3}$ where it is until it is possible to simplify:

$$\frac{1}{3}g(\ \ ) = \frac{1}{3}\left(2(\ \ )^2 + 8\right)$$
$$\frac{1}{3}g(3x) = \frac{1}{3}\left(2(3x)^2 + 8\right)$$
$$= \frac{1}{3}\left(2\left(9x^2\right) + 8\right)$$
$$= \frac{1}{3}\left(18x^2 + 8\right)$$
$$= 6x^2 + \frac{8}{3}$$

Example 10.4.7 If $k(x) = x^2 - 3x$, find and simplify a formula for $k(x - 4)$.

Explanation. This type of exercise is often challenging for algebra students. But let's focus on those parentheses one more time. They encase "$x - 4$," so we are meant to treat "$x - 4$" as the input.

$$k(\quad) = (\quad)^2 - 3(\quad)$$
$$k(x - 4) = (x - 4)^2 - 3(x - 4)$$
$$= x^2 - 8x + 16 - 3x + 12$$
$$= x^2 - 11x + 28$$

Checkpoint 10.4.8. If $q(x) = x + \sqrt{x + 8}$, find and simplify a formula for $q(x + 5)$.

$q(x + 5) = \boxed{}$

Explanation. Starting with the generic formula for q:

$$q(\quad) = (\quad) + \sqrt{(\quad) + 8}$$
$$q(x + 5) = x + 5 + \sqrt{x + 5 + 8}$$
$$= x + 5 + \sqrt{x + 13}$$

Example 10.4.9 If $f(x) = \frac{1}{x}$, find and simplify a formula for $f(x + 3) + 2$.

Explanation. Do not be tempted to add the 3 and the 2. The 3 is added to input *before* the function f does its work. The 2 is added to the result *after* f has done its work.

$$f(\quad) + 2 = \frac{1}{(\quad)} + 2$$
$$f(x + 3) + 2 = \frac{1}{x + 3} + 2$$

This last expression is considered fully simplified. However you might combine the two terms using a technique from Section 13.3.

The tasks we have practiced in this section are the kind of tasks that will make it easier to understand interesting and useful material in college algebra and calculus.

Exercises

Review and Warmup

1. Use the distributive property to write an equivalent expression to $8(p + 5)$ that has no grouping symbols.

2. Use the distributive property to write an equivalent expression to $5(q + 8)$ that has no grouping symbols.

3. Use the distributive property to write an equivalent expression to $-10(y - 6)$ that has no grouping symbols.

4. Use the distributive property to write an equivalent expression to $-5(r + 2)$ that has no grouping symbols.

5. Multiply the polynomials.

$2(y + 4)^2 = \boxed{}$

6. Multiply the polynomials.

$4(y + 10)^2 = \boxed{}$

7. Expand the square of a *binomial*. **8.** Expand the square of a *binomial*.

$(7r + 9)^2 = $ ⬚ $(4r + 3)^2 = $ ⬚

Simplifying Function Expressions

9. Simplify $K(r + 7)$, where $K(r) = 4 + r$.

10. Simplify $G(t + 2)$, where $G(t) = 3 - 4t$.

11. Simplify $g(-t)$, where $g(t) = 3 + 8t$.

12. Simplify $K(-x)$, where $K(x) = 4 + 8x$.

13. Simplify $F(x + 4)$, where $F(x) = 3 - 1.1x$.

14. Simplify $g(y + 8)$, where $g(y) = 2 - 5.5y$.

15. Simplify $H(y - \frac{2}{3})$, where $H(y) = -\frac{8}{3} + \frac{2}{9}y$.

16. Simplify $F(r + \frac{1}{3})$, where $F(r) = -\frac{7}{6} + \frac{2}{5}r$.

17. Simplify $f(r) + 1$, where $f(r) = -3r + 2$.

18. Simplify $H(r) + 5$, where $H(r) = -8r + 2$.

19. Simplify $F(t) + 8$, where $F(t) = 1 + 4.4t$.

20. Simplify $f(t) + 3$, where $f(t) = 1 - 0.1t$.

21. Simplify $H(7x)$, where $H(x) = -5x^2 + x + 8$.

22. Simplify $h(2x)$, where $h(x) = 7x^2 + x - 1$.

23. Simplify $f(-y)$, where $f(y) = y^2 + 3y + 7$.

24. Simplify $G(-y)$, where $G(y) = 8y^2 - 2y - 1$.

25. Simplify $4h(r)$, where $h(r) = -7r^2 + 7r + 8$.

26. Simplify $8f(r)$, where $f(r) = 6r^2 - r - 8$.

27. Simplify $G(r - 6)$, where $G(r) = 0.9r^2 + 7r - 6$.

28. Simplify $h(t + 2)$, where $h(t) = -3.6t^2 - t - 1$.

29. Simplify $K(t) + 2$, where $K(t) = -8t^2 - t + 7$.

30. Simplify $G(x) + 5$, where $G(x) = 4x^2 - x - 1$.

31. Simplify $g(x + 3)$, where $g(x) = \sqrt{-1 - 7x}$.

32. Simplify $h(x + 9)$, where $h(x) = \sqrt{-2 - 2x}$.

33. Simplify $h(x) + 6$, where $h(x) = \sqrt{-2 + 6x}$.

34. Simplify $F(x) + 3$, where $F(x) = \sqrt{-2 + x}$.

35. Simplify $G(x + 8)$, where $G(x) = 8x + \sqrt{-2 - 5x}$. **36.** Simplify $H(x + 5)$, where $H(x) = -2x + \sqrt{-2 - 7x}$.

37. Simplify $g(t + 4)$, where $g(t) = \frac{8}{5t - 3}$.

38. Simplify $HK(t + 8)$, where $HK(t) = -\frac{7}{-3t - 2}$.

39. Simplify $F(-3x)$, where $F(x) = \frac{2x}{-3x^2 + 7}$.

40. Simplify $g(6x)$, where $g(x) = \frac{3x}{-3x^2 - 2}$.

41. Let f be a function given by $f(x) = 4x - 9$. Find and simplify the following:

a. $f(x) + 7 =$ [____]

b. $f(x + 7) =$ [____]

c. $7f(x) =$ [____]

d. $f(7x) =$ [____]

42. Let f be a function given by $f(x) = -5x - 1$. Find and simplify the following:

a. $f(x) + 5 =$ [____]

b. $f(x + 5) =$ [____]

c. $5f(x) =$ [____]

d. $f(5x) =$ [____]

43. Let f be a function given by $f(x) = -4x^2 + 4x$. Find and simplify the following:

a. $f(x) - 5 =$ [____]

b. $f(x - 5) =$ [____]

c. $-5f(x) =$ [____]

d. $f(-5x) =$ [____]

44. Let f be a function given by $f(x) = 4x^2 - 2x$. Find and simplify the following:

a. $f(x) - 2 =$ [____]

b. $f(x - 2) =$ [____]

c. $-2f(x) =$ [____]

d. $f(-2x) =$ [____]

Applications

45. A circular oil slick is expanding with radius, r in feet, at time t in hours given by $r = 18t - 0.3t^2$, for t in hours, $0 \le t \le 10$.

Find a formula for $A = f(t)$, the area of the oil slick as a function of time.

$A = f(t) =$ [____]

46. Suppose $T(t)$ represents the temperature outside, in Fahrenheit, at t hours past noon, and a formula for T is $T(t) = \frac{20t}{t^2+1} + 58$.

If we introduce $F(t)$ as the temperature outside, in Fahrenheit, at t hours past 1:00pm, then $F(t) = T(t + 1)$. Find a simplified formula for $T(t + 1)$.

$T(t + 1) =$ [____]

47. Suppose $G(t)$ represents how many gigabytes of data has been downloaded t minutes after you started a download.

If we introduce $M(t)$ as how many megabytes of data has been downloaded t minutes after you started a download, then $M(t) = 1024G(t)$. Find a simplified formula for $1024G(t)$.

$1024G(t) =$ [____]

10.5 Technical Definition of a Function

In Section 10.1, we discussed a conceptual understanding of functions and Definition 10.1.3. In this section we'll start with a more technical definition of what is a function, consistent with the ideas from Section 10.1.

10.5.1 Formally Defining a Function

Definition 10.5.2 Function (Technical Definition). A **function** is a collection of ordered pairs (x, y) such that any particular value of x is paired with at most one value for y.

How is this definition consistent with the informal Definition 10.1.3, which describes a function as a *process*? Well, if you have a collection of ordered pairs (x, y), you can choose to view the left number as an input, and the right value as the output. If the function's name is f and you want to find $f(x)$ for a particular number x, look in the collection of ordered pairs to see if x appears among the first coordinates. If it does, then $f(x)$ is the (unique) y-value it was paired with. If it does not, then that x is just not in the domain of f, because you have no way to determine what $f(x)$ would be.

> **Example 10.5.3** Using Definition 10.5.1, a function f could be given by
>
> $$\{(1,4), (2,3), (5,3), (6,1)\}.$$
>
> a. What is $f(1)$? Since the ordered pair $(1,4)$ appears in the collection of ordered pairs, we would say that $f(1) = 4$.
>
> b. What is $f(2)$? Since the ordered pair $(2,3)$ appears in the collection of ordered pairs, we would say that $f(2) = 3$.
>
> c. What is $f(3)$? None of the ordered pairs in the collection start with 3, so $f(3)$ is undefined, and we would say that 3 is not in the domain of f.

Let's spend some time seeing how this new definition applies to things that we already understand as functions from Section 10.1.

x	$g(x)$
12	0.16
15	3.2
18	1.4
21	1.4
24	0.98

Consider the function g expressed by Table 10.5.5. How is this "a collection of ordered pairs?" With tables the connection is most easily apparent. Pair off each x-value with its y-value.

Table 10.5.5

> **Example 10.5.4 A Function Given as a Table.** In this case, we can view this function as:
>
> $$\{(12, 0.16), (15, 3.2), (18, 3.2), (21, 1.4), (24, 0.98)\}.$$

Example 10.5.6 A Function Given as a Formula. Consider the function h expressed by the formula $h(x) = x^2$. How is this "a collection of ordered pairs?"

This time, the collection is *really big*. Imagine an x-value, like $x = 2$. We can calculate that $f(2) = 2^2 = 4$. So the input 2 pairs with the output 4 and the ordered pair $(2, 4)$ is part of the collection.

You could move on to *any* x-value, like say $x = 2.1$. We can calculate that $f(2.1) = 2.1^2 = 4.41$. So the input 2.1 pairs with the output 4.41 and the ordered pair $(2.1, 4.41)$ is part of the collection.

The collection is so large that we cannot literally list all the ordered pairs as was done in Example 10.5.3 and Example 10.5.4. We just have to imagine this giant collection of ordered pairs. And if it helps to conceptualize it, we know that the ordered pairs $(2, 4)$ and $(2.1, 4.41)$ are included.

Example 10.5.7 A Function Given as a Graph. Consider the functions p and q expressed in Figure 10.5.8 and Figure 10.5.9. How is each of these "a collection of ordered pairs?"

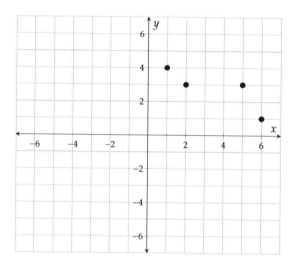

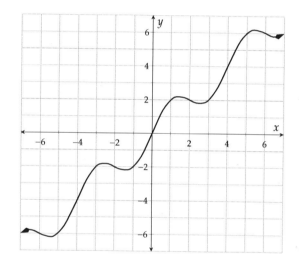

Figure 10.5.8: $y = p(x)$ **Figure 10.5.9:** $y = q(x)$

In Figure 10.5.8, we see that $p(1) = 4$, $p(2) = 3$, $p(5) = 3$, and $p(6) = 1$. The graph *literally is* the collection

$$\{(1, 4), (2, 3), (5, 3), (6, 1)\}.$$

In Figure 10.5.9, we can see a few whole number function values, like $q(0) = 0$ and $q(1) = 2$. But the entire curve has infinitely many points on it and we'd never be able to list them all. We just have to imagine the giant collection of ordered pairs. And if it helps to conceptualize it, we know that the ordered pairs $(0, 0)$ and $(1, 2)$ are included.

Try it yourself in the following exercise.

Checkpoint 10.5.10. The graph below is of $y = f(x)$.

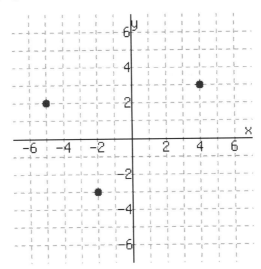

Write the function f as a set of ordered pairs.

Explanation. The function can be expressed as the set $\{(-5, 2), (-2, -3), (4, 3)\}$.

10.5.2 Identifying What is *Not* a Function

Just because you have a set of order pairs, a table, a graph, or an equation, it does not necessarily mean that you have a function. Conceptually, whatever you have needs to give consistent outputs if you feed it the same input. More technically, the set of ordered pairs is not allowed to have two ordered pairs that have the same x-value but different y-values.

Example 10.5.11 Consider each set of ordered pairs. Does it make a function?

a. $\left\{(5, 9), (3, 2), \left(\frac{1}{2}, 0.6\right), (5, 1)\right\}$

b. $\left\{(-5, 12), (3, 7), \left(\sqrt{2}, 1\right), (-0.9, 4)\right\}$

c. $\left\{(5, 9), (3, 9), \left(4.2, \sqrt{2}\right), \left(\frac{4}{3}, \frac{1}{2}\right)\right\}$

d. $\left\{(5, 9), (0.7, 2), \left(\sqrt{25}, 3\right), \left(\frac{2}{3}, \frac{3}{2}\right)\right\}$

Explanation.

a. This set of ordered pairs is *not* a function. The problem is that it has both $(5, 9)$ and $(5, 1)$. It uses the same x-value paired with two different y-values. We have no clear way to turn the input 5 into an output.

b. This set of ordered pairs *is* a function. It is a collection of ordered pairs, and the x-values are never reused.

c. This set of ordered pairs *is* a function. It is a collection of ordered pairs, and the x-values are never reused. You might note that the *output* value 9 appears twice, but that doesn't matter. That just tells us that the function turns 5 into 9 and it also turns 3 into 9.

d. This set of ordered pairs is *not* a function, but it's a little tricky. One of the ordered pairs uses $\sqrt{25}$

as an input value. But that is the same as 5, which is also used as an input value.

Now that we understand how some sets of ordered pairs might not be functions, what about tables, graphs, and equations? If we are handed one of these things, can we tell whether or not it is giving us a function?

Checkpoint 10.5.12 Does This Table Make a Function? Which of these tables make y a function of x?

x	y
2	1
3	1
4	2
5	2
6	2

x	y
8	3
9	2
5	1
2	0
8	1

x	y
5	9
5	9
6	2
6	2
6	2

a. This table (□ does □ does not) make y a function of x.

b. This table (□ does □ does not) make y a function of x.

c. This table (□ does □ does not) make y a function of x.

Explanation.

a. This table does make y a function of x. In the table, no x-value is repeated.

b. This table does not make y a function of x. In the table, the x-value 8 is repeated, and it is paired with two different y-values, 3 and 1.

c. This table does make y a function of x, but you have to think carefully. It's true that the x-value 5 is used more than once in the table. But in both places, the y-value is the same, 9. So there is no conceptual issue with asking for $f(5)$; it would definitely be 9. Similarly, the repeated use of 6 as an x-value is not a problem since it is always paired with output 2.

Example 10.5.13 Does This Graph Make a Function? Which of these graphs make y a function of x?

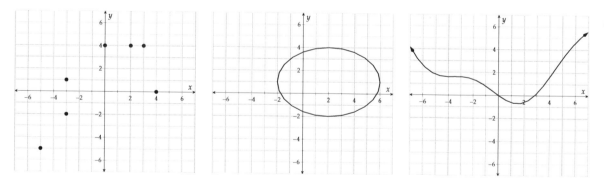

Figure 10.5.14   Figure 10.5.15   Figure 10.5.16

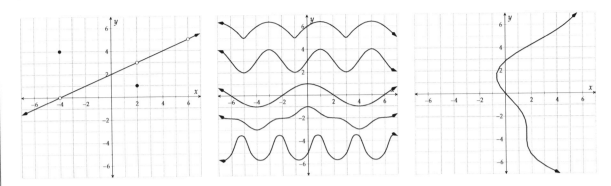

Figure 10.5.17   **Figure 10.5.18**   **Figure 10.5.19**

Explanation. The graph in Figure 10.5.14 does *not* make y a function of x. Two ordered pairs on that graph are $(-3, 1)$ and $(-3, -2)$, so an input value is used twice with different output values.

The graph in Figure 10.5.15 does *not* make y a function of x. There are many ordered pairs with the same input value but different output values. For example, $(2, -2)$ and $(2, 4)$.

The graph in Figure 10.5.16 *does* make y a function of x. It appears that no matter what x-value you choose on the x-axis, there is exactly one y-value paired up with it on the graph.

The graph in Figure 10.5.17 *does* make y a function of x, but we should discuss. The hollow dots on the line indicate that the line goes right up to that point, but never reaches it. We say there is a "hole" in the graph at these places. For two of these holes, there is a separate ordered pair immediately above or below the hole. The graph has the ordered pair $(-4, 4)$. It *also* has ordered pairs like (very close to -4, very close to 0), but it does not have $(-4, 0)$. Overall, there is no x-value that is used twice with different y-values, so this graph does make y a function of x

The graph in Figure 10.5.15 does *not* make y a function of x. There are many ordered pairs with the same input value but different output values. For example, $(0, 1)$, $(0, 3)$, $(0, -1)$, $(0, 5)$, and $(0, -6)$ all use $x = 0$.

The graph in Figure 10.5.15 does *not* make y a function of x. There are many ordered pairs with the same input value but different output values. For example at $x = 2$, there is both a positive and a negative associated y-value. It's hard to say exactly what these y-values are, but we don't have to.

This last set of examples might reveal something to you. For instance in Figure 10.5.15, the issue is that there are places on the graph with the same x-value, but different y-values. Visually, what that means is there are places on the graph that are directly above/below each other. Thinking about this leads to a quick visual "test" to determine if a graph gives y as a function of x.

Fact 10.5.20 Vertical Line Test. *Given a graph in the xy-plane, if a vertical line ever touches it in more than one place, the graph does* not *give y as a function of x. If vertical lines only ever touch the graph once or never at all, then the graph* does *give y as a function of x.*

Example 10.5.21 In each graph from Example 10.5.13, we can apply the Vertical Line Test.

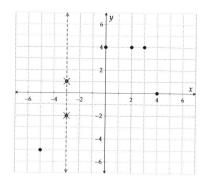

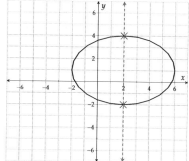

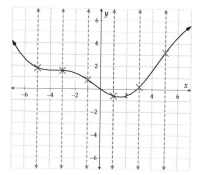

Figure 10.5.22: A vertical line touching the graph twice makes this graph not give y as a function of x.

Figure 10.5.23: A vertical line touching the graph twice makes this graph not give y as a function of x.

Figure 10.5.24: All vertical lines only touch the graph once, so this graph does give y as a function of x.

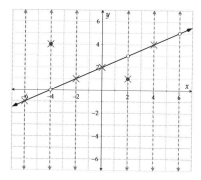

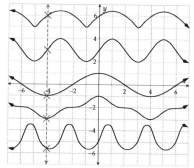

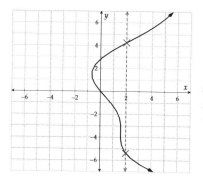

Figure 10.5.25: All vertical lines only touch the graph once, or not at all, so this graph does give y as a function of x.

Figure 10.5.26: A vertical line touching the graph more than once makes this graph not give y as a function of x.

Figure 10.5.27: A vertical line touching the graph more than once makes this graph not give y as a function of x.

Lastly, we come to equations. Certain equations with variables x and y clearly make y a function of x. For example, $y = x^2 + 1$ says that if you have an x-value, all you have to do is substitute it into that equation and you will have determined an output y-value. You could then name the function f and give a formula for it: $f(x) = x^2 + 1$.

With other equations, it may not be immediately clear whether or not they make y a function of x.

Example 10.5.28 Do each of these equations make y a function of x?

    a. $2x + 3y = 5$ b. $y = \pm\sqrt{x + 4}$ c. $x^2 + y^2 = 9$

Explanation.

    a. The equation $2x + 3y = 5$ *does* make y a function of x. Here are three possible explanations.

        i. You recognize that the graph of this equation would be a non-vertical line, and so it would pass the Vertical Line Test.

ii. Imagine that you have a specific value for x and you substitute it in to $2x + 3y = 5$. Will you be able to use algebra to solve for y? All you will need is to simplify, subtract from both sides, and divide on both sides, so you will be able to determine y.

iii. Can you just isolate y in terms of x? Yes, a few steps of algebra can turn $2x + 3y = 5$ into $y = \frac{5-2x}{3}$. Now you have an explicit formula for y in terms of x, so y is a function of x.

b. The equation $y = \pm\sqrt{x+4}$ does *not* make y a function of x. Just having the $\pm$ (plus *or* minus) in the equation immediately tells you that for almost any valid x-value, there would be *two* associated y-values.

c. The equation $x^2 + y^2 = 9$ does *not* make y a function of x. Here are three possible explanations.

i. Imagine that you have a specific value for x and you substitute it in to $x^2 + y^2 = 9$. Will you be able to use algebra to solve for y? For example, if you substitute in $x = 1$, then you have $1 + y^2 = 9$, which simplifies to $y^2 = 8$. Can you really determine what y is? No, because it could be $\sqrt{8}$ or it could be $-\sqrt{8}$. So this equation does not provide you with a way to turn x-values into y-values.

ii. Can you just isolate y in terms of x? You might get started and use algebra to convert $x^2 + y^2 = 9$ into $y^2 = 9 - x^2$. But what now? The best you can do is acknowledge that y is either the positive or the negative square root of $9 - x^2$. You might write $y = \pm\sqrt{9 - x^2}$. But now for almost any valid x-value, there are *two* associated y-values.

iii. You recognize that the graph of this equation would be a circle with radius 3, and so it would not pass the Vertical Line Test.

Checkpoint 10.5.29. Do each of these equations make y a function of x?

a. $5x^2 - 4y = 12$

b. $5x - 4y^2 = 12$

c. $x = \sqrt{y}$

This equation ($\square$ does $\square$ does not) make y a function of x.

This equation ($\square$ does $\square$ does not) make y a function of x.

This equation ($\square$ does $\square$ does not) make y a function of x.

Explanation.

a. The equation $5x^2 - 4y = 12$ *does* make y a function of x. You can isolate y in terms of x. A few steps of algebra can turn $5x^2 - 4y = 12$ into $y = \frac{5x^2 - 12}{4}$. Now you have an explicit formula for y in terms of x, so y is a function of x.

b. The equation $5x - 4y^2 = 12$ does *not* make y a function of x. You cannot isolate y in terms of x. You might get started and use algebra to convert $5x - 4y^2 = 12$ into $y^2 = \frac{5x - 12}{4}$. But what now? The best you can do is acknowledge that y is either the positive or the negative square root of $\frac{5x - 12}{4}$. You might write $y = \pm\sqrt{\frac{5x - 12}{4}}$. But now for almost any valid x-value, there are *two* associated y-values.

c. The equation $x = \sqrt{y}$ *does* make y a function of x. If you try substituting a non-negative x-value, then you can square both sides and you know exactly what the value of y is.

If you try substituting a negative x-value, then you are saying that $\sqrt{y}$ is negative which is impossible. So for negative x, there are no y-values. This is not a problem for the equation giving you a function. This just means that the domain of that function does not include negative numbers. Its domain would be $[0, \infty)$.

Exercises

Determining If Sets of Ordered Pairs Are Functions

1. Do these sets of ordered pairs make functions of x? What are their domains and ranges?

    a. $\left\{(-6,8),(2,9)\right\}$

    This set of ordered pairs ($\square$ describes $\square$ does not describe) a function of x. This set of ordered pairs has domain [   ] and range [   ].

    b. $\left\{(-8,8),(-8,3),(4,4)\right\}$

    This set of ordered pairs ($\square$ describes $\square$ does not describe) a function of x. This set of ordered pairs has domain [   ] and range [   ].

    c. $\left\{(9,7),(-6,7),(6,5),(2,4)\right\}$

    This set of ordered pairs ($\square$ describes $\square$ does not describe) a function of x. This set of ordered pairs has domain [   ] and range [   ].

    d. $\left\{(2,0),(-5,1),(-7,4),(3,1),(-3,1)\right\}$

    This set of ordered pairs ($\square$ describes $\square$ does not describe) a function of x. This set of ordered pairs has domain [   ] and range [   ].

2. Do these sets of ordered pairs make functions of x? What are their domains and ranges?

    a. $\left\{(-5,8),(-8,5)\right\}$

    This set of ordered pairs ($\square$ describes $\square$ does not describe) a function of x. This set of ordered pairs has domain [   ] and range [   ].

    b. $\left\{(-7,6),(2,9),(10,6)\right\}$

    This set of ordered pairs ($\square$ describes $\square$ does not describe) a function of x. This set of ordered pairs has domain [   ] and range [   ].

    c. $\left\{(0,2),(5,9),(-1,7),(5,5)\right\}$

    This set of ordered pairs ($\square$ describes $\square$ does not describe) a function of x. This set of ordered pairs has domain [   ] and range [   ].

    d. $\left\{(7,6),(-9,5),(-6,2),(3,0),(-10,2)\right\}$

    This set of ordered pairs ($\square$ describes $\square$ does not describe) a function of x. This set of ordered pairs has domain [   ] and range [   ].

3. Does the following set of ordered pairs make for a function of x?

$$\left\{(0,4),(-1,1),(-5,0),(-3,1),(-1,9)\right\}$$

This set of ordered pairs   (□ describes   □ does not describe)   a function of x. This set of ordered

pairs has domain [   ] and range [   ].

4. Does the following set of ordered pairs make for a function of x?

$$\left\{(-5,10),(9,8),(-3,10),(-5,8),(-3,5)\right\}$$

This set of ordered pairs   (□ describes   □ does not describe)   a function of x. This set of ordered

pairs has domain [   ] and range [   ].

Domain and Range

5. Below is all of the information that exists about a function H.

$H(3) = 4$ $H(5) = -2$ $H(8) = 4$

Write H as a set of ordered pairs.

H has domain [   ]

and range [   ].

6. Below is all of the information about a function K.

$K(a) = 3$ $K(b) = 1$
$K(c) = 0$ $K(d) = 3$

Write K as a set of ordered pairs.

K has domain [   ]

and range [   ].

Determining If Graphs Are Functions

7. Decide whether each graph shows a relationship where y is a function of x.

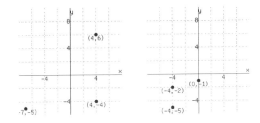

The first graph   (□ does   □ does not)   give a function of x. The second graph   (□ does   □ does not)   give a function of x.

8. Decide whether each graph shows a relationship where y is a function of x.

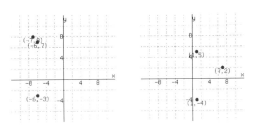

The first graph   (□ does   □ does not)   give a function of x. The second graph   (□ does   □ does not)   give a function of x.

9. The following graphs show two relationships. Decide whether each graph shows a relationship where y is a function of x.

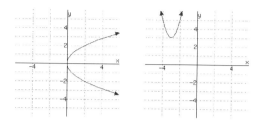

The first graph (☐ does ☐ does not) give a function of x. The second graph (☐ does ☐ does not) give a function of x.

10. The following graphs show two relationships. Decide whether each graph shows a relationship where y is a function of x.

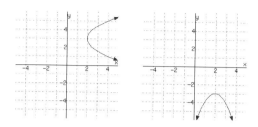

The first graph (☐ does ☐ does not) give a function of x. The second graph (☐ does ☐ does not) give a function of x.

Determining If Tables Are Functions Determine whether or not the following table could be the table of values of a function. If the table can not be the table of values of a function, give an input that has more than one possible output.

11.

Input	Output
2	0
4	14
6	−5
8	−15
−2	−19

Could this be the table of values for a function? (☐ yes ☐ no)

If not, which input has more than one possible output? (☐ -2 ☐ 2 ☐ 4 ☐ 6 ☐ 8 ☐ None, the table represents a function.)

12.

Input	Output
2	5
4	0
6	12
8	15
−2	−20

Could this be the table of values for a function? (☐ yes ☐ no)

If not, which input has more than one possible output? (☐ -2 ☐ 2 ☐ 4 ☐ 6 ☐ 8 ☐ None, the table represents a function.)

13.

Input	Output
−4	13
−3	−4
−2	5
−3	13
−1	−19

Could this be the table of values for a function? (☐ yes ☐ no)

If not, which input has more than one possible output? (☐ -4 ☐ -3 ☐ -2 ☐ -1 ☐ None, the table represents a function.)

14.

Input	Output
−4	−7
−3	−2
−2	−19
−3	19
−1	0

Could this be the table of values for a function? (☐ yes ☐ no)

If not, which input has more than one possible output? (☐ -4 ☐ -3 ☐ -2 ☐ -1 ☐ None, the table represents a function.)

Determining If Equations Are Functions

15. Select all of the following relations that make y a function of x. There are several correct answers.

   □ $x^2 + y^2 = 81$ □ $y = \sqrt[6]{x}$ □ $y = x^3$

   □ $y = \pm\sqrt{1 - x^2}$ □ $|y| = x$ □ $4x + 3y = 1$

   □ $y = \frac{1}{x^2}$ □ $y = \sqrt{1 - x^2}$ □ $x = y^9$

   □ $y = \frac{x+7}{8-x}$ □ $x = y^8$ □ $y = |x|$

16. Select all of the following relations that make y a function of x. There are several correct answers.

   □ $y = \sqrt[4]{x}$ □ $x = y^9$ □ $x^2 + y^2 = 16$

   □ $y = x^2$ □ $|y| = x$ □ $y = \pm\sqrt{64 - x^2}$

   □ $y = \frac{x+2}{4-x}$ □ $4x + 5y = 1$ □ $y = \frac{1}{x^3}$

   □ $y = |x|$ □ $y = \sqrt{64 - x^2}$ □ $x = y^8$

17. Some equations involving x and y define y as a function of x, and others do not. For example, if $x + y = 1$, we can solve for y and obtain $y = 1 - x$. And we can then think of $y = f(x) = 1 - x$. On the other hand, if we have the equation $x = y^2$ then y is not a function of x, since for a given positive value of x, the value of y could equal $\sqrt{x}$ or it could equal $-\sqrt{x}$.

Select all of the following relations that make y a function of x. There are several correct answers.

   □ $y^2 + x^2 = 1$ □ $y - |x| = 0$ □ $y^6 + x = 1$

   □ $x + y = 1$ □ $3x + 8y + 8 = 0$ □ $y + x^2 = 1$

   □ $y^3 + x^4 = 1$ □ $|y| - x = 0$

On the other hand, some equations involving x and y define x as a function of y (the other way round).

Select all of the following relations that make x a function of y. There are several correct answers.

   □ $|y| - x = 0$ □ $y^4 + x^5 = 1$ □ $y - |x| = 0$

   □ $3x + 8y + 8 = 0$ □ $y^2 + x^2 = 1$

18. Some equations involving x and y define y as a function of x, and others do not. For example, if $x + y = 1$, we can solve for y and obtain $y = 1 - x$. And we can then think of $y = f(x) = 1 - x$. On the other hand, if we have the equation $x = y^2$ then y is not a function of x, since for a given positive value of x, the value of y could equal $\sqrt{x}$ or it could equal $-\sqrt{x}$.

Select all of the following relations that make y a function of x. There are several correct answers.

   □ $4x + 5y + 4 = 0$ □ $y + x^2 = 1$ □ $|y| - x = 0$

   □ $y - |x| = 0$ □ $y^2 + x^2 = 1$ □ $x + y = 1$

   □ $y^3 + x^4 = 1$ □ $y^6 + x = 1$

On the other hand, some equations involving x and y define x as a function of y (the other way round).

Select all of the following relations that make x a function of y. There are several correct answers.

   □ $4x + 5y + 4 = 0$ □ $y - |x| = 0$ □ $y^2 + x^2 = 1$

   □ $|y| - x = 0$ □ $y^4 + x^5 = 1$

10.6 Functions and Their Representations Chapter Review

10.6.1 Function Basics

In Section 10.1 we defined functions 10.1.3 informally, as well as function notation 10.1.6. We saw functions in four forms 10.1.31: verbal descriptions, formulas, graphs and tables.

> **Example 10.6.1 Informal Definition of a Function.** Determine whether each example below describes a function.
>
>   a. The area of a circle given its radius.   b. The number you square to get 9.
>
> **Explanation.**
>
>   a. The area of a circle given its radius is a function because there is a set of steps or a formula that changes the radius into the area of the circle. We could write $A(r) = \pi r^2$.
>
>   b. The number you square to get 9 is not a function because the process we would apply to get the result does not give a single answer. There are two different answers, -3 and 3. A function must give a single output for a given input.

> **Example 10.6.2 Tables and Graphs.** Make a table and a graph of the function f, where $f(x) = x^2$.
>
> **Explanation.**
>
> First we will set up a table with negative and positive inputs and calculate the function values. The values are shown in Table 10.6.3, which in turn gives us the graph in Figure 10.6.4.
>
> | input, x | output, $f(x)$ |
> |:---:|:---:|
> | -3 | 9 |
> | -2 | 4 |
> | -1 | 1 |
> | 0 | 0 |
> | 1 | 2 |
> | 2 | 4 |
> | 3 | 9 |
>
>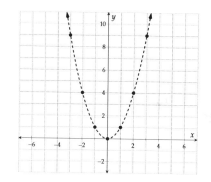
>
> **Table 10.6.3**   **Figure 10.6.4:** $y = f(x) = x^2$

> **Example 10.6.5 Translating between Four Descriptions of the Same Function.** Consider a function f that triples its input and then adds 4. Translate this verbal description of f into a table, a graph, and a formula.

Explanation.

To make a table for f, we'll have to select some input x-values so we will choose some small negative and positive values that are easy to work with. Given the verbal description, we should be able to compute a column of output values. Table 10.6.6 is one possible table that we might end up with.

x	$f(x)$
-2	$3(-2) + 4 = -2$
-1	$3(-1) + 4 = 1$
0	$3(0) + 4 = 4$
1	$3(1) + 4 = 7$
2	$3(2) + 4 = 10$

Table 10.6.6

Once we have a table for f, we can make a graph for f as in Figure 10.6.7, using the table to plot points.

Lastly, we must find a formula for f. This means we need to write an algebraic expression that says the same thing about f as the verbal description, the table, and the graph. For this example, we can focus on the verbal description. Since f takes its input, triples it, and adds 4, we have the formula

$$f(x) = 3x + 4.$$

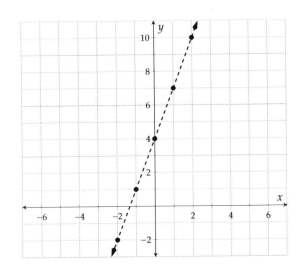

Figure 10.6.7: $y = f(x)$

10.6.2 Domain and Range

In Section 10.2 we saw the definition of domain 10.2.6 and range 10.2.19, and three types of domain restrictions 10.2.18. We also learned how to write the domain and range in interval and set-builder notation.

Example 10.6.8 Domain. Determine the domain of p, where $p(x) = \dfrac{x}{2x - 1}$.

Explanation. This is an example of the first type of domain restriction, when you have a variable in the denominator. The denominator cannot equal 0 so a *bad* value for x would be when

$$2x - 1 = 0$$
$$2x = 1$$
$$x = \frac{1}{2}$$

The domain is all real numbers except $\frac{1}{2}$.

Example 10.6.9 Interval, Set, and Set-Builder Notation. What is the domain of the function C, where $C(x) = \sqrt{2x - 3} - 5$?

Explanation. This is an example of the second type of domain restriction where the value inside the radical cannot be negative. So the *good* values for x would be when

$$2x - 3 \geq 0$$
$$2x \geq 3$$
$$x \geq \frac{3}{2}$$

So on a number line, if we wanted to picture the domain of C, we would make a sketch like:

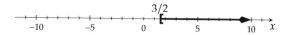

The domain is the interval $\left[\frac{3}{2}, \infty\right)$.

Find the range of the function q using its graph shown in Figure 10.6.11.

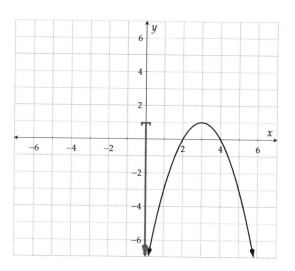

Figure 10.6.11: $y = q(x)$. The range is marked as an interval on the y-axis.

Example 10.6.10 Range. Explanation. The range is the collection of possible numbers that q can give for output. Figure 10.6.11 displays a graph of q, with the range shown as an interval on the y-axis.

The output values are the y-coordinates so we can see that the y-values start from 1 and continue downward forever. Therefore the range is $(-\infty, 1]$.

10.6.3 Using Technology to Explore Functions

In Section 10.3 we covered how to find a good graphing window and use it to identify all of the key features of a function. We also learned how to solve equations and inequalities using a graph. Here are some examples for review.

Example 10.6.12 Finding an Appropriate Window. Graph the function t, where $t(x) = (x + 10)^2 - 15$, using technology and find a good viewing window.

Explanation.

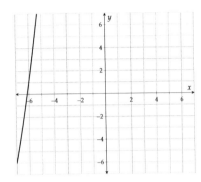

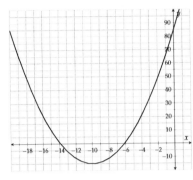

After some trial and error we found this window that goes from -20 to 2 on the x-axis and -20 to 100 on the y-axis.

Figure 10.6.13: $y = t(x)$ in the viewing window of -7 to 7 on the x and y axes. We need to zoom out and move our window to the left.

Figure 10.6.14: $y = t(x)$ in a good viewing window.

Now we can see the vertex and all of the intercepts and we will identify them in the next example.

Example 10.6.15 Using Technology to Determine Key Features of a Graph. Use the previous graph in figure 10.6.14 to identify the intercepts, minimum or maximum function value, and the domain and range of the function t, where $t(x) = (x + 10)^2 - 15$.

Explanation.

From our graph we can now identify the vertex at $(-10, -15)$, the y-intercept at $(0, 85)$, and the x-intercepts at approximately $(-13.9, 0)$ and $(-6.13, 0)$.

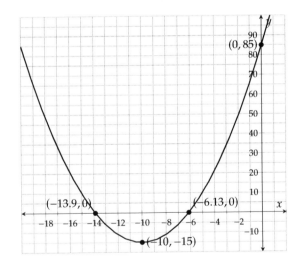

Figure 10.6.16: $y = t(x) = (x + 10)^2 - 15$.

Example 10.6.17 Solving Equations and Inequalities Graphically Using Technology. Use graphing technology to solve the equation $t(x) = 40$, where $t(x) = (x + 10)^2 - 15$.

Explanation.

To solve the equation $t(x) = 40$, we need to graph $y = t(x)$ and $y = 40$ on the same axes and find the x-values where they intersect.

From the graph we can see that the intersection points are approximately $(-17.4, 40)$ and $(-2.58, 40)$. The solution set is $\{-17.4, -2.58\}$.

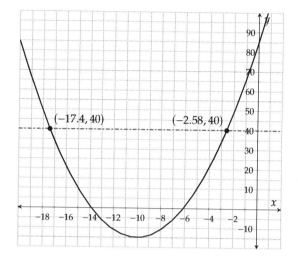

Figure 10.6.18: $y = t(x)$ where $t(x) = (x + 10)^2 - 15$ and $y = 40$.

10.6.4 Simplifying Expressions with Function Notation

In Section 10.4 we learned about the difference between $f(-x)$ and $-f(x)$ and how to simplify them. We also learned how to simplify other changes to the input and output like $f(3x)$ and $\frac{1}{3}f(x)$. Here are some examples.

Example 10.6.19 Negative Signs in and out of Function Notation. Find and simplify a formula for $f(-x)$ and $-f(x)$, where $f(x) = -3x^2 - 7x + 1$.

Explanation. To find $f(-x)$, we use an input of $-x$ in our function f and simplify to get:

$$f(-x) = -3(-x)^2 - 7(-x) + 1$$
$$= -3x^2 + 7x + 1$$

To find $-f(x)$, we take the opposite of the function f and simplify to get:

$$-f(x) = -(-3x^2 - 7x + 1)$$
$$= 3x^2 + 7x - 1$$

Example 10.6.20 Other Nontrivial Simplifications. If $g(x) = 2x^2 - 3x - 5$, find and simplify a formula for $g(x - 1)$.

Explanation. To find $g(x - 1)$, we put in $(x - 1)$ for the input. It is important to keep the parentheses because we have exponents and negative signs in the function.

$$g(x - 1) = 2(x - 1)^2 - 3(x - 1) - 5$$
$$= 2(x^2 - 2x + 1) - 3x + 3 - 5$$
$$= 2x^2 - 4x + 2 - 3x - 2$$
$$= 2x^2 - 7x$$

10.6.5 Technical Definition of a Function

In Section 10.5 we gave a formal definition of a function 10.5.2 and learned to identify what is and is not a function with sets or ordered pairs, tables and graphs. We also used the vertical line test 10.5.20.

Example 10.6.21 Formally Defining a Function. We learned that sets of ordered pairs, tables and graphs can meet the formal definition of a function. Here is an example that shows a function in all three forms. We can verify that each input has at most one output.

$\{(1, 4), (2, 4), (3, 3), (4, 6), (5, -2)\}$

x	$f(x)$
1	4
2	4
3	3
4	6
5	-2

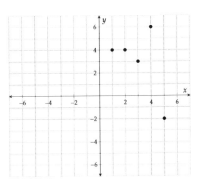

Figure 10.6.22: The function f represented as a collection of ordered pairs.

Table 10.6.23: The function f represented as a table.

Figure 10.6.24: The function f represented as a graph.

Example 10.6.25 Identifying What is *Not* a Function. Identify whether each graph represents a function using the vertical line test 10.5.20.

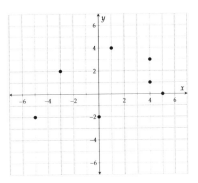

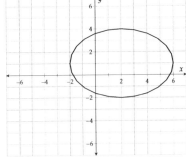

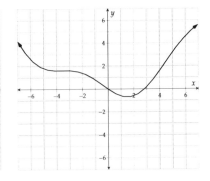

Figure 10.6.26

Figure 10.6.27

Figure 10.6.28

Explanation.

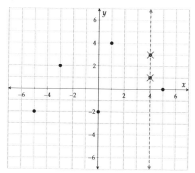
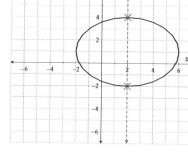
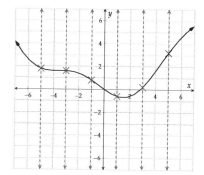

Figure 10.6.29: A vertical line touching the graph twice makes this graph not give y as a function of x.

Figure 10.6.30: A vertical line touching the graph twice makes this graph not give y as a function of x.

Figure 10.6.31: All vertical lines only touch the graph once, so this graph does give y as a function of x.

Exercises

Function Basics

1. Samantha will spend $240 to purchase some bowls and some plates. Each plate costs $1, and each bowl costs $6. The function $q(x) = -\frac{1}{6}x + 40$ models the number of bowls Samantha will purchase, where x represents the number of plates to be purchased.

Interpret the meaning of $q(36) = 34$.

⊙ A. 34 plates and 36 bowls can be purchased.

⊙ B. 36 plates and 34 bowls can be purchased.

⊙ C. $34 will be used to purchase bowls, and $36 will be used to purchase plates.

⊙ D. $36 will be used to purchase bowls, and $34 will be used to purchase plates.

2. Fabrienne will spend $140 to purchase some bowls and some plates. Each plate costs $8, and each bowl costs $7. The function $q(x) = -\frac{8}{7}x + 20$ models the number of bowls Fabrienne will purchase, where x represents the number of plates to be purchased.

Interpret the meaning of $q(14) = 4$.

⊙ A. 4 plates and 14 bowls can be purchased.

⊙ B. $4 will be used to purchase bowls, and $14 will be used to purchase plates.

⊙ C. $14 will be used to purchase bowls, and $4 will be used to purchase plates.

⊙ D. 14 plates and 4 bowls can be purchased.

The following figure has the graph $y = d(t)$, which models a particle's distance from the starting line in feet, where t stands for time in seconds since timing started.

3.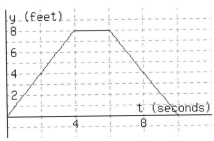

4.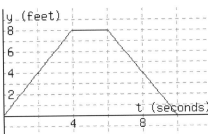

a. $d(8) =$ ⬚

b. Interpret the meaning of $d(8)$:

⊙ *A.* The particle was 8 feet away from the starting line 4 seconds since timing started.

⊙ *B.* In the first 4 seconds, the particle moved a total of 8 feet.

⊙ *C.* The particle was 4 feet away from the starting line 8 seconds since timing started.

⊙ *D.* In the first 8 seconds, the particle moved a total of 4 feet.

c. Solve $d(t) = 6$ for t. $t =$ ⬚

d. Interpret the meaning of part c's solution(s):

⊙ *A.* The article was 6 feet from the starting line 3 seconds since timing started.

⊙ *B.* The article was 6 feet from the starting line 7 seconds since timing started.

⊙ *C.* The article was 6 feet from the starting line 3 seconds since timing started, or 7 seconds since timing started.

⊙ *D.* The article was 6 feet from the starting line 3 seconds since timing started, and again 7 seconds since timing started.

a. $d(4) =$ ⬚

b. Interpret the meaning of $d(4)$:

⊙ *A.* The particle was 4 feet away from the starting line 8 seconds since timing started.

⊙ *B.* In the first 4 seconds, the particle moved a total of 8 feet.

⊙ *C.* In the first 8 seconds, the particle moved a total of 4 feet.

⊙ *D.* The particle was 8 feet away from the starting line 4 seconds since timing started.

c. Solve $d(t) = 4$ for t. $t =$ ⬚

d. Interpret the meaning of part c's solution(s):

⊙ *A.* The article was 4 feet from the starting line 2 seconds since timing started, or 8 seconds since timing started.

⊙ *B.* The article was 4 feet from the starting line 2 seconds since timing started, and again 8 seconds since timing started.

⊙ *C.* The article was 4 feet from the starting line 8 seconds since timing started.

⊙ *D.* The article was 4 feet from the starting line 2 seconds since timing started.

5. Evaluate the function at the given values.

$$H(x) = -\frac{12}{x+6}.$$

   a. $H(-9) =$ [_____].

   b. $H(-6) =$ [_____].

6. Evaluate the function at the given values.

$$K(x) = -\frac{24}{x+1}.$$

   a. $K(5) =$ [_____].

   b. $K(-1) =$ [_____].

7. Use the graph of f below to evaluate the given expressions. (Estimates are OK.)

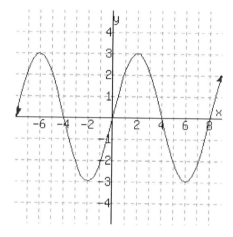

$f(-6) =$ [_____]

$f(6) =$ [_____]

8. Use the graph of g below to evaluate the given expressions. (Estimates are OK.)

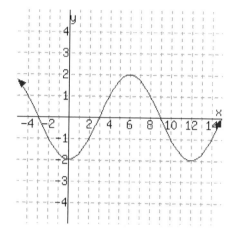

$g(-3) =$ [_____]

$g(12) =$ [_____]

9. Use the table of values for g below to evaluate the given expressions.

x	-3	-1	1	3	5
$g(x)$	6.4	7.9	7.5	4.9	1.8

$g(-1) =$ [_____]

$g(3) =$ [_____]

10. Use the table of values for h below to evaluate the given expressions.

x	-4	1	6	11	16
$h(x)$	3.1	7.7	1.4	6	6.8

$h(1) =$ [_____]

$h(16) =$ [_____]

11. Make a table of values for the function h, defined by $h(x) = 2x^2$. Based on values in the table, sketch a graph of h.

x	$h(x)$

12. Make a table of values for the function H, defined by $H(x) = \dfrac{2^x + 2}{x^2 + 2}$. Based on values in the table, sketch a graph of H.

x	$H(x)$

Domain and Range   A function is graphed.

13.

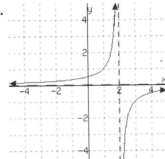

This function has domain ☐ and range ☐.

14.

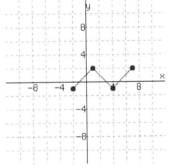

This function has domain ☐ and range ☐.

15.

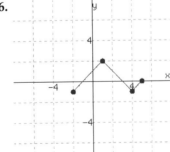

The function has domain ☐ and range ☐.

16.

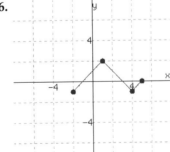

The function has domain ☐ and range ☐.

17.

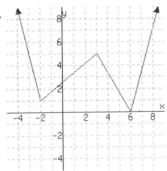

The function has domain ☐ and range ☐.

18.

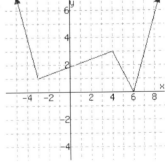

The function has domain ☐ and range ☐.

19. Find the domain of r where $r(x) = \dfrac{\sqrt{8+x}}{9-x}$.

20. Find the domain of B where $B(x) = \dfrac{\sqrt{10+x}}{5-x}$.

21. An object was shot up into the air at an initial vertical speed of 512 feet per second. Its height as time passes can be modeled by the quadratic function f, where $f(t) = -16t^2 + 512t$. Here t represents the number of seconds since the object's release, and $f(t)$ represents the object's height in feet.

Find the function's domain and range in this context.

The function's domain in this context is $\boxed{}$.

The function's range in this context is $\boxed{}$.

22. An object was shot up into the air at an initial vertical speed of 544 feet per second. Its height as time passes can be modeled by the quadratic function f, where $f(t) = -16t^2 + 544t$. Here t represents the number of seconds since the object's release, and $f(t)$ represents the object's height in feet.

Find the function's domain and range in this context.

The function's domain in this context is $\boxed{}$.

The function's range in this context is $\boxed{}$.

Using Technology to Explore Functions

23. Use technology to make a table of values for the function H defined by $H(x) = -4x^2 + 4x + 3$.

x	$H(x)$
___	___
___	___
___	___
___	___
___	___
___	___
___	___

24. Use technology to make a table of values for the function K defined by $K(x) = -2x^2 + 16x - 1$.

x	$K(x)$
___	___
___	___
___	___
___	___
___	___
___	___
___	___

25. Choose an appropriate window for graphing the function f defined by $f(x) = 1456x - 7423$ that shows its key features.

The x-interval could be $\boxed{}$ and the y-interval could be $\boxed{}$.

26. Choose an appropriate window for graphing the function f defined by $f(x) = -169x + 139$ that shows its key features.

The x-interval could be $\boxed{}$ and the y-interval could be $\boxed{}$.

27. Use technology to determine how many times the equations $y = -4x^3 - 3x^2 + x$ and $y = 4x + 2$ intersect. They intersect (☐ zero times ☐ one time ☐ two times ☐ three times) .

28. Use technology to determine how many times the equations $y = -2x^3 + x^2 + 9x$ and $y = -x + 1$ intersect. They intersect (☐ zero times ☐ one time ☐ two times ☐ three times) .

29. For the function L defined by

$$L(x) = 3000x^2 + 10x + 4,$$

use technology to determine the following. Round answers as necessary.

   a. Any intercepts.

   b. The vertex.

   c. The domain.

   d. The range.

30. For the function M defined by

$$M(x) = -(300x - 2950)^2,$$

use technology to determine the following. Round answers as necessary.

   a. Any intercepts.

   b. The vertex.

   c. The domain.

   d. The range.

31. Let $f(x) = 4x^2 + 5x - 1$ and $g(x) = 5$. Use graphing technology to determine the following.

   a. What are the points of intersection for these two functions?

   b. Solve $f(x) = g(x)$.

   c. Solve $f(x) < g(x)$.

   d. Solve $f(x) \geq g(x)$.

32. Let $p(x) = 6x^2 - 3x + 4$ and $k(x) = 7$. Use graphing technology to determine the following.

   a. What are the points of intersection for these two functions?

   b. Solve $p(x) = k(x)$.

   c. Solve $p(x) < k(x)$.

   d. Solve $p(x) \geq k(x)$.

33. Use graphing technology to solve the equation $-0.02x^2 + 1.97x - 51.5 = 0.05(x - 50)^2 - 0.03(x - 50)$. Approximate the solution(s) if necessary.

34. Use graphing technology to solve the equation $-200x^2 + 60x - 55 = -20x - 40$. Approximate the solution(s) if necessary.

35. Use graphing technology to solve the inequality $-15x^2 - 6 \leq 10x - 4$. State the solution set using interval notation, and approximate if necessary.

36. Use graphing technology to solve the inequality $\frac{1}{2}x^2 + \frac{3}{2}x \geq \frac{1}{2}x - \frac{3}{2}$. State the solution set using interval notation, and approximate if necessary.

Simplifying Expressions with Function Notation

37. Let f be a function given by $f(x) = 3x^2 + 2x$. Find and simplify the following:

    a. $f(x) - 2 =$ [   ]

    b. $f(x - 2) =$ [   ]

    c. $-2f(x) =$ [   ]

    d. $f(-2x) =$ [   ]

38. Let f be a function given by $f(x) = -3x^2 - 4x$. Find and simplify the following:

    a. $f(x) - 3 =$ [   ]

    b. $f(x - 3) =$ [   ]

    c. $-3f(x) =$ [   ]

    d. $f(-3x) =$ [   ]

39. Simplify $H(r) + 5$, where $H(r) = -1 - 1.8r$.

40. Simplify $F(r) + 9$, where $F(r) = -1 - 6.3r$.

Technical Definition of a Function

41. Does the following set of ordered pairs make for a function of x?

    $$\{(9, 2), (5, 8), (8, 6), (-3, 3), (-5, 9)\}$$

    This set of ordered pairs ($\square$ describes $\square$ does not describe) a function of x. This set of ordered pairs has domain [   ] and range [   ].

42. Does the following set of ordered pairs make for a function of x?

    $$\{(5, 8), (-6, 5), (10, 4), (-6, 10), (-7, 5)\}$$

    This set of ordered pairs ($\square$ describes $\square$ does not describe) a function of x. This set of ordered pairs has domain [   ] and range [   ].

43. Below is all of the information that exists about a function f.

    $$f(0) = 2 \quad f(2) = 2 \quad f(3) = 2$$

    Write f as a set of ordered pairs.

    f has domain [   ] and range [   ].

44. Below is all of the information about a function g.

    $$g(a) = 1 \quad g(b) = 5$$
    $$g(c) = -5 \quad g(d) = 5$$

    Write g as a set of ordered pairs.

    g has domain [   ] and range [   ].

45. The following graphs show two relationships. Decide whether each graph shows a relationship where y is a function of x.

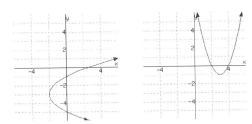

The first graph ($\square$ does $\square$ does not) give a function of x. The second graph ($\square$ does $\square$ does not) give a function of x.

46. The following graphs show two relationships. Decide whether each graph shows a relationship where y is a function of x.

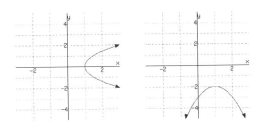

The first graph ($\square$ does $\square$ does not) give a function of x. The second graph ($\square$ does $\square$ does not) give a function of x.

47. Some equations involving x and y define y as a function of x, and others do not. For example, if $x + y = 1$, we can solve for y and obtain $y = 1 - x$. And we can then think of $y = f(x) = 1 - x$. On the other hand, if we have the equation $x = y^2$ then y is not a function of x, since for a given positive value of x, the value of y could equal $\sqrt{x}$ or it could equal $-\sqrt{x}$.

Select all of the following relations that make y a function of x. There are several correct answers.

$\square \, |y| - x = 0$ $\square \, y + x^2 = 1$ $\square \, y^2 + x^2 = 1$
$\square \, y - |x| = 0$ $\square \, 5x + 2y + 9 = 0$ $\square \, x + y = 1$
$\square \, y^6 + x = 1$ $\square \, y^3 + x^4 = 1$

On the other hand, some equations involving x and y define x as a function of y (the other way round).

Select all of the following relations that make x a function of y. There are several correct answers.

$\square \, y - |x| = 0$ $\square \, 5x + 2y + 9 = 0$ $\square \, |y| - x = 0$
$\square \, y^2 + x^2 = 1$ $\square \, y^4 + x^5 = 1$

48. Some equations involving x and y define y as a function of x, and others do not. For example, if $x + y = 1$, we can solve for y and obtain $y = 1 - x$. And we can then think of $y = f(x) = 1 - x$. On the other hand, if we have the equation $x = y^2$ then y is not a function of x, since for a given positive value of x, the value of y could equal $\sqrt{x}$ or it could equal $-\sqrt{x}$.

Select all of the following relations that make y a function of x. There are several correct answers.

$\square \, y^6 + x = 1$ $\square \, x + y = 1$ $\square \, 6x + 7y + 4 = 0$
$\square \, y^3 + x^4 = 1$ $\square \, y + x^2 = 1$ $\square \, y^2 + x^2 = 1$
$\square \, y - |x| = 0$ $\square \, |y| - x = 0$

On the other hand, some equations involving x and y define x as a function of y (the other way round).

Select all of the following relations that make x a function of y. There are several correct answers.

$\square \, y^4 + x^5 = 1$ $\square \, y - |x| = 0$ $\square \, y^2 + x^2 = 1$
$\square \, |y| - x = 0$ $\square \, 6x + 7y + 4 = 0$

49. Determine whether or not the following table could be the table of values of a function. If the table can not be the table of values of a function, give an input that has more than one possible output.

Input	Output
2	9
4	−5
6	9
8	5
−2	−8

Could this be the table of values for a function? (□ yes   □ no)

If not, which input has more than one possible output?   (□ -2   □ 2   □ 4   □ 6   □ 8   □ None, the table represents a function.)

50. Determine whether or not the following table could be the table of values of a function. If the table can not be the table of values of a function, give an input that has more than one possible output.

Input	Output
2	13
4	−19
6	−15
8	−7
−2	−9

Could this be the table of values for a function? (□ yes   □ no)

If not, which input has more than one possible output?   (□ -2   □ 2   □ 4   □ 6   □ 8   □ None, the table represents a function.)

51. Determine whether or not the following table could be the table of values of a function. If the table can not be the table of values of a function, give an input that has more than one possible output.

Input	Output
−4	7
−3	−4
−2	−5
−3	13
−1	−3

Could this be the table of values for a function? (□ yes   □ no)

If not, which input has more than one possible output?   (□ -4   □ -3   □ -2   □ -1   □ None, the table represents a function.)

52. Determine whether or not the following table could be the table of values of a function. If the table can not be the table of values of a function, give an input that has more than one possible output.

Input	Output
−4	−14
−3	−2
−2	12
−3	19
−1	15

Could this be the table of values for a function? (□ yes   □ no)

If not, which input has more than one possible output?   (□ -4   □ -3   □ -2   □ -1   □ None, the table represents a function.)

CHAPTER 11

Absolute Value Functions

11.1 Introduction to Absolute Value Functions

This section will introduce the basic concepts behind absolute value functions and their graphs. This information will also be useful at the end of this chapter when we solve absolute value equations and inequalities.

11.1.1 Definition of Absolute Value

Recall that in Section 1.3, we defined the **absolute value** of a number to be the distance between that number and 0 on a number line. Also recall that this causes the output of the absolute value function to never be a negative number since we are under the presumption that "distance" is always positive (or zero).

> **Example 11.1.2**
>
> a. Since the number 5 is 5 units from 0, then $|5| = 5$.
>
> b. Since the number -3 is 3 units from 0, then $|-3| = 3$.

> **Example 11.1.3** Yonas takes a 5-block walk north from his home to a food cart. After enjoying dinner, he then walks 9 blocks south of the food cart to his favorite movie theater.
>
> a. How many blocks has Yonas walked in total when he reaches the theater?
>
> b. How many blocks is Yonas from home when he reaches the theater?
>
> **Explanation.**
>
> a. Since we only care about total distance, we can ignore the "signs" on the distances walked (either north or south) and simply add the two values together. Mathematically, if we think of north as positive values and south as having negative values, this situation is the same as
>
> $$|5| + |-9| = 5 + 9$$
> $$= 14$$
>
> Yonas has walked a total of 14 blocks when he reaches the theater.
>
> b. When he reaches the theater, Yonas's actual position could be thought of as $5 + (-9)$. But the actual distance from the theater to his home is better thought of as:
>
> $$|5 + (-9)| = |-4|$$
> $$= 4$$

Yonas was 4 blocks from home when he reached the theater.

11.1.2 Evaluating Absolute Value Functions

The formula $f(x) = |x|$ *does* satisfy the requirements for f to be a function because no matter what number you put in for x, there is only one measured distance from 0 to that value x.

Example 11.1.4 Let $f(x) = |x|$ and $g(x) = |2x - 5|$. Evaluate the following expressions.

a. $f(34)$ b. $f(-63)$ c. $f(0)$ d. $g(13)$ e. $g(1)$

Explanation.

a. $f(34) = |34|$
$ = 34$

b. $f(-63) = |-63|$
$ = 63$

c. $f(0) = |0|$
$ = 0$

d. $g(13) = |2 \cdot 13 - 5|$
$ = |21|$
$ = 21$

e. $g(1) = |2 \cdot 1 - 5|$
$ = |-3|$
$ = 3$

Checkpoint 11.1.5. Mark each equation as True or False.

a.   ($\square$ True $\square$ False) $|10| = 10$.

b.   ($\square$ True $\square$ False) $|-3|$ is both 3 and -3.

c.   ($\square$ True $\square$ False) $|x + 4| = x + 4$.

d.   ($\square$ True $\square$ False) $|-6| = |6|$.

e.   ($\square$ True $\square$ False) $|x - 3| = |x + 3|$.

Explanation.   Remember that to be "false" when there is a variable in the equation, all that has to occur is a single input number that makes the equation false.

a. True: $|10| = 10$.

b. False: $|-3|$ is only 3.

c. False: $|x + 4| \neq x + 4$ for many values of x. When you have to decide whether or not an equation is true, one good method to help you decide is to plug in a few numbers to see if each number makes the equation true or false. Be sure to pick a variety and input at least two numbers, if not three. In this case, we will choose 10 and -20.

When $x = 10$:

$$|x + 4| \overset{?}{=} x + 4$$
$$|10 + 4| \overset{?}{=} 10 + 4$$
$$|14| \overset{\checkmark}{=} 14$$

When $x = -20$:

$$|x + 4| \overset{?}{=} x + 4$$
$$|-20 + 4| \overset{?}{=} -20 + 4$$
$$|-16| \overset{\text{no}}{=} -16$$

When we input -20, the equation was false, which indicates that the equation $|x + 4| = x + 4$ is false for general x.

d. True: $|-6| = |6|$. Both $|-6|$ and $|6|$ are equal to 6.

e. False: $|x - 3| \neq |x + 3|$. Again we should choose some numbers to check the validity of the equation. We will choose -12 and 15.

When $x = -12$:

$$|x - 3| \overset{?}{=} |x + 3|$$
$$|-12 - 3| \overset{?}{=} |-12 + 3|$$
$$|-15| \overset{?}{=} |-9|$$
$$15 \overset{\text{no}}{=} 9$$

Since we had a false equation for our first value, we don't need to check the second input. The original equation is simply false.

11.1.3 Graphs of Absolute Value Functions

Absolute value functions have generally the same shape. They are usually described as "V"-shaped graphs and the tip of the "V" is called the **vertex**. A few graphs of various absolute value functions are shown in Figure 11.1.5. In general, the domain of an absolute value function (where there is a polynomial inside the absolute value) is $(-\infty, \infty)$.

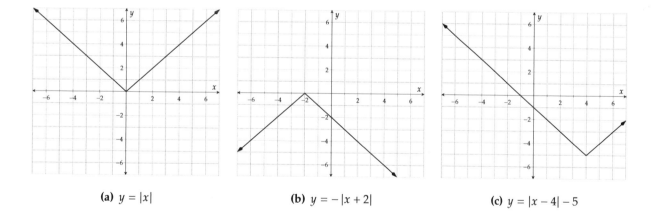

(a) $y = |x|$ (b) $y = -|x + 2|$ (c) $y = |x - 4| - 5$

Figure 11.1.5

Example 11.1.6 Let $h(x) = -2|x - 3| + 5$. Using technology, create table of values with x-values from -3 to 3, using an increment of 1. Then sketch a graph of $y = h(x)$. State the domain and range of h.

Explanation.

x	y
-3	-7
-2	-5
-1	-3
0	-1
1	1
2	3
3	5

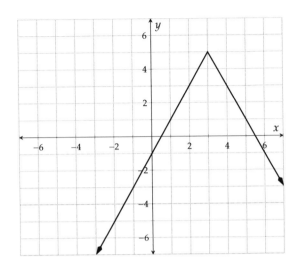

Table 11.1.7: Table for $y = h(x)$.

Figure 11.1.8: Graph of $y = h(x)$

The graph indicates that the domain is (∞, ∞) as it goes to the right and left indefinitely. The range is $(-\infty, 5]$.

Example 11.1.9 Let $j(x) = \left||x + 1| - 2\right| - 1$. Using technology, create table of values with x-values from -5 to 5, using an increment of 1 and sketch a graph of $y = j(x)$. State the domain and range of j.

Explanation. This is a strange one because it has an absolute value within an absolute value.

x	y
-5	1
-4	0
-3	-1
-2	0
-1	1
0	0
1	-1
2	0
3	1
4	2
5	3

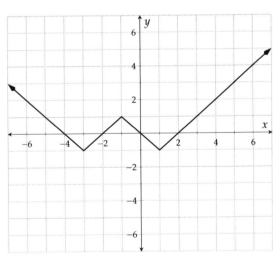

Table 11.1.10: A table of values for $y = j(x)$.

Figure 11.1.11: $y = \left||x + 1| - 2\right| - 1$

The graph indicates that the domain is (∞, ∞) as it goes to the right and left indefinitely. The range is $[-1, \infty)$.

11.1.4 Another Definition for Absolute Value

How many definitions do we really need? Bear with us because this one is important.

Example 11.1.12 Consider the function f defined by $f(x) = \sqrt{x^2}$. First, we will evaluate this function at a few arbitrary values: 3, 0, and -5.

$$f(3) = \sqrt{3^2} \qquad\qquad f(0) = \sqrt{0^2} \qquad\qquad f(-5) = \sqrt{(-5)^2}$$
$$= \sqrt{9} \qquad\qquad\qquad = \sqrt{0} \qquad\qquad\qquad = \sqrt{25}$$
$$= 3 \qquad\qquad\qquad\quad = 0 \qquad\qquad\qquad = 5$$

These results should seem familiar: $f(3) = 3$, $f(0) = 0$, and $f(-5) = 5$. The outputs are the same as the inputs, except for a missing negative sign on 5. It seems like we've seen a function that does that exact same thing already …

Make a quick graph using technology to see what the graph of $y = f(x)$ looks like.

Explanation. Figure 11.1.13 shows a graph of f where $f(x) = \sqrt{x^2}$. It looks just like that of $y = |x|$.

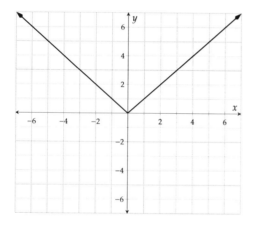

Since the graphs of $y = \sqrt{x^2}$ and $y = |x|$ match up exactly, that must mean that

$$|x| = \sqrt{x^2}$$

This fact will be used later in this chapter and it will continue to pop up in subsequent math courses here and there as important.

Figure 11.1.13: $y = \sqrt{x^2}$

Example 11.1.14 Simplify the following expressions using the fact that $|x| = \sqrt{x^2}$.

a. $\sqrt{(x-2)^2}$ b. $\sqrt{x^6}$ c. $\sqrt{x^2 + 10x + 25}$ d. $\sqrt{x^4}$

Explanation.

a. $\sqrt{(x-2)^2} = |x-2|$. Note that $x - 2$ might be a negative number depending on the value of x, so the absolute value will change those negative numbers to be positive values.

b. $\sqrt{x^6} = \sqrt{(x^3)^2}$

   $= |x^3|$

We know from exponent rules that $x^6 = (x^3)^2$. Note that x^3 will be negative whenever x is a negative number, so the absolute value bars must remain.

c. $\sqrt{x^2 + 10x + 25} = \sqrt{(x+5)^2}$

   $= |x + 5|$

Note again that $x + 5$ can be negative for certain values of x, so the absolute value bars must remain.

d. $\sqrt{x^4} = \sqrt{(x^2)^2}$

   $= |x^2|$

   $= x^2$

Note here that x^2 is never negative. No matter what number you substitute in for x in x^2, you always either get a positive result or zero. So the absolute value around x^2 doesn't have any effect. Absolute values change negative numbers to positive values but leave positive values alone. Thus, it is OK in this case to drop the absolute value bars.

11.1.5 Applications Involving Absolute Values

Absolute values are quite useful as models in a variety of real world applications. One example is the path of a billiards (pool) ball: when the ball bounces off one of the side rails, its path is mirrored and creates a "V" shape. The game gets more complicated when more than the rail is hit, but the fundamental mathematics doesn't change: absolute values model the bounces each time.

Here are some more examples. The first one we'll explore involves light reflecting off of a mirror.

When light reflects off of a mirror, the path it takes is in the shape of an absolute value graph. Khenbish was playing with a laser pointer in his bedroom mirror. He set up the laser pointer on his windowsill and the light hit the center of the mirror and reflected onto the corner of his room. He declared that the laser pointer is sitting at the origin, and x should stand for the horizontal distance from the left wall to the light beam. Shown is a birds-eye view of the situation.

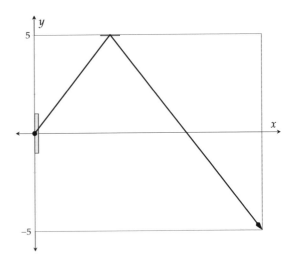

Figure 11.1.16: Birds-Eye View of Khenbish's Room with Laser

Example 11.1.15 After a little bit of work, Khenbish was able to come up with a formula for the light's path:

$$p(x) = 5 - \frac{5}{4} |x - 4|$$

where $p(x)$ stands for the position, in ft, above (for positive values) or below (for negatives) the center line through his room that represents the x-axis, where x is also measured in ft. Use technology and a graph of this formula to answer the following questions.

a. Khenbish's room is 10 ft wide according to Figure 11.1.16 (in the vertical direction in the figure). What is the room's length (in the horizontal direction in the figure)?

b. How far along the wall is the mirror centered?

c. If you stood 9 ft from the left wall, how far above or below the room's center line (x-axis) should you stand to have the laser pointer hit you?

Explanation.

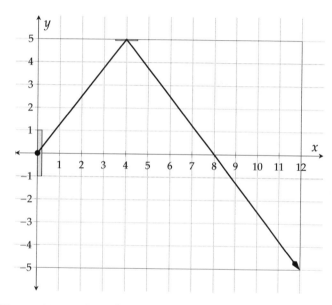

Figure 11.1.17: Detailed Birds-Eye View of Khenbish's Room

a. To find the room's length, first note that since the laser hits the corner of the room, the x coordinate of the lasers position would tell us the room's width. According to the detailed graph, the x-coordinate when $y = -5$ is 12. So the room must be 12 feet wide.

b. The mirror is centered exactly where the laser hits the wall. This is the vertex of the absolute value graph which, according to the graph, is at the point $(4, 5)$. This tells us that the mirror is centered 4 feet from the left wall.

c. If you are standing 9 feet from the left wall, the laser's position will be a bit more than one foot behind the rooms center line, by the diagram. While technology can tell us the exact answer, here is how to do this problem algebraically.

$$d(9) = 5 - \frac{5}{4}|9 - 4|$$
$$= 5 - \frac{5}{4}|5|$$
$$= 5 - \frac{5}{4} \cdot 5$$
$$= \frac{20}{4} - \frac{25}{4}$$
$$= -\frac{5}{4}$$
$$= -1.25$$

So, it looks like if you stand 9 feet from the left wall, you need to stand 1.25 feet behind the center line (which would be 6.25 feet from the wall with the mirror on it) to be hit by the laser.

Absolute value functions are also used when a value must be within a certain distance or tolerance. For example, a person's body temperature is considered "normal" if it is within 0.5 degrees of 98.6 °F, so their temperature could be up to 0.5 degrees less than or greater than that temperature. To be within normal range, the difference between the two values must be less than or equal to 0.5, and it does not matter whether it is positive or negative. We will introduce a function for measuring this in the next example.

Example 11.1.18 The function D defined by $D(T) = |T - 98.6|$ represents the difference between a person's temperature, T, in Fahrenheit, and 98.6 °F. A person's temperature is considered "normal" if $D(T)$ is less than or equal to 0.5. Use $D(T)$ to determine whether each person's temperature is within the normal range.

a. LaShonda has a temperature of 98.3 °F.

b. Castel has a temperature of 99.3 °F.

c. Daniel has a temperature of 97.3 °F.

Explanation.

a. LaShonda has a temperature of 98.3 °F, so we have:

$$D(98.3) = |98.3 - 98.6|$$
$$= |-0.3|$$
$$= 0.3$$

Since the value of $D(98.3)$ is a number smaller than 0.5, her temperature of 98.3 °F is within the normal range.

b. If Castel has a temperature of 99.3 °F, then we have:

$$D(99.3) = |99.3 - 98.6|$$
$$= |0.7|$$
$$= 0.7$$

Since the value of $D(99.3)$ is a number bigger than 0.5, their temperature of 99.3 °F is *not* within the normal range.

c. Daniel's temperature is 97.3 °F, so we have:

$$D(97.3) = |97.3 - 98.6|$$
$$= |-1.3|$$
$$= 1.3$$

Since the value of $D(97.3)$ is a number bigger than 0.5, his temperature of 97.3 °F is *not* within the normal range.

Example 11.1.19 The entryway to the Louvre Museum in Paris is through I. M. Pei's metal and glass Louvre Pyramid. This pyramid has a square base and is 71 feet high and 112 feet wide. The formula $h(x) = 71 - \frac{71}{56}|x - 56|$ gives the height above ground level of the pyramid at a distance of x from the left side of the pyramid base.

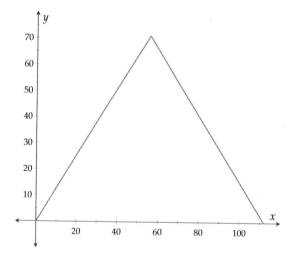

a. If you are 20 feet from the left edge, how high will the pyramid rise in front of you? Round your result to the nearest tenth of an inch.

b. How far from the left edge is the center of the pyramid?

c. Using your previous answer, check that the formula gives you the correct height at the center.

Figure 11.1.20: A Diagram of the Front of the Louvre Pyramid

Explanation.

a. If you are 20 feet from the left edge, then x is 20. Substituting 20 for x we have

$$h(20) = 71 - \frac{71}{56}|20 - 56|$$
$$= 71 - \frac{71}{56}|-36|$$
$$= 71 - \frac{71}{56} \cdot 36$$
$$\approx 25.4$$

The pyramid is about 25.4 feet high at the position 20 feet from the left edge.

b. The center of the pyramid is 56 feet from the either edge since it's half of 112 feet.

c. Putting $x = 56$ into the formula for h gives us

$$h(56) = 71 - \frac{71}{56}\,|56 - 56|$$
$$= 71 - \frac{71}{56}\,|0|$$
$$= 71 - \frac{71}{56} \cdot 0$$
$$= 71$$

And so the formula does give us the correct maximum height of 71 feet at the center of the pyramid.

Exercises

Review and Warmup   Evaluate the following.

1.
   a. $|3| = \boxed{}$

   b. $|-4| = \boxed{}$

   c. $|0| = \boxed{}$

   d. $|18 + (-8)| = \boxed{}$

   e. $|-9 - (-4)| = \boxed{}$

2.
   a. $|4| = \boxed{}$

   b. $|-8| = \boxed{}$

   c. $|0| = \boxed{}$

   d. $|12 + (-1)| = \boxed{}$

   e. $|-9 - (-2)| = \boxed{}$

3.
   a. $-|5 - 9| = \boxed{}$

   b. $|-5 - 9| = \boxed{}$

   c. $-3|9 - 5| = \boxed{}$

4.
   a. $-|3 - 6| = \boxed{}$

   b. $|-3 - 6| = \boxed{}$

   c. $-3|6 - 3| = \boxed{}$

5.
   a. $-3 - 3|7 - 1| = \boxed{}$

   b. $-3 - 3|1 - 7| = \boxed{}$

6.
   a. $-2 - 8|9 - 4| = \boxed{}$

   b. $-2 - 8|4 - 9| = \boxed{}$

7.
   a. $-1 - 5|6 - 3| = \boxed{}$

   b. $-1 - 5|3 - 6| = \boxed{}$

8.
   a. $-10 - 3|8 - 2| = \boxed{}$

   b. $-10 - 3|2 - 8| = \boxed{}$

9. $1 - 8|1 - 2| + 2 = \boxed{}$

10. $2 - 6|3 - 8| + 2 = \boxed{}$

11. $4 - 2\left|-1 + (3 - 5)^3\right| = \boxed{}$

12. $5 - 8\left|-7 + (2 - 5)^3\right| = \boxed{}$

Function Notation with Absolute Value

13. Given $H(t) = |t - 296|$, find and simplify $H(159)$.

$H(159) = \boxed{}$

14. Given $g(r) = |r - 261|$, find and simplify $g(170)$.

$g(170) = \boxed{}$

15. Given $h(x) = |x + 16|$, find and simplify $h(18)$.

$h(18) = \boxed{}$

16. Given $f(x) = |-4x - 5|$, find and simplify $f(20)$.

$f(20) = \boxed{}$

17. Given $f(x) = 10 - |3x - 26|$, find and simplify $f(10)$.

$f(10) = \boxed{}$

18. Given $f(r) = 15 - |4r + 14|$, find and simplify $f(11)$.

$f(11) = \boxed{}$

19. Given $g(t) = t + |-2t - 7|$, find and simplify $g(12)$.

$g(12) = \boxed{}$

20. Given $K(t) = t + |2t - 29|$, find and simplify $K(14)$.

$K(14) = \boxed{}$

21. Given $G(t) = |t^2 - 16|$, find and simplify $G(-5)$.

$G(-5) = \boxed{}$

22. Given $h(t) = |t^2 - 25|$, find and simplify $h(1)$.

$h(1) = \boxed{}$

23. Given $f(t) = |t^2 - 2t - 15|$, find and simplify $f(7)$.

$f(7) = \boxed{}$

24. Given $H(t) = |t^2 - 2t - 24|$, find and simplify $H(-2)$.

$H(-2) = \boxed{}$

Domain

25. Find the domain of K where $K(x) = |-4x + 6|$.

26. Find the domain of f where $f(x) = |9x - 6|$.

27. Find the domain of f where $f(x) = 2x - |2x + 3|$.

28. Find the domain of g where $g(x) = 8x - |-5x - 9|$.

Tables

29. Make a table of values for the function h defined by $h(x) = |3x - 2|$.

x	$h(x)$

30. Make a table of values for the function F defined by $F(x) = |2x - 3|$.

x	$F(x)$

31. Make a table of values for the function G defined by $G(x) = |x^2 - 2x - 1|$.

x	$G(x)$

32. Make a table of values for the function G defined by $G(x) = |x^2 - x - 2|$.

x	$G(x)$

33. Make a table of values for the function H defined by $H(x) = -|2x - 3| + 1$.

x	$H(x)$

34. Make a table of values for the function K defined by $K(x) = -3|2x - 3| + 2$.

x	$K(x)$

35. Make a table of values for the function f defined by $f(x) = \left|-2|2x - 3| - 1\right|$.

x	$f(x)$

36. Make a table of values for the function f defined by $f(x) = \left||-x + 2| + 2\right|$.

x	$f(x)$

Graphs

37. Graph $y = f(x)$, where $f(x) = |2x - 1|$.

38. Graph $y = f(x)$, where $f(x) = |x - 2|$.

39. Graph $y = f(x)$, where $f(x) = |x^2 - 2x - 1|$.

40. Graph $y = f(x)$, where $f(x) = |x^2 + 3x - 2|$.

41. Graph $y = f(x)$, where $f(x) = \frac{1}{2}|4x - 5| - 3$.

42. Graph $y = f(x)$, where $f(x) = \frac{3}{4}|6 + x| + 2$.

43. Graph $y = f(x)$, where $f(x) = |2|3 - x| - 2|$.

44. Graph $y = f(x)$, where $f(x) = |3 - 2|2x - 3||$.

Absolute Value and Square Roots

45. Simplify the expression. Do not assume the variables take only positive values.

$\sqrt{9z^2}$

46. Simplify the expression. Do not assume the variables take only positive values.

$\sqrt{64t^2}$

47. Simplify the expression.

$\sqrt{(r - 43)^2}$

48. Simplify the expression.

$\sqrt{(m - 8)^2}$

49. Simplify the expression.

$\sqrt{(-13698)^2}$

50. Simplify the expression.

$\sqrt{(-15696)^2}$

51. Simplify the expression.

$\sqrt{x^2 + 20x + 100}$

52. Simplify the expression.

$\sqrt{n^2 + 2n + 1}$

Applications

53. The height inside a camping tent when you are d feet from the edge of the tent is given by

$$h = -0.5|d - 4.4| + 5.5$$

where h stands for height in feet.

Determine the height when you are:

a. 5.9 ft from the edge.

The height inside a camping tent when you 5.9 ft from the edge of the tent is [   ]

b. 3.5 ft from the edge.

The height inside a camping tent when you 3.5 ft from the edge of the tent is [   ]

54. The height inside a camping tent when you are d feet from the edge of the tent is given by

$$h = -0.5|d - 6.6| + 6.5$$

where h stands for height in feet.

Determine the height when you are:

    a. 11.5 ft from the edge.

        The height inside a camping tent when you 11.5 ft from the edge of the tent is

    b. 1.2 ft from the edge.

        The height inside a camping tent when you 1.2 ft from the edge of the tent is

Challenge

    55. Write two numbers so that

        • The first number is less than the second number, and

        • The absolute value of the first number is greater than the absolute value of the second number

        and

11.2 Compound Inequalities

On the newest version of the SAT (an exam that often qualifies students for colleges) the minimum score that you can earn is 400 and the maximum score that you can earn is 1600. This means that only numbers between 400 and 1600, including these endpoints, are possible scores. To plot all of these values on a number line would look something like:

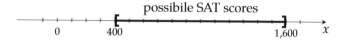

Figure 11.2.1: Possible SAT Scores

Going back to the original statement, "the minimum score that you can earn is 400 and the maximum score that you can earn is 1600," this really says two things. First, it says that (a SAT score) $\geq$ 400, and second, that (a SAT score) $\leq$ 1600. When we combine two inequalities like this into a single problem, it becomes a **compound inequality**.

Our lives are often constrained by the compound inequalities of reality: you need to buy enough materials to complete your project, but you can only fit so much into your vehicle; you would like to finish your degree early, but only have so much money and time to put toward your courses; you would like a vegetable garden big enough to supply you with veggies all summer long, but your yard or balcony only gets so much sun. In the rest of the section we hope to illuminate how to think mathematically about problems like these.

Before continuing, a review on how notation for intervals works may be useful, and you may benefit from revisiting Section 1.6. Then a refresher on solving linear inequalities may also benefit you, which you can revisit in Section 3.2 and Section 3.3.

11.2.1 Unions of Intervals

Definition 11.2.3. The **union** of two sets, A and B, is the set of all elements contained in either A or B (or both). We write $A \cup B$ to indicate the union of the two sets.

In other words, the union of two sets is what you get if you toss every number in both sets into a bigger set.

Example 11.2.4 The union of sets $\{1, 2, 3, 4\}$ and $\{3, 4, 5, 6\}$ is the set of all elements from either set. So $\{1, 2, 3, 4\} \cup \{3, 4, 5, 6\} = \{1, 2, 3, 4, 5, 6\}$. Note that we don't write duplicates.

Example 11.2.5 Let's visualize the union of the sets $(-\infty, 4)$ and $[7, \infty)$. First we make a number line with both intervals drawn to understand what both sets mean.

Figure 11.2.6: A number line sketch of $(-\infty, 4)$ as well as $[7, \infty)$

The two intervals should be viewed as a single object when stating the union, so here is the picture of the union. It looks the same, but now it is a graph of a single set.

Figure 11.2.7: A number line sketch of $(-\infty, 4) \cup [7, \infty)$

11.2.2 "Or" Compound Inequalities

Definition 11.2.8. A **compound inequality** is a grouping of two or more inequalities into a larger inequality statement. These usually come in two flavors: "or" and "and" inequalities. For an example of an "or" compound inequality, you might get a discount at the movie theater if your age is less than 13 *or* greater than 64. In general, compound inequalities of the "and" variety currently are beyond the scope of this book. However a special type of the "and" variety is covered later in Subsection 11.2.3.

In math, the technical term **or** means "either or both." So, mathematically, if we asked if you would like "chocolate cake *or* apple pie" for dessert, your choices are either "chocolate cake," "apple pie," or "both chocolate cake and apple pie." This is slightly different than the English "or" which usually means "one or the other but not both."

"Or" shows up in math between equations (as in when solving a quadratic equation, you might end up with "$x = 2$ or $x = -3$") or between inequalities (which is what we're about to discuss).

Remark 11.2.9. The definition of "or" is very close to the definition of a union where you combine elements from either or both sets together. In fact, when you have an "or" between inequalities in a compound inequality, to find the solution set of the compound inequality, you find the union of the the solutions sets of each of the pieces.

Example 11.2.10 Solve the compound inequality.

$$x \leq 1 \text{ or } x > 4$$

Explanation.

Writing the solution set to this compound inequality doesn't require any algebra beforehand because each of the inequalities is already solved for x. The first thing we should do is understand what each inequality is saying using a graph.

Figure 11.2.11: A number line sketch of solutions to $x \leq 1$ as well as to $x > 4$

An "or" statement becomes a union of solution sets, so the solution set to the compound inequality must be:

$$(-\infty, 1] \cup (4, \infty).$$

Example 11.2.12 Solve the compound inequality.

$$3 - 5x > -7 \text{ or } 2 - x \leq -3$$

Explanation.   First we need to do some algebra to isolate x in each piece. Note that we are going to do algebra on both pieces simultaneously. Also note that the mathematical symbol "or" should be written on each line.

$$3 - 5x > -7 \qquad\qquad \text{or} \qquad\qquad 2 - x \le -3$$
$$3 - 5x - 3 > -7 - 3 \qquad\qquad \text{or} \qquad\qquad 2 - x - 2 \le -3 - 2$$
$$-5x > -10 \qquad\qquad \text{or} \qquad\qquad -x \le -5$$
$$\frac{-5x}{-5} < \frac{-10}{-5} \qquad\qquad \text{or} \qquad\qquad \frac{-x}{-1} \ge \frac{-5}{-1}$$
$$x < 2 \qquad\qquad \text{or} \qquad\qquad x \ge 5$$

The solution set for the compound inequality $x < 2$ is $(-\infty, 2)$ and the solution set to $x \ge 5$ is $[5, \infty)$. To do the "or" portion of the problem, we need to take the union of these two sets. Let's first make a graph of the solution sets to visualize the problem.

Figure 11.2.13: A number line sketch of $(-\infty, 2)$ as well as $[5, \infty)$

The union combines both solution sets into one, and so

$$(-\infty, 2) \cup [5, \infty)$$

We have finished the problem, but for the sake of completeness, let's try to verify that our answer is reasonable.

- First, let's choose a number that is *not* in our proposed solution set. We will arbitrarily choose 3.

$$3 - 5x > -7 \qquad\qquad \text{or} \qquad\qquad 2 - x \le -3$$
$$3 - 5(3) \overset{?}{>} -7 \qquad\qquad \text{or} \qquad\qquad 2 - (3) \overset{?}{\le} -3$$
$$-9 \overset{\text{no}}{>} -7 \qquad\qquad \text{or} \qquad\qquad -1 \overset{\text{no}}{\le} -3$$

  This value made *both* inequalities false which is why 3 isn't in our solution set.

- Next, let's choose a number that *is* in our solution region. We will arbitrarily choose 1.

$$3 - 5x > -7 \qquad\qquad \text{or} \qquad\qquad 2 - x \le -3$$
$$3 - 5(1) \overset{?}{>} -7 \qquad\qquad \text{or} \qquad\qquad 2 - (1) \overset{?}{\le} -3$$
$$-12 \overset{\checkmark}{<} -7 \qquad\qquad \text{or} \qquad\qquad -1 \overset{\text{no}}{\le} -3$$

This value made *one* of the inequalities true. Since this is an "or" statement, only one *or* the other piece has to be true to make the compound inequality true.

- Last, what will happen if we choose a value that was in the other solution region in Figure 11.2.13, like the number 6?

$$3 - 5x > -7 \qquad \text{or} \qquad 2 - x \le -3$$

$$3 - 5(6) \overset{?}{>} -7 \qquad \text{or} \qquad 2 - (6) \overset{?}{\le} -3$$

$$-27 \overset{no}{>} -7 \qquad \text{or} \qquad -4 \overset{\checkmark}{\le} -3$$

This solution made the *other* inequality piece true.

This completes the check. Numbers from within the solution region make the compound inequality true and numbers outside the solution region make the compound inequality false.

Example 11.2.14 Solve the compound inequality.

$$\frac{3}{4}t + 2 \le \frac{5}{2} \text{ or } -\frac{1}{2}(t - 3) < -2$$

Explanation. First we will solve each inequality for t. Recall that we usually try to clear denominators by multiplying both sides by the least common denominator.

$$\frac{3}{4}t + 2 \le \frac{5}{2} \qquad \text{or} \qquad -\frac{1}{2}(t - 3) < -2$$

$$4 \cdot \left(\frac{3}{4}t + 2\right) \le 4 \cdot \frac{5}{2} \qquad \text{or} \qquad 2 \cdot \left(-\frac{1}{2}(t - 3)\right) < 2 \cdot (-2)$$

$$3t + 8 \le 10 \qquad \text{or} \qquad -t + 3 < -4$$

$$3t + 8 - 8 \le 10 - 8 \qquad \text{or} \qquad -t + 3 - 3 < -4 - 3$$

$$3t \le 2 \qquad \text{or} \qquad -t < -7$$

$$\frac{3t}{3} \le \frac{2}{3} \qquad \text{or} \qquad \frac{-t}{-1} > \frac{-7}{-1}$$

$$t \le \frac{2}{3} \qquad \text{or} \qquad t > 7$$

The solution set to $t \le \frac{2}{3}$ is $\left(-\infty, \frac{2}{3}\right]$ and the solution set to $t > 7$ is $(7, \infty)$. Figure 11.2.15 shows these two sets.

Figure 11.2.15: A number line sketch of $\left(-\infty, \frac{2}{3}\right]$ and also $(7, \infty)$

Note that the two sets do not overlap so there will be no way to simplify the union. Thus the solution set to the compound inequality is:

$$\left(-\infty, \frac{2}{3}\right] \cup (7, \infty)$$

Example 11.2.16 Solve the compound inequality.

$$3y - 15 > 6 \text{ or } 7 - 4y \geq y - 3$$

Explanation. First we solve each inequality for y.

$$
\begin{array}{lcl}
3y - 15 > 6 & \text{or} & 7 - 4y \geq y - 3 \\
3y - 15 + 15 > 6 + 15 & \text{or} & 7 - 4y - 7 - y \geq y - 3 - 7 - y \\
3y > 21 & \text{or} & -5y \geq -10 \\
\dfrac{3y}{3} > \dfrac{21}{3} & \text{or} & \dfrac{-5y}{-5} \leq \dfrac{-10}{-5} \\
y > 7 & \text{or} & y \leq 2
\end{array}
$$

The solution set to $y > 7$ is $(7, \infty)$ and the solution set to $y \leq 2$ is $(-\infty, 2]$. Figure 11.2.17 shows these two sets.

Figure 11.2.17: A number line sketch of $(7, \infty)$ as well as $(-\infty, 2]$

So the solution set to the compound inequality is:

$$(-\infty, 2] \cup (7, \infty)$$

11.2.3 Three-Part Inequalities

The inequality $1 \leq 2 < 3$ says a lot more than you might think. It actually says four different single inequalities which are highlighted for you to see.

$$1 \leq 2 < 3 \quad 1 \leq 2 < 3 \quad 1 \leq 2 < 3 \quad 1 \leq 2 < 3$$

This might seem trivial at first, but if you are presented with an inequality like $-1 < 3 \geq 2$, at first it might look sensible; however, in reality, you need to check that *all four* linear inequalities make sense. Those are highlighted here.

$$-1 < 3 \geq 2 \quad -1 < 3 \geq 2 \quad -1 < 3 \geq 2 \quad -1 < 3 \geq 2$$

One of these inequalities is false: $-1 \not\geq 2$. This implies that the entire original inequality, $-1 < 3 \geq 2$, is nonsense.

Example 11.2.18 Decide whether or not the following inequalities are true or false.

a. True or False: $-5 < 7 \leq 12$?

b. True or False: $-7 \leq -10 < 4$?

c. True or False: $-2 \leq 0 \geq 1$?

d. True or False: $5 > -3 \geq -9$?

e. True or False: $3 < 3 \leq 5$?

f. True or False: $9 > 1 < 5$?

g. True or False: $3 < 8 \leq -2$?

h. True or False: $-9 < -4 \leq -2$?

Explanation. We need to go through all four single inequalities for each. If the inequality is false, for simplicity's sake, we will only highlight the one single inequality that makes the inequality false.

a. True: $-5 < 7 \leq 12$.

b. False: $-7 \overset{no}{\leq} -10 < 4$.

c. False: $-2 \leq 0 \overset{no}{\geq} 1$.

d. True: $5 > -3 \geq -9$.

e. False: $3 \overset{no}{<} 3 \leq 5$.

f. False: $9 > 1 \overset{no}{<} 5$.

g. False: $3 < 8 \overset{no}{\leq} -2$.

h. True: $-9 < -4 \leq -2$.

As a general hint, no (nontrivial) three-part inequality can ever be true if the inequality signs are not pointing in the same direction. So no matter what numbers a, b, and c are, both $a < b \geq c$ and $a \geq b < c$ cannot be true! Soon you will be writing inequalities like $2 < x \leq 4$ and you need to be sure to check that your answer is feasible. You will know that if you get $2 > x \leq 4$ or $2 < x \geq 4$ that something went wrong in the solving process. The only exception is that something like $1 \leq 1 \geq 1$ is true because $1 = 1 = 1$, although this shouldn't come up very often!

Example 11.2.19 Write the solution set to the compound inequality.

$$-7 < x \leq 5$$

Explanation. The solutions to the three-part inequality $-7 < x \leq 5$ are those numbers that are trapped between -7 and 5, including 5 but not -7. Keep in mind that there are infinitely many decimal numbers and irrational numbers that satisfy this inequality like -2.781828 and π. We will write these numbers in interval notation as $(-7, 5]$ or in set builder notation as $\{x \mid -7 < x \leq 5\}$.

Example 11.2.20 Solve the compound inequality.

$$4 \leq 9x + 13 < 20$$

Explanation.

This is a three-part inequality which we can treat just as a regular inequality with three "sides." The goal is to isolate x in the middle and whatever you do to one "side," you have to do to the other two "sides."

The solutions to the three-part inequality $-1 \leq x < \frac{7}{9}$ are those numbers that are trapped between -1 and $\frac{7}{9}$, including -1 but not $\frac{7}{9}$. The solution set in interval notation is $\left[-1, \frac{7}{9}\right)$.

$$4 \leq 9x + 13 < 20$$
$$4 - 13 \leq 9x + 13 - 13 < 20 - 13$$
$$-9 \leq 9x < 7$$
$$\frac{-9}{9} \leq \frac{9x}{9} < \frac{7}{9}$$
$$-1 \leq x < \frac{7}{9}$$

Example 11.2.21 Solve the compound inequality.

$$-13 < 7 - \frac{4}{3}x \leq 15$$

Explanation.

This is a three-part inequality which we can treat just as a regular inequality with three "sides." The goal is to isolate x in the middle and whatever you do to one "side," you have to do to the other two "sides." We will begin by canceling the fraction by multiplying each part by the least common denominator.

At the end we reverse the entire statement to go from smallest to largest. The solution set is $(-6, 15]$.

$$-13 < 7 - \frac{4}{3}x \le 15$$

$$-13 \cdot 3 < \left(7 - \frac{4}{3}x\right) \cdot 3 \le 15 \cdot 3$$

$$-39 < 21 - 4x \le 45$$

$$-39 - 21 < 21 - 4x - 21 \le 45 - 21$$

$$-60 < -4x \le 24$$

$$\frac{-60}{-4} > \frac{-4x}{-4} \ge \frac{24}{-4}$$

$$15 > x \ge -6$$

$$-6 \le x < 15$$

11.2.4 Solving Compound Inequalities Graphically

So far we have focused on solving inequalities algebraically. Next, we will describe how to solve compound inequalities graphically.

Example 11.2.22 Figure 11.2.23 shows a graph of $y = f(x)$. Use the graph to solve the inequality $2 \le f(x) < 6$.

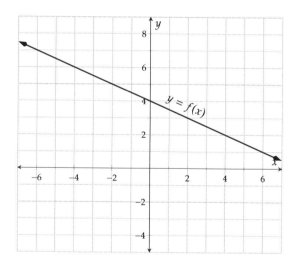

To solve the inequality $2 \le f(x) < 6$ means to find the x-values that give function values between 2 and 6, not including 6. We draw the horizontal lines $y = 2$ and $y = 6$. Then we look for the points of intersection and find their x-values. We see that when x is between -4 and 4, not including -4, the inequality will be true.

Figure 11.2.23: Graph of $y = f(x)$

We have drawn the interval $(-4, 4]$ along the x-axis, which is the solution set.

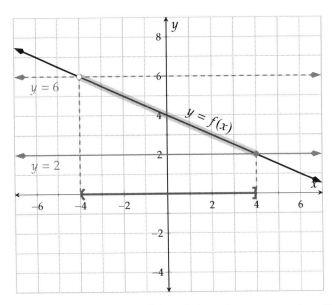

Figure 11.2.24: Graph of $y = f(x)$ and the solution set to $2 \le f(x) < 6$

Example 11.2.25 Figure 11.2.26 shows a graph of $y = g(x)$. Use the graph to solve the inequality $-4 < g(x) \le 3$.

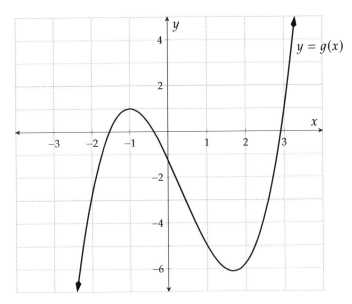

Figure 11.2.26: Graph of $y = g(x)$

Explanation. To solve $-4 < g(x) \le 3$, we first draw the horizontal lines $y = -4$ and $y = 3$. To solve this inequality we notice that there are two pieces of the function g that are trapped between the y-values -4 and 3.

The solution set is the compound inequality $(-2.1, 0.7) \cup (2.4, 3.2]$.

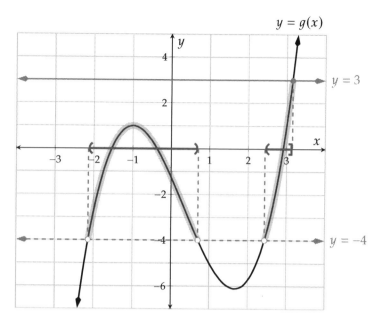

Figure 11.2.27: Graph of $y = g(x)$ and solution set to $-4 < g(x) \le 3$

11.2.5 Applications of Compound inequalities

Example 11.2.28 Raphael's friend is getting married and he's decided to give them some dishes from their registry. Raphael doesn't want to seem cheap but isn't a wealthy man either, so he wants to buy "enough" but not "too many." He's decided that he definitely wants to spend at least $150 on his friend, but less than $250. Each dish is $21.70 and shipping on an order of any size is going to be $19.99. Given his budget, set up and algebraically solve a compound inequality to find out what his different options are for the number of dishes that he can buy.

Explanation. First, we should define our variable. Let x represent the number of dishes that Raphael can afford. Next we should write a compound inequality that describes this situation. In this case, Raphael wants to spend between $150 and $250 and, since he's buying x dishes, the price that he will pay is $21.70x + 19.99$. All of this translates to a triple inequality

$$150 < 21.70x + 19.99 < 250$$

Now we have to solve this inequality in the usual way.

$$150 < 21.70x + 19.99 < 250$$
$$150 - 19.99 < 21.70x + 19.99 - 19.99 < 250 - 19.99$$
$$130.01 < 21.70x < 230.01$$
$$\frac{130.01}{21.70} < \frac{21.70x}{21.70} < \frac{230.01}{21.70}$$
$$5.991 < x < 10.6 \qquad \text{(note: these values are approximate)}$$

The interpretation of this inequality is a little tricky. Remember that x represents the number of dishes Raphael can afford. Since you cannot buy 5.991 dishes (manufacturers will typically only ship whole number amounts of tableware) his minimum purchase must be 6 dishes. We have a similar problem with his maximum purchase: clearly he cannot buy 10.6 dishes. So, should we round up or down? If we rounded up, that would be 11 dishes and that would cost $21.70 \cdot 11 + \$19.99 = \$258.69$, which is outside his price range. Therefore, we should actually round *down* in this case.

In conclusion, Raphael should buy somewhere between 6 and 10 dishes for his friend to stay within his budget.

Example 11.2.29 Oak Ridge National Laboratory, a renowned scientific research facility, compiled some data in table 4.28[a] on fuel efficiency of a mid-size hybrid car versus the speed that the car was driven. A model for the fuel efficiency $e(x)$ (in miles per gallon, mpg) at a speed x (in miles per hour, mph) is $e(x) = 88 - 0.7x$.

    a. Evaluate and interpret $e(60)$ in the context of the problem.

    b. Note that this model only applies between certain speeds. The maximum fuel efficiency for which this formula applies is 55 mpg and the minimum fuel efficiency for which it applies is 33 mpg. Set up and algebraically solve a compound inequality to find the range of speeds for which this model applies.

Explanation.

    a. Let's evaluate $e(60)$ first.

$$e(x) = 88 - 0.7x$$
$$e(60) = 88 - 0.7(60)$$
$$= 46$$

So, when the hybrid car travels at a speed of 60 mph, it has a fuel efficiency of 46 mpg.

    b. In this case, the minimum efficiency is 33 mpg and the maximum efficiency is 55 mpg. We need to trap our formula between these two values to solve for the respective speeds.

$$33 < 88 - 0.7x < 55$$
$$33 - 88 < 88 - 0.7x - 88 < 55 - 88$$
$$-55 < -0.7x < -33$$
$$\frac{-55}{-0.7} > \frac{-0.7x}{-0.7} > \frac{-33}{-0.7}$$
$$78.57 > x > 47.14 \qquad \text{(note: these values are approximate)}$$

This inequality says that our model is applicable when the car's speed is between about 47 mph and about 79 mph.

———————————

[a]cta.ornl.gov/data/chapter4.shtml

Exercises

Review and Warmup

1. Here is an interval:

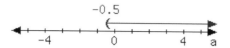

   Write the interval using set-builder notation.

   Write the interval using interval notation.

2. Here is an interval:

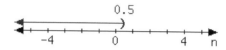

   Write the interval using set-builder notation.

   Write the interval using interval notation.

3. Here is an interval:

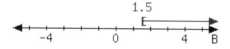

   Write the interval using set-builder notation.

   Write the interval using interval notation.

4. Here is an interval:

   Write the interval using set-builder notation.

   Write the interval using interval notation.

5. Solve this inequality.

   $5 > x + 10$

   In set-builder notation, the solution set is ☐.

   In interval notation, the solution set is ☐.

6. Solve this inequality.

   $1 > x + 8$

   In set-builder notation, the solution set is ☐.

   In interval notation, the solution set is ☐.

7. Solve this inequality.

   $-2x \geq 4$

   In set-builder notation, the solution set is ☐.

   In interval notation, the solution set is ☐.

8. Solve this inequality.

   $-2x \geq 8$

   In set-builder notation, the solution set is ☐.

   In interval notation, the solution set is ☐.

9. Solve this inequality.

   $4 \geq -5x + 4$

   In set-builder notation, the solution set is ☐.

   In interval notation, the solution set is ☐.

10. Solve this inequality.

    $2 \geq -6x + 2$

    In set-builder notation, the solution set is ☐.

    In interval notation, the solution set is ☐.

11. Solve this inequality.

$8t + 9 < 3t + 34$

In set-builder notation, the solution set is ☐.

In interval notation, the solution set is ☐.

12. Solve this inequality.

$9t + 6 < 5t + 18$

In set-builder notation, the solution set is ☐.

In interval notation, the solution set is ☐.

Check Solutions   Decide whether the given value for the variable is a solution.

13. a. $x > 8$ and $x \le 2$ $x = 4$

The given value   (☐ is   ☐ is not)   a solution.

b. $x < 8$ or $x \ge 5$ $x = 4$

The given value   (☐ is   ☐ is not)   a solution.

c. $x \ge -2$ and $x \le 6$ $x = 8$

The given value   (☐ is   ☐ is not)   a solution.

d. $-2 \le x \le 3$ $x = 1$

The given value   (☐ is   ☐ is not)   a solution.

14. a. $x > 9$ and $x \le 7$ $x = 8$

The given value   (☐ is   ☐ is not)   a solution.

b. $x < 6$ or $x \ge 5$ $x = 9$

The given value   (☐ is   ☐ is not)   a solution.

c. $x \ge -1$ and $x \le 9$ $x = -5$

The given value   (☐ is   ☐ is not)   a solution.

d. $-1 \le x \le 2$ $x = 1$

The given value   (☐ is   ☐ is not)   a solution.

Compound Inequalities and Interval Notation

15. Solve the compound inequality. Write the solution set in interval notation.

$-10 < x \le 5$

x is in ☐

16. Solve the compound inequality. Write the solution set in interval notation.

$-9 < x \le 1$

x is in ☐

17. Solve the compound inequality. Write the solution set in interval notation.

$-8 > x$ or $x \ge 8$

x is in ☐

18. Solve the compound inequality. Write the solution set in interval notation.

$-7 > x$ or $x \ge 4$

x is in ☐

19. Express the following inequality using interval notation.

$x < -6$ or $x \le 1$

x is in ☐

20. Express the following inequality using interval notation.

$x < -5$ or $x \le 7$

x is in ☐

Solving a Compound Inequality Algebraically   Solve the compound inequality algebraically.

21. $-14 < 7 - x \leq -9$

    x is in $\boxed{}$

22. $-16 < 20 - x \leq -11$

    x is in $\boxed{}$

23. $19 \leq x + 13 < 24$

    x is in $\boxed{}$

24. $1 \leq x + 7 < 6$

    x is in $\boxed{}$

25. $4 \leq \dfrac{5}{9}(F - 32) \leq 50$

    F is in $\boxed{}$

26. $8 \leq \dfrac{5}{9}(F - 32) \leq 43$

    F is in $\boxed{}$

27. $-10x - 11 \leq -20$ and $-2x - 18 < -10$

28. $17x + 6 \leq 9$ and $18x - 14 \leq 7$

29. $-9x - 8 \leq -5$ or $-4x - 15 \geq -13$

30. $-12x - 13 \geq -11$ or $-7x - 1 < -17$

31. $13x + 13 \leq -13$ or $-5x + 11 \leq -4$

32. $-9x + 13 < 2$ or $16x - 2 < -1$

33. $-13x + 10 < -7$ and $-3x + 4 < 8$

34. $13x - 14 < -18$ and $18x + 8 \leq -16$

35. $6 < \dfrac{2}{5}x < 20$

    In set-builder notation, the solution set is $\boxed{}$.

    In interval notation, the solution set is $\boxed{}$.

36. $15 < \dfrac{3}{2}x < 54$

    In set-builder notation, the solution set is $\boxed{}$.

    In interval notation, the solution set is $\boxed{}$.

37. $5 > -1 - \dfrac{3}{7}x \geq -10$

    In set-builder notation, the solution set is $\boxed{}$.

    In interval notation, the solution set is $\boxed{}$.

38. $20 > 4 - \dfrac{4}{5}x \geq -8$

    In set-builder notation, the solution set is $\boxed{}$.

    In interval notation, the solution set is $\boxed{}$.

Solving a Compound Inequality Graphically A graph of f is given. Use the graph alone to solve the compound inequalities.

39.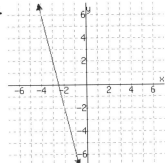

a. $f(x) > 2$

b. $-2 < f(x) \le 2$

40.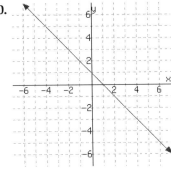

a. $f(x) > -1$

b. $-1 < f(x) \le 3$

41.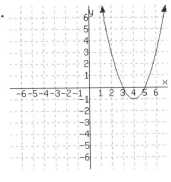

a. $f(x) > 0$

b. $f(x) \le 0$

42.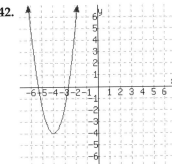

a. $f(x) > 5$

b. $f(x) \le 5$

43.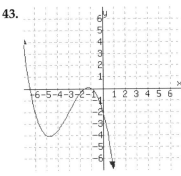

a. $f(x) > -2$

b. $f(x) \le -2$

44.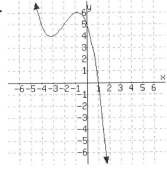

a. $f(x) > 5$

b. $f(x) \le 5$

Applications

45. As dry air moves upward, it expands. In so doing, it cools at a rate of about 1°C for every 100 m rise, up to about 12 km.

   a. If the ground temperature is 18°C, write a formula for the temperature at height x km. $T(x) = $

   b. What range of temperature will a plane be exposed to if it takes off and reaches a maximum height of 5 km? Write answer in interval notation.

      The range is

11.3 Absolute Value Equations and Inequalities

Whether it's a washer, nut, bolt, or gear, when a machine part is made, it must be made to fit with all of the other parts of the system. Since no manufacturing process is perfect, there are small deviations from the norm when each piece is made. In fact, manufacturers have a *range* of acceptable values for each measurement of every screw, bolt, etc.

Let's say we were examining some new bolts just out of the factory. The manufacturer specifies that each bolt should be within a *tolerance* of 0.04 mm to 10 mm in diameter. So the lowest diameter that the bolt could be to make it through quality assurance is 0.04 mm smaller than 10 mm, which is 9.96 mm. Similarly, the largest diameter that the bolt could be is 0.04 mm larger than 10 mm, which is 10.04 mm.

Summarizing, we want the difference between the actual diameter and the specification to be less than or equal to 0.04 mm. Since absolute values are used to describe distances, we can summarize our thoughts mathematically as $|x - 10| \le 0.04$, where x represents the diameter of an acceptably sized bolt, in millimeters. Since the minimum value is 9.96 mm and the maximum value is 10.04 mm, our range of acceptable values should be $9.96 \le x \le 10.04$.

In this section we will examine a variety of problems and applications that relate to this sort of math with absolute values.

11.3.1 Solving Absolute Value Equations

Recall in Section 11.1 that we learned that graphs of absolute value function are in general shaped like "V"s. We can now solve some absolute value equations graphically.

Example 11.3.2 Solve the equations graphically using the graphs provided.

a. $|x| = 3$

b. $|2x + 3| = 5$

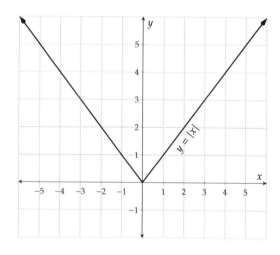
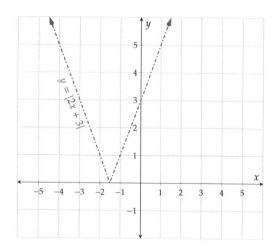

Explanation. To solve the equations graphically, first we need to graph the right sides of the equations also.

a. $|x| = 3$

b. $|2x + 3| = 5$

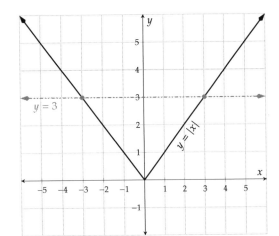

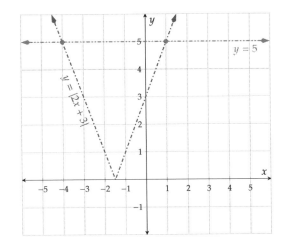

Since the graph of $y = |x|$ crosses $y = 3$ at the x-values -3 and 3, the solution set to the equation $|x| = 3$ must be $\{-3, 3\}$.

Since the graph of $y = |2x + 3|$ crosses $y = 5$ at the x-values -4 and 1, the solution set to the equation $|2x + 3| = 5$ must be $\{-4, 1\}$.

Remark 11.3.3. At this point, please note that there is a big difference between the expression $|3|$ and the equation $|x| = 3$.

1. The expression $|3|$ is describing the distance from 0 to the number 3. The distance is just 3. So $|3| = 3$.

2. The equation $|x| = 3$ is asking you to find the numbers that are a distance of 3 from 0. We saw in Explanation 11.3.2.1 that these two numbers are 3 and -3.

Example 11.3.4

a. Verify that the value 4 is a solution to the absolute value equation $|2x - 3| = 5$.

b. Verify that the value $\frac{3}{2}$ is a solution to the absolute value equation $\left|\frac{1}{6}x - \frac{1}{2}\right| = \frac{1}{4}$.

Explanation.

a. We will substitute the value 4 into the absolute value equation $|2x - 3| = 5$. We get:

$$|2x - 3| = 5$$
$$|2 \cdot 4 - 3| \stackrel{?}{=} 5$$
$$|8 - 3| \stackrel{?}{=} 5$$
$$|5| \stackrel{\checkmark}{=} 5$$

b. We will substitute the value $\frac{3}{2}$ into the absolute value equation $\left|\frac{1}{6}x - \frac{1}{2}\right| = \frac{1}{4}$. We get:

$$\left|\frac{1}{6}x - \frac{1}{2}\right| = \frac{1}{4}$$
$$\left|\frac{1}{6} \cdot \frac{3}{2} - \frac{1}{2}\right| \stackrel{?}{=} \frac{1}{4}$$

$$\left|\frac{1}{4} - \frac{1}{2}\right| \overset{?}{=} \frac{1}{4}$$

$$\left|-\frac{1}{4}\right| \overset{\checkmark}{=} \frac{1}{4}$$

Now we will learn to solve absolute value equations algebraically. To motivate this, we will think about what an absolute value equation means in terms of the "distance from zero" definition of absolute value. If

$$|X| = n,$$

where $n \geq 0$, then this means that we want all of the numbers, X, that are a distance n from 0. Since we can only go left or right along the number line, this is describing both $X = n$ as well as $X = -n$.

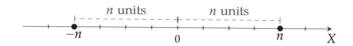

Figure 11.3.5: A Numberline with Points a Distance n from 0

Let's summarize this with a fact.

Fact 11.3.6 Equations with an Absolute Value Expression. *Let n be a non-negative number and X be an algebraic expression. Then the equation*

$$|X| = n$$

has the same solutions as

$$X = n \text{ or } X = -n.$$

Example 11.3.7 Solve the absolute value equations using Fact 11.3.6. Write solutions in a solution set.

a. $|x| = 6$

b. $|x| = -4$

c. $|5x - 7| = 23$

d. $|14 - 3x| = 8$

e. $|3 - 4x| = 0$

Explanation.

a. Fact 11.3.6 says that the equation $|x| = 6$ is the same as

$$x = 6 \text{ or } x = -6.$$

Thus the solution set is $\{6, -6\}$.

b. Fact 11.3.6 doesn't actually apply to the equation $|x| = -4$ because the value on the right side is *negative*. How often is an absolute value of a number negative? Never! Thus, there are no solutions and the solution set is the empty set, denoted $\emptyset$.

c. The equation $|5x - 7| = 23$ breaks into two pieces, each of which needs to be solved independently.

$$
\begin{array}{ccc}
5x - 7 = 23 & \text{or} & 5x - 7 = -23 \\
5x = 30 & \text{or} & 5x = -16 \\
x = 6 & \text{or} & x = -\frac{16}{5}
\end{array}
$$

Thus the solution set is $\left\{6, -\frac{16}{5}\right\}$.

d. The equation $|14 - 3x| = 8$ breaks into two pieces, each of which needs to be solved independently.

$$14 - 3x = 8 \qquad\qquad \text{or} \qquad\qquad 14 - 3x = -8$$
$$-3x = -6 \qquad\qquad \text{or} \qquad\qquad -3x = -22$$
$$x = 2 \qquad\qquad \text{or} \qquad\qquad x = \frac{22}{3}$$

Thus the solution set is $\left\{2, \frac{22}{3}\right\}$.

e. The equation $|3 - 4x| = 0$ breaks into two pieces, each of which needs to be solved independently.

$$3 - 4x = 0 \qquad\qquad \text{or} \qquad\qquad 3 - 4x = -0$$

Since these are identical equations, all we have to do is solve one equation.

$$3 - 4x = 0$$
$$-4x = -3$$
$$x = \frac{3}{4}$$

Thus, the equation $|3 - 4x| = 0$ only has one solution, and the solution set is $\left\{\frac{3}{4}\right\}$.

Now we will look at an equation with an absolute value expression on each side, such as $|x| = |2x + 6|$. Since $|x| = 5$ has two solutions, you might be wondering how many solutions $|x| = |2x + 6|$ will have. Let's look at a graph to find out.

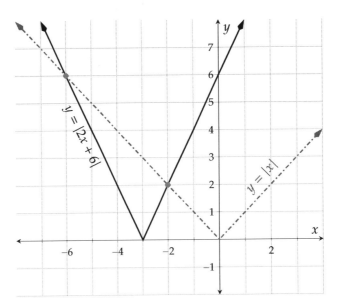

Figure 11.3.8: $y = |x|$ and $y = |2x + 6|$

Figure 11.3.8 shows that there are also two points of intersection between the graphs of $y = |x|$ and $y = |2x + 6|$. The solutions to the equation $|x| = |2x + 6|$ are the x-values where the graphs cross. So, the solution set is $\{-6, -2\}$.

Example 11.3.9 Solve the equation $|x + 1| = |2x - 4|$ graphically.

Explanation.

First break up the equation into the left side and the right side and graph each separately, as in $y = |x + 1|$ and $y = |2x - 4|$. We can see in the graph that the graphs intersect twice. The x-values of those intersections are 1 and 5 so the solution set to the equation $|x + 1| = |2x - 4|$ is $\{1, 5\}$.

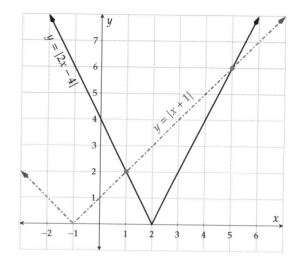

Figure 11.3.10: $y = |x + 1|$ and $y = |2x - 4|$

Fact 11.3.11 Equations with Two Absolute Value Expressions. *Let X and Y be linear algebraic expressions. Then, the equation*

$$|X| = |Y|$$

has the same solutions as

$$X = Y \text{ or } X = -Y.$$

Remark 11.3.12. You might be confused as to why the negative sign *has* to go on the right side of the equation in $X = -Y$. Well, it doesn't: it can go on either side of the equation. The equations $X = -Y$ and $-X = Y$ are equivalent. Similarly, $-X = -Y$ is equivalent to $X = Y$. That's why we only need to solve two of the four possible equations.

Example 11.3.13 Solve the equations using Fact 11.3.11.

    a. $|x - 4| = |3x - 2|$ c. $|x - 2| = |x + 1|$

    b. $\left|\frac{1}{2}x + 1\right| = \left|\frac{1}{3}x + 2\right|$ d. $|x - 1| = |1 - x|$

Explanation.

    a. The equation $|x - 4| = |3x - 2|$ breaks down into two pieces:

$$
\begin{array}{lcl}
x - 4 = 3x - 2 & \text{or} & x - 4 = -(3x - 2) \\
x - 4 = 3x - 2 & \text{or} & x - 4 = -3x + 2 \\
-2 = 2x & \text{or} & 4x = 6 \\
\dfrac{-2}{2} = \dfrac{2x}{2} & \text{or} & \dfrac{4x}{4} = \dfrac{6}{4} \\
-1 = x & \text{or} & x = \dfrac{3}{2}
\end{array}
$$

So, the solution set is $\left\{-1, \frac{3}{2}\right\}$.

b. The equation $\left|\frac{1}{2}x + 1\right| = \left|\frac{1}{3}x + 2\right|$ breaks down into two pieces:

$$\frac{1}{2}x + 1 = \frac{1}{3}x + 2 \qquad \text{or} \qquad \frac{1}{2}x + 1 = -\left(\frac{1}{3}x + 2\right)$$

$$\frac{1}{2}x + 1 = \frac{1}{3}x + 2 \qquad \text{or} \qquad \frac{1}{2}x + 1 = -\frac{1}{3}x - 2$$

$$6 \cdot \left(\frac{1}{2}x + 1\right) = 6 \cdot \left(\frac{1}{3}x + 2\right) \qquad \text{or} \qquad 6 \cdot \left(\frac{1}{2}x + 1\right) = 6 \cdot \left(-\frac{1}{3}x - 2\right)$$

$$3x + 6 = 2x + 12 \qquad \text{or} \qquad 3x + 6 = -2x - 12$$

$$x = 6 \qquad \text{or} \qquad 5x = -18$$

$$x = 6 \qquad \text{or} \qquad x = -\frac{18}{5}$$

So, the solution set is $\left\{6, -\frac{18}{5}\right\}$.

c. The equation $|x - 2| = |x + 1|$ breaks down into two pieces:

$$x - 2 = x + 1 \qquad \text{or} \qquad x - 2 = -(x + 1)$$

$$x - 2 = x + 1 \qquad \text{or} \qquad x - 2 = -x - 1$$

$$x = x + 3 \qquad \text{or} \qquad 2x = 1$$

$$0 = 3 \qquad \text{or} \qquad x = \frac{1}{2}$$

Note that one of the two pieces gives us an equation with no solutions. Since $0 \neq 3$, we can safely ignore this piece. Thus the only solution is $\frac{1}{2}$.

We should visualize this equation graphically because our previous assumption was that two absolute value graphs would cross twice. The graph shows why there is only one crossing: the left and right sides of each "V" are parallel.

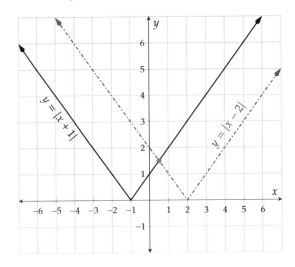

d. The equation $|x - 1| = |1 - x|$ breaks down into two pieces:

$$x - 1 = 1 - x \qquad \text{or} \qquad x - 1 = -(1 - x)$$
$$x - 1 = 1 - x \qquad \text{or} \qquad x - 1 = -1 + x$$
$$2x = 2 \qquad \text{or} \qquad x = 0 + x$$
$$x = 1 \qquad \text{or} \qquad 0 = 0$$

Note that our second equation is an identity so recall from Section 3.6 that the solution set is "all real numbers."

So, our two pieces have solutions 1 and "all real numbers." Since 1 *is* a real number and we have an *or* statement, our overall solution set is $(-\infty, \infty)$. The graph confirms our answer since the two "V" graphs are coinciding.

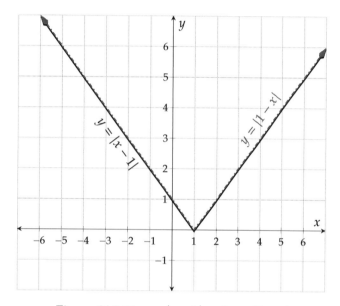

Figure 11.3.14: $y = |x - 1|$ and $y = |1 - x|$

11.3.2 Solving Absolute Value Inequalities

Now we turn our attention away from equations and onto absolute value inequalities. Don't dismiss this topic as it will actually be used in some capacity in many subsequent math courses. So let's give these the full treatment. We start with a graphical interpretation of what $|2x - 1| \leq 5$ means.

Graphically solving the inequality $|2x - 1| \leq 5$ means looking for the x-values where the graph of $y = |2x - 1|$ is below (or touching) the line $y = 5$. On the graph the highlighted region of $y = |2x - 1|$ is the portion that is below the line $y = 5$, and the x-values in that region are $[-2, 5]$.

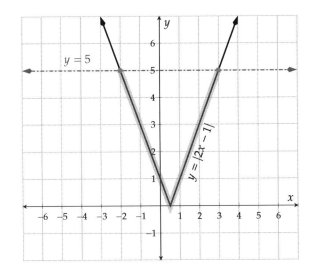

Figure 11.3.15: $y = |2x - 1|$ and $y = 5$

Example 11.3.16 Solve the inequality $\left|\frac{2}{3}x + 1\right| < 3$ graphically.

Explanation. To solve the inequality $\left|\frac{2}{3}x + 1\right| < 3$, we will start by making a graph with both $y = \left|\frac{2}{3}x + 1\right|$ and $y = 3$.

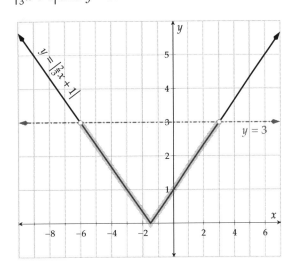

Figure 11.3.17: $y = \left|\frac{2}{3}x + 1\right|$ and $y = 3$

The portion of the graph of $y = \left|\frac{2}{3}x + 1\right|$ that is below $y = 3$ is highlighted and the x-values of that highlighted region are trapped between -6 and 3: $-6 < x < 3$. That means that the solution set is $(-6, 3)$. Note that we shouldn't include the endpoints of the interval because at those values, the two graphs are *equal* whereas the original inequality was only *less than* and not equal.

For a more verbal approach to understanding the concept, let's try to describe "values that are less than 4 units from 0." We would say that those are "numbers between -4 and 4." Let's translate each sentence into math. "Values that are less than 4 units from 0" translates to "$|x| < 4$," and the piece "numbers between -4 and 4" translates to be "$-4 < x < 4$."

For a graphical interpretation, let's think in terms of the "distance from zero" definition of absolute value. If

$$|X| \leq n,$$

where $n \geq 0$, then we want all of the numbers, X, that are a distance n or less from 0. Since we can only go left or right along the number line, this is describing all numbers from $-n$ to n.

Figure 11.3.18: A Numberline with Points a Distance n or less from 0

Fact 11.3.19 An Absolute Value Expression Less Than a Value. *Let n be a non-negative number and X be a linear algebraic expression.*

Then, the inequality $|X| < n$ has the same solutions as the compound inequality $-n < X < n$.

Likewise, the inequality $|X| \le n$ has the same solutions as the compound inequality $-n \le X \le n$.

Example 11.3.20 Solve the absolute value inequalities using Fact 11.3.19.

a. $|x| \le 9$

b. $|x| < -6$

c. $|4x + 3| < 9$

d. $3 \cdot |3 - x| + 1 \le 13$

Explanation.

a. The inequality $|x| < 9$ breaks down into a triple inequality:

$$-9 \le x \le 9$$

This inequality is already written in simplest form and all that remains for us to do is to write the solution set in interval notation: $[-9, 9]$.

b. Fact 11.3.19 doesn't apply to the inequality $|x| < -6$ because the right side is a negative number. Let's translate the meaning of the inequality into English. It says, "The distance from 0 to what numbers is less than -6?" Since we define distance to be non-negative, there are no possible numbers that are less than -6 units distance from 0. Thus, the solution set is the empty set, denoted $\emptyset$.

c. The inequality $|4x + 3| < 9$ breaks down into a triple inequality that we can then solve:

$$-9 < 4x + 3 < 9$$
$$-9 - 3 < 4x + 3 - 3 < 9 - 3$$
$$-12 < 4x < 6$$
$$\frac{-12}{4} < \frac{4x}{4} < \frac{6}{4}$$
$$-3 < x < \frac{3}{2}$$

So, the solution set to the inequality is $\left(-3, \frac{3}{2}\right)$.

d. The inequality $3 \cdot |3 - x| + 1 \le 13$ *must* be simplified into the form that matches Fact 11.3.19, so we will first isolate the absolute value expression on the left side of the inequality:

$$3 \cdot |3 - x| + 1 \le 13$$
$$3 \cdot |3 - x| \le 12$$

$$|3 - x| \leq 4$$

Now that we have the absolute value isolated, we can split it into a triple inequality that we can finish solving:

$$-4 \leq 3 - x \leq 4$$
$$-4 - 3 \leq 3 - x - 3 \leq 4 - 3$$
$$-7 \leq -x \leq 1$$
$$\frac{-7}{-1} \geq \frac{-x}{-1} \geq \frac{1}{-1}$$
$$7 \geq x \geq -1$$

So, the solution set to the inequality is $[-1, 7]$.

Example 11.3.21 If a machined circular washer must have a circumference that is within 0.2 mm of 36 mm, then what is the acceptable range for the radius of the washer? Round your answers to the nearest hundredth of a millimeter.

Explanation. We will first define the radius of the washer to be r, measured in millimeters. The formula $C = 2\pi r$ gives us the circumference, C, of a circle with radius r. Now we know that "distance" between the circumference and our preferred circumference of 36 mm must be less than or equal to 0.2 mm. In math, this translates to

$$|C - 36| \leq 0.2$$

Now we can substitute our formula for circumference and solve for r.

$$|C - 36| \leq 0.2$$
$$|2\pi r - 36| \leq 0.2$$

To solve this we will use Fact 11.3.19 to break the absolute value inequality into a triple inequality:

$$-0.2 \leq 2\pi r - 36 \leq 0.2$$
$$-0.2 + 36 \leq 2\pi r - 36 + 36 \leq 0.2 + 36$$
$$35.8 \leq 2\pi r \leq 36.2$$
$$\frac{35.8}{2\pi} \leq \frac{2\pi r}{2\pi} \leq \frac{36.2}{2\pi}$$
$$5.70 \leq r \leq 5.76 \qquad \text{(note: these values are rounded)}$$

This shows that the radius must be somewhere between 5.70 mm and 5.76 mm, inclusive.

The last few examples have all revolved around absolute values being *less than* some value. We now need to investigate what happens when we have an absolute value that is *greater than* a value. We will again start with a graphical interpretation.

Example 11.3.22 To graphically solve the inequality $|x - 1| > 3$ would mean looking for the x-values where the graph of $y = |x - 1|$ is *above* the line $y = 3$.

On the graph the highlighted region of $y = |x - 1|$ is the portion that is above the line $y = 3$ and the x-values in that region can be represented by $(-\infty, -2) \cup (4, \infty)$.

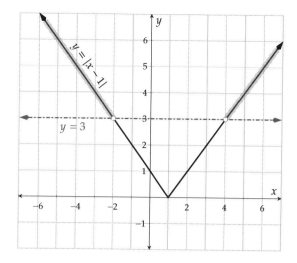

Figure 11.3.23: $y = |x - 1|$ and $y = 3$

Example 11.3.24 Solve the inequality $\left|\frac{1}{3}x + 2\right| \geq 6$ graphically.

Explanation. To solve the inequality $\left|\frac{1}{3}x + 2\right| \geq 6$, we will start by making a graph with both $y = \left|\frac{1}{3}x + 2\right|$ and $y = 6$.

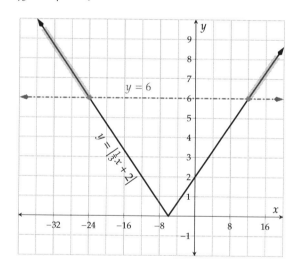

The portion of the graph of $y = \left|\frac{1}{3}x + 2\right|$ that is above $y = 6$ is highlighted and the x-values of that highlighted region are those below (or equal to) -24 and those above (or equal to) 12: $x \leq -24$ or $x \geq 12$. That means that the solution set is $(-\infty, -24) \cup (12, \infty)$.

Figure 11.3.25: $y = \left|\frac{1}{3}x + 2\right|$ and $y = 3$

Again, for a more verbal approach to understanding the concept, lets try to describe "values that are more than 4 units from 0." We would say that those are "numbers below −4 as well as numbers above 4." We will again translate each sentence into math. "Values that are more than 4 units from 0" translates to "$|x| > 4$," and the piece "numbers below −4 as well as numbers above 4" translates to be "$x < -4$ or $x > 4$."

For a graphical interpretation, let's think in terms of the "distance from zero" definition of absolute value. If

$$|X| \geq n,$$

where $n \geq 0$, then we want all of the numbers, X, that are a distance n or more from 0. Since we can only go left or right along the number line, this is describing all numbers below $-n$ as well as those above n.

Figure 11.3.26: A Numberline with Points a Distance n or less from 0

Fact 11.3.27 An Absolute Value Expression Greater Than a Value. *Let n be a non-negative number and X be a linear algebraic expression.*

Then, the inequality $|X| > n$ has the same solutions as the compound inequality $X < -n$ or $X > n$.

Likewise, the inequality $|X| \geq n$ has the same solutions as the compound inequality $X \leq -n$ or $X \geq n$.

Remark 11.3.28. Since Fact 11.3.27 specifies that an "absolute value greater than a number"-type inequality breaks down into an *or* statement, we will therefore need to find the *union* of the solution sets of the pieces.

Example 11.3.29 Solve the absolute value inequalities using Fact 11.3.27.

a. $|x| \geq 4$

b. $|x| > -2$

c. $|5x - 7| > 7$

d. $2 \cdot |3 - 2x| - 5 \geq 13$

Explanation.

a. The inequality $|x| \geq 4$ breaks down into a compound inequality:

$$x \leq -4 \qquad \text{or} \qquad x \geq 4$$

So, the solution set is $(-\infty, -4] \cup [4, \infty)$.

b. Fact 11.3.27 doesn't apply to the inequality $|x| > -2$ because the right side is negative. Instead, we will make sense of it logically. This is asking, "When is an absolute value greater than a negative number?" The answer is that absolute values are *always* bigger than negative numbers! So, our solution set is $(-\infty, \infty)$.

c. The inequality $|5x - 7| > 7$ breaks down into a compound inequality:

$$
\begin{array}{ccc}
5x - 7 < -7 & \text{or} & 5x - 7 > 7 \\
5x < 0 & \text{or} & 5x > 14 \\
x < 0 & \text{or} & x > \dfrac{14}{5}
\end{array}
$$

We will write the solution set as $(-\infty, 0) \cup \left(\frac{14}{5}, \infty\right)$.

d. Before we break up the inequality $2 \cdot |3 - 2x| - 5 \geq 13$ into an "or" statement, we *must* isolate the absolute value expression:

$$
\begin{aligned}
2 \cdot |3 - 2x| - 5 &\geq 13 \\
2 \cdot |3 - 2x| &\geq 18 \\
|3 - 2x| &\geq 9
\end{aligned}
$$

Now that the absolute value expression has been isolated on the left side, we can use Fact 11.3.27 to break it into an "or" statement:

$$3 - 2x \le -9 \qquad \text{or} \qquad 3 - 2x \ge 9$$
$$-2x \le -12 \qquad \text{or} \qquad -2x \ge 6$$
$$x \ge 6 \qquad \text{or} \qquad x \le -3$$

Our final simplified solution set is $(-\infty, 3] \cup [6, \infty)$.

Example 11.3.30 Phuong is taking the standard climbing route on Mount Hood from Timberline Lodge up the Southside Hogsback and back down. Her altitude can be very closely modeled by an absolute value function since the angle of ascent is nearly constant. Let x represent the number of miles walked from Timberline Lodge, and let $f(x)$ represent the altitude, in miles, after walking for a distance x. The altitude can be modeled by $f(x) = 2.1 - 0.3077 \cdot |x - 3.25|$. Note that below Timberline Lodge this model fails to be accurate.

  a. Solve the equation $f(x) = 1.1$ and interpret the results in the context of the problem.

  b. Altitude sickness can occur at altitudes above 1.5 miles. Set up and solve an inequality to find out how far Phuong can walk the trail and still be under 1.5 miles of elevation.

Explanation.

  a. First, we substitute the formula for $f(x)$ and simplify the equation.

$$f(x) = 1.1$$
$$2.1 - 0.3077 \cdot |x - 3.25| = 1.1$$
$$-0.3077 \cdot |x - 3.25| = -1$$
$$\frac{-0.3077 \cdot |x - 3.25|}{-0.3077} = \frac{-1}{-0.3077}$$
$$|x - 3.25| \approx 3.25$$

At this point, we can use Fact 11.3.6 to split apart the equation:

$$x - 3.25 \approx 3.25 \qquad \text{or} \qquad x - 3.25 \approx -3.25$$
$$x \approx 6.5 \qquad \text{or} \qquad x \approx 0$$

According to the model, Phuong will be at 1.1 miles of elevation after walking about 0 miles and about 6.5 miles along the trail. This seems to imply that Timberline Lodge is very close to 1.1 miles of elevation. In addition, it implies that the entire hike is 6.5 miles round trip, ending at Timberline Lodge again.

  b. The inequality we are looking for will describe when the altitude is below 1.5 miles. Since $f(x)$ is the altitude, the inequality we need is:

$$f(x) < 1.5$$

To solve this, we need to input the formula and simplify before using one of the absolute value inequality rules.

$$f(x) < 1.5$$
$$2.1 - 0.3077 \cdot |x - 3.25| < 1.5$$
$$-0.3077 \cdot |x - 3.25| < -0.6$$

$$\frac{-0.3077 \cdot |x - 3.25|}{-0.3077} > \frac{-0.6}{-0.3077}$$

$$|x - 3.25| > 1.95 \qquad \text{(note: this value is rounded)}$$

At this point, we can use Fact 11.3.27 to split apart the inequality:

$x - 3.25 < -1.95$	or	$x - 3.25 > 1.95$
$x < 1.3$	or	$x > 5.2$

The image only shows the portion of the graph that is above Timberline Lodge, which we learned was at 1.1 miles in elevation in the previous part. The highlighted portions of the graph are those indicated by $x > 5.2$ or $x < 1.3$.

Figure 11.3.31: $y = f(x)$, the Graph of the Mt Hood Ascent and Descent

In conclusion, based both on our math and the reality of the situation, regions of the trail that are below 1.5 miles are those that are from Timberline Lodge (at 0 miles on the trail), to 1.3 miles along the trail and then also from 5.2 miles along the trail (and by now we are on our way back down) to 6.5 miles along the trail (back at Timberline Lodge). If we wanted to write this in interval notation, we might write $[0, 1.3) \cup (5.2, 6.5]$. There is a big portion along the trail (from 1.3 miles to 5.2 miles) that Phuong will be above the 1.5 mile altitude and should watch for signs of altitude sickness.

Exercises

Review and Warmup

1. Solve the equation.

$$\frac{c}{5} - 6 = \frac{c}{7}$$

2. Solve the equation.

$$\frac{A}{3} - 10 = \frac{A}{8}$$

3. Solve the equation.

$$-30 = -10(C + 10)$$

4. Solve the equation.

$$-98 = -7(m + 5)$$

5. Solve the equation.

$$4p + 9 = 7p + 10$$

6. Solve the equation.

$$10x + 4 = 5x + 10$$

7. Solve this inequality.

$17 \geq 3x - 4$

In set-builder notation, the solution set is [ ].

In interval notation, the solution set is [ ].

8. Solve this inequality.

$6 \geq 4x - 2$

In set-builder notation, the solution set is [ ].

In interval notation, the solution set is [ ].

9. Solve this inequality.

$-5x - 9 < -39$

In set-builder notation, the solution set is [ ].

In interval notation, the solution set is [ ].

10. Solve this inequality.

$-6x - 6 < -66$

In set-builder notation, the solution set is [ ].

In interval notation, the solution set is [ ].

11. Solve this inequality.

$-3 > 3 - x$

In set-builder notation, the solution set is [ ].

In interval notation, the solution set is [ ].

12. Solve this inequality.

$-6 > 4 - x$

In set-builder notation, the solution set is [ ].

In interval notation, the solution set is [ ].

Solving Absolute Value Equations Algebraically

13. a. Write the equation $6 = |3x| - 5$ as two separate equations. Neither of your equations should use absolute value.

[ ]

[ ]

b. Solve both equations above.

14. a. Write the equation $7 = |5x| - 3$ as two separate equations. Neither of your equations should use absolute value.

[ ]

[ ]

b. Solve both equations above.

15. a. Write the equation $\left| 8 - \dfrac{r}{7} \right| = 3$ as two separate equations. Neither of your equations should use absolute value.

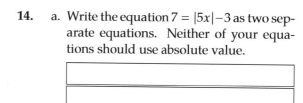

b. Solve both equations above.

16. a. Write the equation $\left| 2 - \dfrac{r}{5} \right| = 7$ as two separate equations. Neither of your equations should use absolute value.

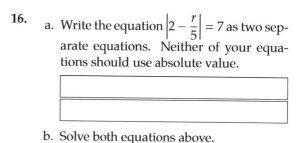

b. Solve both equations above.

17. (a) Verify that the value -1 is a solution to the absolute value equation $\left| \dfrac{x-3}{2} \right| = 2$.

(b) Verify that the value $\dfrac{2}{3}$ is a solution to the absolute value equation $|6x - 5| < 4$.

18. (a) Verify that the value 8 is a solution to the absolute value equation $\left| \dfrac{1}{2}x - 2 \right| = 2$.

(b) Verify that the value 6 is a solution to the absolute value equation $|7 - 2x| \geq 5$.

19. Solve the following equation.

$$|3x - 9| = 9$$

20. Solve the following equation.

$$|4x + 5| = 3$$

21. Solve the equation $|3x - 1| = 17$.

22. Solve the equation $|4x - 4| = 10$.

23. Solve: $|x| = 9$

24. Solve: $|x| = 5$

25. Solve: $|y - 1| = 11$

26. Solve: $|y - 5| = 15$

27. Solve: $|2a + 3| = 9$

28. Solve: $|2b + 7| = 13$

29. Solve: $\left|\dfrac{2b - 5}{9}\right| = 3$

30. Solve: $\left|\dfrac{2t - 3}{5}\right| = 1$

31. Solve: $|t| = -4$

32. Solve: $|x| = -6$

33. Solve: $|x + 2| = 0$

34. Solve: $|y + 4| = 0$

35. Solve: $|4 - 3y| = 9$

36. Solve: $|2 - 3a| = 14$

37. Solve: $\left|\frac{1}{4}b + 3\right| = 1$

38. Solve: $\left|\frac{1}{2}b + 5\right| = 1$

39. Solve: $|0.2 - 0.1t| = 4$

40. Solve: $|0.8 - 0.4t| = 3$

41. Solve: $|x + 5| - 2 = 2$

42. Solve: $|x + 1| - 4 = 6$

43. Solve: $|4y - 20| + 6 = 6$

44. Solve: $|3y - 6| + 4 = 4$

45. Solve: $|a + 1| + 7 = 6$

46. Solve: $|b + 7| + 7 = 2$

47. Solve: $|4b + 3| + 7 = 4$

48. Solve: $|4t + 1| + 8 = 6$

49. Solve the equation *by inspection* (meaning in your head).

$$|6x + 18| = 0$$

50. Solve the equation *by inspection* (meaning in your head).

$$|6x + 12| = 0$$

51. The equation $|x| = |y|$ is satisfied if $x = y$ or $x = -y$. Use this fact to solve the following equation.

$$|2x + 4| = |-3x - 1|$$

52. The equation $|x| = |y|$ is satisfied if $x = y$ or $x = -y$. Use this fact to solve the following equation.

$$|3x + 1| = |x - 3|$$

53. The equation $|x| = |y|$ is satisfied if $x = y$ or $x = -y$. Use this fact to solve the following equation.

$$|x + 6| = |x - 5|$$

54. The equation $|x| = |y|$ is satisfied if $x = y$ or $x = -y$. Use this fact to solve the following equation.

$$|x + 2| = |x - 1|$$

55. Solve the equation: $|2x - 6| = |9x + 4|$

56. Solve the equation: $|4x - 3| = |5x + 2|$

57. Solve the following equation.

$$|3x + 8| = |9x + 6|$$

58. Solve the following equation.

$$|3x + 1| = |6x - 4|$$

Testing Possible Solutions   Decide whether the given value for the variable is a solution.

59. a. $|x - 6| \leq 2$ $x = 5$

The given value   ($\square$ is $\square$ is not)   a solution.

b. $\left|\frac{2}{3}x - 1\right| \geq 7$ $x = 6$

The given value   ($\square$ is $\square$ is not)   a solution.

c. $|8t - 5| > 6$ $t = 7$

The given value   ($\square$ is $\square$ is not)   a solution.

d. $|3(z - 3)| < 8$ $z = \pi$

The given value   ($\square$ is $\square$ is not)   a solution.

60. a. $|x - 7| \leq 8$ $x = 9$

The given value   ($\square$ is $\square$ is not)   a solution.

b. $\left|3x - \frac{3}{8}\right| \geq 2$ $x = -4$

The given value   ($\square$ is $\square$ is not)   a solution.

c. $|4t - 5| > 5$ $t = 6$

The given value   ($\square$ is $\square$ is not)   a solution.

d. $|8(z - 6)| < 6$ $z = \pi$

The given value   ($\square$ is $\square$ is not)   a solution.

Solving Absolute Value Equations Graphically

61. Solve the equations and inequalities graphically. Use interval notation when applicable.

a. $\left|\frac{2}{3}x + 2\right| = 4$

b. $\left|\frac{2}{3}x + 2\right| > 4$

c. $\left|\frac{2}{3}x + 2\right| \leq 4$

62. Solve the equations and inequalities graphically. Use interval notation when applicable.

a. $\left|\frac{11 - 2x}{5}\right| = 4$

b. $\left|\frac{11 - 2x}{5}\right| > 4$

c. $\left|\frac{11 - 2x}{5}\right| \leq 4$

Solving Absolute Value Inequalities Algebraically Solve the inequality.

63. $\left|\dfrac{7-x}{6}\right| \geq 7$

64. $\left|\dfrac{8-x}{3}\right| \geq 12$

65. $|9 - x| \geq 5$

66. $|6 - 2x| \geq 10$

67. $|3x - 2| < 3$

68. $|4x - 8| < 8$

69. $\left|\dfrac{x+5}{5}\right| \leq 13$

70. $\left|\dfrac{x+6}{2}\right| \leq 6$

71. $|x - 7| > 13$

72. $|x - 7| > 10$

73. $|2 - 8x| < 9$

74. $|7 - x| < 15$

75. $20 - |3x + 1| \leq 6$

76. $11 - |4x + 7| \leq 1$

Challenge

77. Algebraically, solve for x in the equation:

$$5 = |x - 5| + |x - 10|$$

11.4 Absolute Value Functions Chapter Review

11.4.1 Introduction to Absolute Value Functions

In Section 11.1 we covered the definition of absolute value, what the graphs of absolute value functions look like, the fact that $\sqrt{x^2} = |x|$, and applications of absolute values.

> **Example 11.4.1 Evaluating Absolute Value Functions.** Given that $h(x) = |9 - 4x|$, evaluate the following expressions.
>
> a. $h(-1)$.
>
> b. $h(4)$.
>
> **Explanation.**
>
> a. $h(-1) = |9 - 4(-1)|$
>
> $ = |9 + 4|$
>
> $ = |13|$
>
> $ = 13$
>
> b. $h(4) = |9 - 4(4)|$
>
> $ = |9 - 16|$
>
> $ = |-7|$
>
> $ = 7$

> **Example 11.4.2 Graphing Absolute Value Functions.** Absolute value functions always make "V" shaped graphs. We usually use technology to make graphs to help speed up the process. Use technology to make a graph of $y = |2x - 6| - 4$.
>
> **Explanation.** To make a graph of a function, we often use technology to generate a table of values for that function. Then we use the graph that the technology creates to thoughtfully connect the points.
>
> | x | $y = |2x - 6| - 4$ |
> |---|---|
> | -1 | 4 |
> | 0 | 2 |
> | 1 | 0 |
> | 2 | -2 |
> | 2 | -4 |
> | 2 | -2 |
> | 2 | 0 |
>
>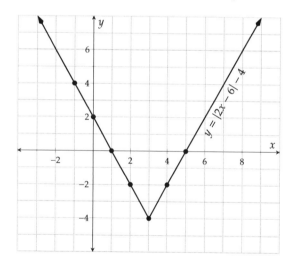
>
> **Table 11.4.3:** A table of values for $y = |2x - 6| - 4$
>
> **Figure 11.4.4:** A graph of $y = |2x - 6| - 4$

Example 11.4.5 The Alternate Definition of Absolute Value: $|x| = \sqrt{x^2}$. Simplify the following expressions using the fact that $|x| = \sqrt{x^2}$.

a. $\sqrt{x^{14}}$

b. $\sqrt{x^2 - 12x + 36}$

Explanation.

a. $\sqrt{x^{14}} = \sqrt{(x^7)^2}$
$\phantom{\sqrt{x^{14}}} = |x^7|$

We know from exponent rules that $(x^7)^2 = x^{14}$. Note that x^7 will be negative whenever x is a negative number, so the absolute value is meaningful.

b. $\sqrt{x^2 - 12x + 36} = \sqrt{(x - 6)^2}$
$\phantom{\sqrt{x^2 - 12x + 36}} = |x - 6|$

Note that $x - 6$ can be negative for certain values of x, so the absolute value is meaningful.

Example 11.4.6 An Application of Absolute Value. Mariam arrived at school one day only to realize that she had left her favorite pencil on her porch at home. She hopped on her bicycle and headed back to get it. Her distance from her home, $d(t)$ in yards, can be modeled as a function of the time, t in seconds, since she left school:

$$d(t) = |5t - 300|$$

Use this function to answer the following questions.

a. Find and interpret the meaning of $d(0)$.

b. Using technology, make a graph of $y = d(t)$.

c. Using your graph, find out how long it took Mariam to get to her home to get her pencil and get back to school.

Explanation.

a. $d(0) = |5(0) - 300|$
$ = |-300|$
$ = 300$

This means that just as Mariam was leaving her school, she was 300 yards from her home.

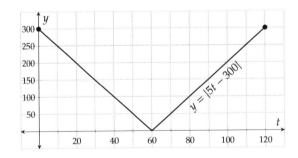

b.

c. Mariam was back at a y-value of 300 at $t = 120$. We should assume that she is back at her school again here. So it took her 120 seconds, which is 2 minutes.

11.4.2 Compound Inequalities

In Section 11.2 we defined the union of intervals, what compound inequalities are, and how to solve both "or" inequalities and triple inequalities.

Example 11.4.7 Unions of Intervals. Draw a representation of the union of the sets $(-\infty, -1]$ and $(2, \infty)$.

Explanation. First we make a number line with both intervals drawn to understand what both sets mean.

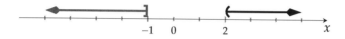

Figure 11.4.8: A number line sketch of $(-\infty, -1]$ as well as $(2, \infty)$

The two intervals should be viewed as a single object when stating the union, so here is the picture of the union. It looks the same, but now it is a graph of a single set.

Figure 11.4.9: A number line sketch of $(-\infty, -1] \cup (2, \infty)$

Example 11.4.10 "Or" Compound Inequalities. Solve the compound inequality.

$$5z + 12 \leq 7 \text{ or } 3 - 9z < -2$$

Explanation. First we will solve each inequality for z.

$$
\begin{array}{ccc}
5z + 12 \leq 7 & \text{or} & 3 - 9z < -2 \\
5z \leq -5 & \text{or} & -9z < -5 \\
z \leq -1 & \text{or} & z > \dfrac{5}{9}
\end{array}
$$

The solution set to the compound inequality is:

$$\left(-\infty, -1\right] \cup \left(\frac{5}{9}, \infty\right)$$

Example 11.4.11 Three-Part Inequalities. Solve the three-part inequality $-4 \leq 20 - 6x < 32$.

Explanation. This is a three-part inequality. The goal is to isolate x in the middle and whatever you do to one "side," you have to do to the other two "sides."

$$-4 \leq 20 - 6x < 32$$

$$-4 - 20 \le 20 - 6x - 20 < 32 - 20$$
$$-24 \le -6x < 12$$
$$\frac{-24}{-6} \ge \frac{-6x}{-6} > \frac{12}{-6}$$
$$4 \ge x > -2$$

The solutions to the three-part inequality $4 \ge x > -2$ are those numbers that are trapped between -2 and 4, including 4 but not -2. The solution set in interval notation is $(-2, 4]$.

Example 11.4.12 Solving Compound Inequalities Graphically. Figure 11.4.13 shows a graph of $y = G(x)$. Use the graph do the following.

    a. Solve $G(x) < -2$.   b. Solve $G(x) \ge 1$.   c. Solve $-1 \le G(x) < 1$.

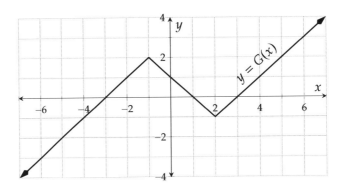

Figure 11.4.13: Graph of $y = G(x)$

Explanation.

    a. To solve $G(x) < -2$, we first draw a dotted line (since it's a less-than, not a less-than-or-equal) at $y = -2$. Then we examine the graph to find out where the graph of $y = G(x)$ is underneath the line $y = -2$. Our graph is below the line $y = -2$ for x-values less than -5. So the solution set is $(-\infty, -5)$.

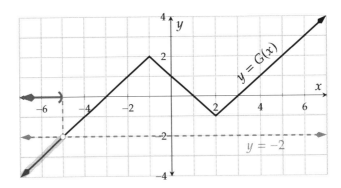

Figure 11.4.14: Graph of $y = G(x)$ and solution set to $G(x) < -2$

b. To solve $G(x) \geq 1$, we first draw a solid line (since it's a greater-than-or-equal) at $y = 1$. Then we examine the graph to find out what parts of the graph of $y = G(x)$ are above the line $y = 1$. Our graph is above (or on) the line $y = 1$ for x-values between -2 and 0 as well as x-values bigger than 4. So the solution set is $[-2, 0] \cup [4, \infty)$.

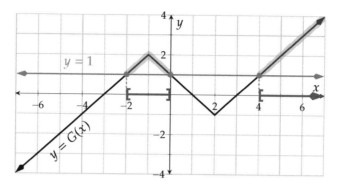

Figure 11.4.15: Graph of $y = G(x)$ and solution set to $G(x) \geq 1$

c. To solve $-1 < G(x) \leq 1$, we first draw a solid line at $y = 1$ and dotted line at $y = -1$. Then we examine the graph to find out what parts of the graph of $y = G(x)$ are trapped between the two lines we just drew. Our graph is between those values for x-values between -4 and -2 as well as x-values between 0 and 2 as well as well as x-values between 2 and 4. We use the solid and hollow dots on the graph to decide whether or not to include those values. So the solution set is $(-4, -2] \cup [0, 2) \cup (2, 4]$.

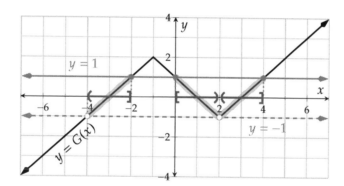

Figure 11.4.16: Graph of $y = G(x)$ and solution set to $-1 < G(x) \leq 1$

Example 11.4.17 Application of Compound Inequalities. Mishel wanted to buy some mulch for their spring garden. Each cubic yard of mulch cost \$27 and delivery for any size load was \$40. If they wanted to spend between \$200 and \$300, set up and solve a compound inequality to solve for the number of cubic yards, x, that they could buy.

Explanation. Since the mulch costs \$27 per cubic yard and delivery is \$40, the formula for the cost of x yards of mulch is $27x + 40$. Since Mishel wants to spend between \$200 and \$300, we just trap their cost between these two values.

$$200 < 27x + 40 < 300$$
$$200 - 40 < 27x + 40 - 40 < 300 - 40$$

$$160 < 27x < 260$$

$$\frac{160}{27} < \frac{27x}{27} < \frac{260}{27}$$

$$5.93 < x < 9.63$$

Note: these values are approximate

Most companies will only sell whole number cubic yards of mulch, so we have to round appropriately. Since Mishel wants to spend more than $200, we have to round our lower value from 5.93 up to 6 cubic yards.

If we round the 9.63 up to 10, then the total cost will be $27 \cdot 10 + 40 = 310$ (which represents $310), which is more than Mishel wanted to spend. So we actually have to round down to 9cubic yards to stay below the $300 maximum.

In conclusion, Mishel could buy 6, 7, 8, or 9 cubic yards of mulch to stay between $200 and $300.

11.4.3 Absolute Value Equations and Inequalities

In Section 11.3 we covered how to solve equations when an absolute value is equal to a number and when an absolute value is equal to an absolute value. We also covered how to solve inequalities when an absolute value is less than a number and when an absolute value is greater than a number.

Example 11.4.18 Solving an Equation with an Absolute Value. Solve the absolute value equation $|9 - 4x| = 17$ using Fact 11.3.6.

Explanation. The equation $|9 - 4x| = 17$ breaks into two pieces, each of which needs to be solved independently.

$9 - 4x = 17$	or	$9 - 4x = -17$
$-4x = 8$	or	$-4x = -26$
$\dfrac{-4x}{-4} = \dfrac{8}{-4}$	or	$\dfrac{-4x}{-4} = \dfrac{-26}{-4}$
$x = -2$	or	$x = \dfrac{13}{2}$

The solution set is $\left\{-2, \frac{13}{2}\right\}$.

Example 11.4.19 Solving an Equation with Two Absolute Values. Solve the absolute value equation $|7 - 3x| = |6x - 5|$ using Fact 11.3.11.

Explanation. The equation $|7 - 3x| = |6x - 5|$ breaks into two pieces, each of which needs to be solved independently.

$7 - 3x = 6x - 5$	or	$7 - 3x = -(6x - 5)$
$7 - 3x = 6x - 5$	or	$7 - 3x = -6x + 5$
$12 - 3x = 6x$	or	$2 - 3x = -6x$
$12 = 9x$	or	$2 = -3x$
$\dfrac{12}{9} = \dfrac{9x}{9}$	or	$\dfrac{2}{-3} = \dfrac{-3x}{-3}$

$$\frac{4}{3} = x \qquad\qquad \text{or} \qquad\qquad -\frac{2}{3} = x$$

The solution set is $\left\{\frac{4}{3}, -\frac{2}{3}\right\}$.

Example 11.4.20 Solving an Absolute Value Less-Than Inequality. Solve the absolute value inequality $4 \cdot |7 - 2x| + 1 < 25$ using Fact 11.3.19.

Explanation. The inequality $4 \cdot |7 - 2x| + 1 < 25$ *must* be simplified into the form that matches Fact 11.3.19, so we will first isolate the absolute value expression on the left side of the equation:

$$4 \cdot |7 - 2x| + 1 < 25$$
$$4 \cdot |7 - 2x| < 24$$
$$|7 - 2x| < 6$$

Now that we have the absolute value isolated, we can us Fact 11.3.19 to split it into a triple inequality that we can finish solving:

$$-6 < 7 - 2x < 6$$
$$-6 - 7 < 7 - 2x - 7 < 6 - 7$$
$$-13 < -2x < -1$$
$$\frac{-13}{-2} > \frac{-2x}{-2} > \frac{-1}{-2}$$
$$\frac{13}{2} > x > \frac{1}{2}$$

So, the solution set to the inequality is $\left(\frac{1}{2}, \frac{13}{2}\right)$.

Example 11.4.21 Solving an Absolute Value Greater-Than Inequality. To solve the absolute value inequality $\left|13 - \frac{3}{2}x\right| \geq 15$ using Fact 11.3.27.

Explanation. Using Fact 11.3.27, the inequality $\left|13 - \frac{3}{2}x\right| \geq 15$ breaks down into a compound inequality:

$$13 - \frac{3}{2}x \leq -15 \qquad \text{or} \qquad 13 - \frac{3}{2}x \geq 15$$
$$-\frac{3}{2}x \leq -28 \qquad \text{or} \qquad -\frac{3}{2}x \geq 2$$
$$-\frac{2}{3}\cdot\left(-\frac{3}{2}x\right) \geq -\frac{2}{3}\cdot(-28) \qquad \text{or} \qquad -\frac{2}{3}\cdot\left(-\frac{3}{2}x\right) \leq -\frac{2}{3}\cdot(2)$$
$$x \geq \frac{56}{3} \qquad \text{or} \qquad x \leq -\frac{4}{3}$$

We will write the solution set as $\left(-\infty, -\frac{4}{3}\right] \cup \left[\frac{56}{3}, \infty\right)$.

Exercises

Introduction to Absolute Value Functions

1. Evaluate the following.

$$3 - 5|3 - 7| + 1 = \boxed{}$$

2. Evaluate the following.

$$4 - 9|1 - 3| + 1 = \boxed{}$$

3. Given $f(x) = 20 - |-x + 5|$, find and simplify $f(18)$.

$$f(18) = \boxed{}$$

4. Given $f(r) = 17 - |3r - 16|$, find and simplify $f(19)$.

$$f(19) = \boxed{}$$

5. Find the domain of K where $K(x) = |8x - 5|$.

6. Find the domain of f where $f(x) = |x + 4|$.

7. Make a table of values for the function g defined by $g(x) = |2x - 3|$.

x	$g(x)$

8. Make a table of values for the function h defined by $h(x) = |-2x + 1|$.

x	$h(x)$

9. Graph $y = f(x)$, where $f(x) = \frac{1}{2}|4x - 5| - 3$.

10. Graph $y = f(x)$, where $f(x) = \frac{3}{4}|6 + x| + 2$.

11. Simplify the expression. Do not assume the variables take only positive values.

$$\sqrt{36r^2}$$

12. Simplify the expression. Do not assume the variables take only positive values.

$$\sqrt{9m^2}$$

13. Simplify the expression.

$$\sqrt{a^2 + 14a + 49}$$

14. Simplify the expression.

$$\sqrt{r^2 + 16r + 64}$$

15. The height inside a camping tent when you are d feet from the edge of the tent is given by

$$h = -0.7|d - 6.6| + 7$$

where h stands for height in feet.

Determine the height when you are:

 a. 7.6 ft from the edge.

 The height inside a camping tent when you 7.6 ft from the edge of the tent is ☐

 b. 3.1 ft from the edge.

 The height inside a camping tent when you 3.1 ft from the edge of the tent is ☐

16. The height inside a camping tent when you are d feet from the edge of the tent is given by

$$h = -0.5|d - 6| + 4.5$$

where h stands for height in feet.

Determine the height when you are:

 a. 8.3 ft from the edge.

 The height inside a camping tent when you 8.3 ft from the edge of the tent is ☐

 b. 1.3 ft from the edge.

 The height inside a camping tent when you 1.3 ft from the edge of the tent is ☐

Compound Inequalities   Solve the compound inequality algebraically.

17. $-4x - 3 \geq 5$ and $-14x - 7 \geq -5$

18. $-18x + 14 \geq -6$ and $6x + 4 > 9$

19. $9x - 10 \geq -17$ and $-15x + 7 \geq -15$

20. $-6x + 7 \geq 12$ and $6x + 11 \leq 2$

21. $7x + 15 > 11$ or $x + 15 < -20$

22. $12x - 2 \geq 10$ or $6x + 6 < 6$

23. $13x + 3 < -20$ or $10x + 1 \geq 8$

24. $19x - 6 \geq 8$ or $5x - 16 \leq -14$

A graph of f is given. Use the graph alone to solve the compound inequalities.

25.

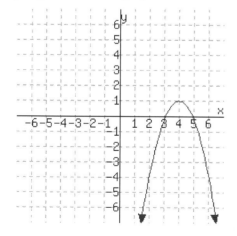

 a. $f(x) > -4$

 b. $f(x) \leq -4$

26.

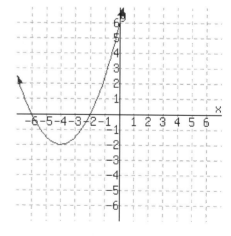

 a. $f(x) > 0$

 b. $f(x) \leq 0$

27.

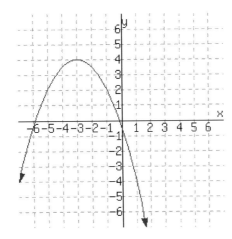

a. $f(x) > 2$

b. $f(x) \le 2$

28.

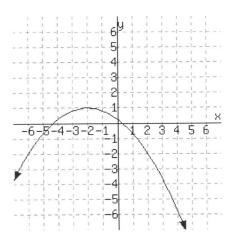

a. $f(x) > -2$

b. $f(x) \le -2$

Absolute Value Equations and Inequalities

29. Solve the following equation.

$|5x - 6| = 5$

30. Solve the following equation.

$|6x + 8| = 9$

31. Solve the equation $|4x - 2| = 18$.

32. Solve the equation $|4x - 3| = 17$.

33. Solve: $\left|\dfrac{2y - 7}{3}\right| = 1$

34. Solve: $\left|\dfrac{2y - 3}{7}\right| = 3$

35. Solve: $\left|\dfrac{1}{2}a + 7\right| = 3$

36. Solve: $\left|\dfrac{1}{4}a + 5\right| = 3$

37. Solve: $|b + 5| - 8 = 2$

38. Solve: $|t + 1| - 2 = 6$

39. Solve: $|5t - 20| + 5 = 5$

40. Solve: $|3x - 3| + 3 = 3$

The equation $|x| = |y|$ is satisfied if $x = y$ or $x = -y$. Use this fact to solve the following equation.

41. $|x + 6| = |x - 5|$

42. $|x + 6| = |x - 1|$

43. Solve the equation: $|8x - 4| = |5x + 5|$

44. Solve the equation: $|2x - 2| = |7x + 3|$

45. Solve the following equation.

$|x + 4| = |6x + 8|$

46. Solve the following equation.

$|2x - 4| = |8x + 2|$

Solve the inequality.

47. $|10 - 4x| \ge 15$

48. $|7 - 5x| \ge 7$

49. $|5x - 4| < 12$

50. $|6x - 10| < 5$

CHAPTER 12

More on Quadratic Functions

12.1 Graphs and Vertex Form

In this section, we will explore quadratic functions using graphing technology and learn the vertex and factored forms of a quadratic function's formula. We will also see how parabola graphs can be shifted.

12.1.1 Exploring Quadratic Functions with Graphing Technology

Graphing technology is very important and useful for applications and for finding points quickly. Let's explore some quadratic functions with graphing technology.

Example 12.1.2 Use technology to graph and make a table of the quadratic function f defined by $f(x) = 2x^2 + 4x - 3$ and find each of the key points or features.

a. Find the vertex.

b. Find the vertical intercept (i.e. y-intercept).

c. Find the horizontal or (i.e. x-intercept(s)).

d. Find $f(-2)$.

e. Solve $f(x) = 3$ using the graph.

f. Solve $f(x) \leq 3$ using the graph.

g. State the domain and range of the function.

Explanation.

The specifics of how to use any one particular technology tool vary. Whether you use an app, a physical calculator, or something else, a table and graph should look like:

x	$f(x)$
-2	-3
-1	-5
0	-3
1	3
2	13

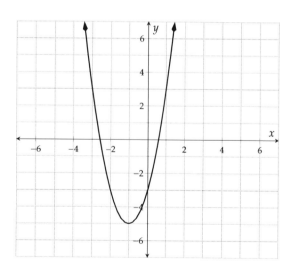

Additional features of your technology tool can enhance the graph to help answer these questions. You may be able to make the graph appear like:

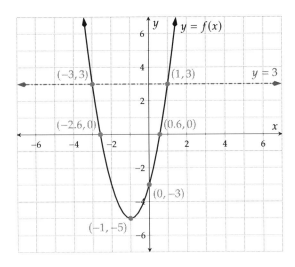

a. The vertex is $(-1, -5)$.

b. The vertical intercept is $(0, -3)$.

c. The horizontal intercepts are approximately $(-2.6, 0)$ and $(0.6, 0)$.

d. When $x = -2$, $y = -3$, so $f(-2) = -3$.

e. The solutions to $f(x) = 3$ are the x-values where $y = 3$. We graph the horizontal line $y = 3$ and find the x-values where the graphs intersect. The solution set is $\{-3, 1\}$.

f. The solutions are all of the x-values where the function's graph is below (or touching) the line $y = 3$. The interval is $[-3, 1]$.

g. The domain is $(-\infty, \infty)$ and the range is $[-5, \infty)$.

Now we will look at an application with graphing technology and put the points of interest in context.

Example 12.1.3 A reduced-gravity aircraft[a] is a fixed-wing airplane that astronauts use for training. The airplane flies up and then down in a parabolic path to simulate the feeling of weightlessness. In one training flight, the pilot will fly 40 to 60 parabolic maneuvers.

For the first parabolic maneuver, the altitude of the plane, in feet, at time t, in seconds since the maneuver began, is given by $H(t) = -16t^2 + 400t + 30500$.

a. Determine the starting altitude of the plane for the first maneuver.

b. What is the altitude of the plane 10 seconds into the maneuver?

c. Determine the maximum altitude of the plane and how long it takes to reach that altitude.

d. The zero-gravity effect is experienced when the plane begins the parabolic path until it gets back down to 30,500 feet. Write an inequality to express this and solve it using the graph. Write the times of the zero-gravity effect as an interval and determine how long the astronauts experience weightlessness during each cycle.

e. Use technology to make a table for H with t-values from 0 to 25 seconds. Use an increment of 5 seconds and then use the table to solve $H(t) = 32100$.

f. State the domain and range for this context.

Explanation. We can answer the questions based on the information in the graph.

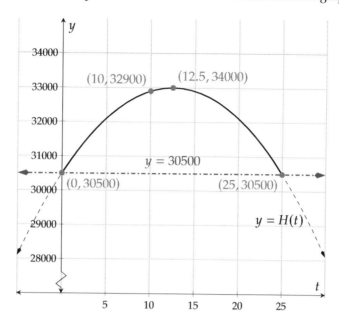

Figure 12.1.4: Graph of $H(t) = -16t^2 + 400t + 30500$ with $y = 30500$

a. The starting altitude can be read from the vertical intercept, which is $(0, 30500)$. The feeling of weightlessness begins at 30,500 feet.

b. After 10 seconds, the altitude of the plane is 32,900 feet.

c. For the maximum altitude of the plane we look at the vertex, which is approximately $(12.5, 33000)$. This tells us that after 12.5 seconds the plane will be at its maximum altitude of 33,000 feet.

d. We can write an inequality to describe when the plane is at or above 30,500 feet and solve it graphically.

$$H(t) \geq 30500$$
$$-16t^2 + 400t + 30500 \geq 30500.$$

We graph the line $y = 30500$ and find the points of intersection with the parabola. The astronauts experience weightlessness from 0 seconds to 25 seconds into the maneuver, or $[0, 25]$ seconds. They experience weightlessness for 25 seconds in each cycle.

e. To solve $H(t) = 32100$ using the table, we look for where the H-values are equal to 32100.

t	0	5	10	15	20	25
$H(t)$	30500	32100	32900	32900	32100	30500

There are two solutions, 5 seconds and 20 seconds. The solution set is $\{5, 20\}$.

f. When we use technology we see the entire function but in this context the plane is only on a parabolic path from $t = 0$ to $t = 25$ seconds. So the domain is $[0, 25]$, and the range is the set of corresponding y-values which is $[30500, 33000]$ feet.

---

[a]en.wikipedia.org/wiki/Reduced-gravity_aircraft

Let's look at the remote-controlled airplane dive from Example 9.3.18. This time we will use technology to answer the questions.

Example 12.1.5 Maia has a remote-controlled airplane and she is going to do a stunt dive where the plane dives toward the ground and back up along a parabolic path. The altitude or height of the plane is given by the function H where $H(t) = 0.7t^2 - 23t + 200$, for $0 \le t \le 30$. The height is measured in feet and the time, t, is measured in seconds since the stunt began.

a. Determine the starting height of the plane as the dive begins.

b. Determine the height of the plane after 5 seconds.

c. Will the plane hit the ground, and if so, at what time?

d. If the plane does not hit the ground, what is the closest it gets to the ground, and at what time?

e. At what time(s) will the plane have an altitude of 50 feet?

f. State the domain and the range of the function (in context).

Explanation. We have graphed the function and we will find the key information and put it in context.

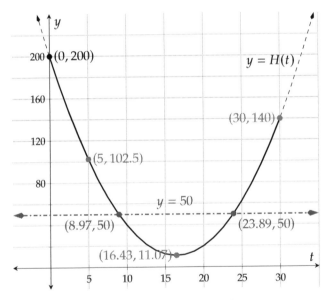

Figure 12.1.6: Graph of $H(t) = 0.7t^2 - 23t + 200$

a. The starting altitude can be read from the vertical intercept, which is $(0, 200)$. When the stunt begins, the plane has a altitude of 200 feet.

b. When $x = 5$, the y-value is 102.5. So $H(5) = 102.5$. This means that after 5 seconds, the plane is 102.5 feet above the ground.

c. From the graph we can see that the parabola does not touch or cross the x-axis, which represents the ground. This means the plane does not hit the ground and there are no real solutions to the equation $H(t) = 0$.

d. The lowest point is the vertex, which is approximately $(16.43, 11.07)$. The minimum altitude of the plane is about 11 feet, which occurs after about 16.4 seconds.

e. We graph the horizontal line $y = 50$ and look for the points of intersection. The plane will be 50 feet above the ground about 9 seconds after the plane begins the stunt, and again at about 24 seconds.

f. The domain for this function is given in the problem statement because only part of the parabola represents the path of the plane. The domain is $[0, 30]$. For the range we look at the possible altitudes of the plane and see that it is $[11.07\ldots, 200]$. The plane is doing this stunt from 0 to 30 seconds and its height ranges from about 11 to 200 feet above the ground.

12.1.2 The Vertex Form of a Parabola

We have learned the standard form of a quadratic function's formula, which is $f(x) = ax^2 + bx + c$. In this subsection, we will learn another form called the **vertex form**.

Using graphing technology, consider the graphs of $f(x) = x^2 - 6x + 7$ and $g(x) = (x-3)^2 - 2$ on the same axes.

We see only one parabola because these are two different forms of the same function. Indeed, if we convert $g(x)$ into standard form:

$$g(x) = (x-3)^2 - 2$$
$$g(x) = x^2 - 6x + 9 - 2$$
$$g(x) = x^2 - 6x + 7$$

it is clear that f and g are the same function.

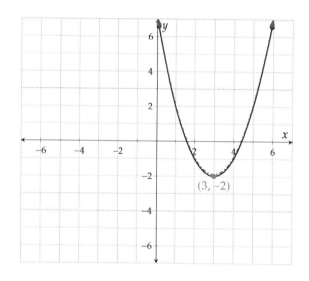

Figure 12.1.7: Graph of $f(x) = x^2 - 6x + 7$ and $g(x) = (x-3)^2 - 2$

The formula given for g is written in vertex form which is very useful because it allows us to read the vertex without doing any calculations. The vertex of the parabola is $(3, -2)$. We can see those numbers in $g(x) = (x-3)^2 - 2$. The x-value is the solution to $(x-3) = 0$, and the y-value is the constant *added* at the end.

Here are the graphs of three more functions with formulas in vertex form. Compare each function with the vertex of its graph.

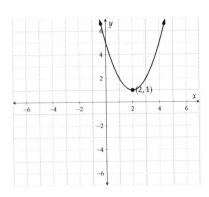
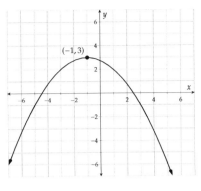
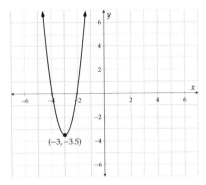

Figure 12.1.8: $r(x) = (x - 2)^2 + 1$ **Figure 12.1.9:** $s(x) = -\frac{1}{4}(x+1)^2 + 3$ **Figure 12.1.10:** $t(x) = 4(x + 3)^2 - 3.5$

Notice that the x-coordinate of the vertex has the opposite sign as the value in the function formula. On the other hand, the y-coordinate of the vertex has the same sign as the value in the function formula. Let's look at an example to understand why. We will evaluate $r(2)$.

$$r(2) = (2 - 2)^2 + 1$$
$$= 1$$

The x-value is the solution to $(x - 2) = 0$, which is positive 2. When we substitute 2 for x we get the value $y = 1$. Note that these coordinates create the vertex at $(2, 1)$. Now we can define the vertex form of a quadratic function.

Fact 12.1.11 Vertex Form of a Quadratic Function. *A quadratic function with the vertex at the point (h, k) is given by $f(x) = a(x - h)^2 + k$.*

Checkpoint 12.1.12. Find the vertex of each quadratic function.

a. $r(x) = -2(x + 4)^2 + 10$

b. $s(x) = 5(x - 1)^2 + 2$

c. $t(x) = (x - 10)^2 - 5$

d. $u(x) = 3(x + 7)^2 - 13$

Explanation.

a. The vertex of $r(x) = -2(x + 4)^2 + 10$ is $(-4, 10)$.

b. The vertex of $s(x) = 5(x - 1)^2 + 2$ is $(1, 2)$.

c. The vertex of $t(x) = (x - 10)^2 - 5$ is $(10, -5)$.

d. The vertex of $u(x) = 3(x + 7)^2 - 13$ is $(-7, -13)$.

Now let's do the reverse. When given the vertex and the value of a, we can write the function in vertex form.

Example 12.1.13 Write a formula for the quadratic function f with the given vertex and value of a.

a. Vertex $(-2, 8)$, $a = 1$

b. Vertex $(4, -9)$, $a = -4$

c. Vertex $(-3, -1)$, $a = 2$

d. Vertex $(5, 12)$, $a = -3$

Explanation.

a. The vertex form is $f(x) = (x + 2)^2 + 8$.

b. The vertex form is $f(x) = -4(x - 4)^2 - 9$.

c. The vertex form is $f(x) = 2(x + 3)^2 - 1$.

d. The vertex form is $f(x) = -3(x - 5)^2 + 12$.

Once we read the vertex we can also state the domain and range. All quadratic functions have a domain of $(-\infty, \infty)$ because we can put in any value to a quadratic function. The range, however, depends on the y-value of the vertex and whether the parabola opens upward or downward. When we have a quadratic function in vertex form we can read the range from the formula. Let's look at the graph of f, where $f(x) = 2(x - 3)^2 - 5$, as an example.

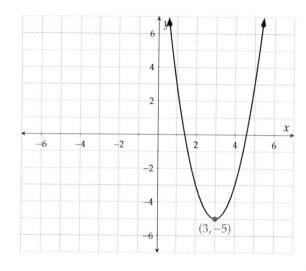

Figure 12.1.14: The graph of $f(x) = 2(x - 3)^2 - 5$

The domain is $(-\infty, \infty)$. The graph of f opens upward (which we know because $a = 2 > 0$) so the vertex is the minimum point. The y-value of -5 is the minimum. The range is $[-5, \infty)$.

Example 12.1.15 Identify the domain and range of g, where $g(x) = -3(x + 1)^2 + 6$.

Explanation.

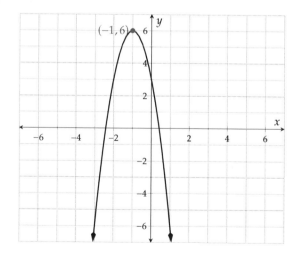

Figure 12.1.16: $g(x) = -3(x + 1)^2 + 6$

The domain is $(-\infty, \infty)$. The graph of g opens downward (which we know because $a = -3 < 0$) so the vertex is the maximum point. The y-value of 6 is the maximum. The range is $(-\infty, 6]$.

Checkpoint 12.1.17. Identify the domain and range of each quadratic function.

a. $w(x) = -3(x + 10)^2 - 11$

     The domain is [⎯⎯⎯⎯] and the range is [⎯⎯⎯⎯] .

b. $u(x) = 4(x - 7)^2 + 20$

     The domain is [⎯⎯⎯⎯] and the range is [⎯⎯⎯⎯] .

c. $y(x) = -(x - 1)^2$

     The domain is [⎯⎯⎯⎯] and the range is [⎯⎯⎯⎯] .

d. $z(x) = 3(x + 9)^2 - 4$

     The domain is [⎯⎯⎯⎯] and the range is [⎯⎯⎯⎯] .

Explanation.

a. The domain of w is $(-\infty, \infty)$. The parabola opens downward so the range is $(-\infty, -11]$.

b. The domain of u is $(-\infty, \infty)$. The parabola opens upward so the range is $[20, \infty)$.

c. The domain of y is $(-\infty, \infty)$. The parabola opens downward so the range is $(-\infty, 0]$.

d. The domain of z is $(-\infty, \infty)$. The parabola opens upward so the range is $[-4, \infty)$.

12.1.3 Horizontal and Vertical Shifts

Let $f(x) = x^2$ and $g(x) = (x - 4)^2 + 1$. The graph of $y = f(x)$ has its vertex at the point $(0, 0)$. Now we will compare this with the graph of $y = g(x)$ on the same axes.

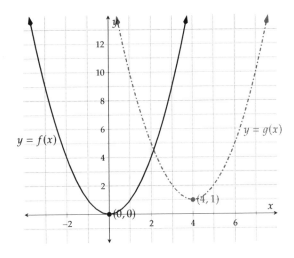

Both graphs open upward and have the same shape. Notice that the graph of g is the same as the graph of f but is shifted to the right by 4 units and up by 1 units because its vertex is $(4, 1)$.

Figure 12.1.18: The graph of f and g

Let's look at another graph. Let $h(x) = -x^2$ and let $j(x) = -(x + 3)^2 + 4$.

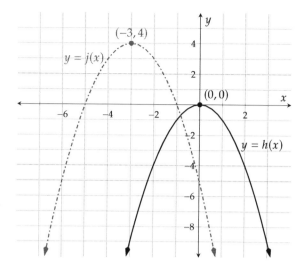

Both parabolas open downward and have the same shape. The graph of j is the same as the graph of h but it has been shifted to the left by 3 units and up by 4 units making its vertex $(-3, 4)$.

Figure 12.1.19: The graph of h and j

To summarize this, when a quadratic function is written in vertex form, the h-value is the horizontal shift and the k-value is the vertical shift.

Example 12.1.20 Identify the horizontal and vertical shifts compared with $f(x) = x^2$.

a. $m(x) = (x + 7)^2 + 3$

b. $n(x) = (x - 1)^2 + 6$

c. $o(x) = (x - 5)^2 - 1$

d. $p(x) = (x + 3)^2 - 11$

Explanation.

a. The graph of $y = m(x)$ has vertex at $(-7, 3)$. Therefore the graph is the same as $y = f(x)$ shifted to the left 7 units and up 3 units.

b. The graph of $y = n(x)$ has vertex at $(1, 6)$. Therefore the graph is the same as $y = f(x)$ shifted to the right 1 unit and up 6 units.

c. The graph of $y = o(x)$ has vertex at $(5, -1)$. Therefore the graph is the same as $y = f(x)$ shifted to the right 5 units and down 1 unit.

d. The graph of $y = p(x)$ has vertex at $(-3, -11)$. Therefore the graph is the same as $y = f(x)$ shifted to the left 3 units and down 11 units.

12.1.4 The Factored Form of a Parabola

There is another form of a quadratic function's formula, called **factored form**, which we will explore next. Let's consider the two functions $q(x) = -x^2 + 3x + 4$ and $s(x) = -(x - 4)(x + 1)$. Using graphing technology, we will graph $y = q(x)$ and $y = s(x)$ on the same axes.

These graphs coincide because the functions are actually the same. We can tell by multiplying out the formula for g to get back to the formula for f.

$$g(x) = -(x - 4)(x + 1)$$
$$g(x) = -(x^2 - 3x - 4)$$
$$g(x) = -x^2 + 3x + 4$$

Now we can see that f and g are really the same function.

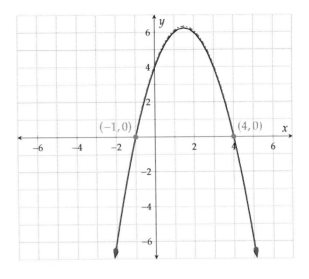

Figure 12.1.21: Graph of q and s

Factored form is very useful because we can read the x-intercepts directly from the function, which in this case are $(4, 0)$ and $(-1, 0)$. We find these by looking for the values that make the factors equal to 0, so the x-values have the opposite signs as are shown in the formula. To demonstrate this, we will find the roots by solving $g(x) = 0$.

$$g(x) = -(x - 4)(x + 1)$$
$$0 = -(x - 4)(x + 1)$$

$x - 4 = 0$	or	$x + 1 = 0$
$x = 4$	or	$x = -1$

This shows us that the x-intercepts are $(4, 0)$ and $(-1, 0)$.

The x-values of the x-intercepts are also called **zeros** or **roots**. The zeros or roots of the function g are -1 and 4.

Fact 12.1.22 Factored Form of a Quadratic Function. *A quadratic function with horizontal intercepts at $(r, 0)$ and $(s, 0)$ has the formula $f(x) = a(x - r)(x - s)$.*

Checkpoint 12.1.23. Write the horizontal intercepts of each function.

a. $t(x) = -(x + 2)(x - 4)$

b. $u(x) = 6(x - 7)(x - 5)$

c. $v(x) = -2(x + 1)(x + 4)$

d. $w(x) = 10(x - 8)(x + 3)$

Explanation.

a. The horizontal intercepts of t are $(-2, 0)$ and $(4, 0)$.

b. The horizontal intercepts of u are $(7, 0)$ and $(5, 0)$.

c. The horizontal intercepts of v are $(-1, 0)$ and $(-4, 0)$.

d. The horizontal intercepts of w are $(8, 0)$ and $(-3, 0)$.

Let's summarize the three forms of a quadratic function's formula:

Standard Form $f(x) = ax^2 + bx + c$, with y-intercept $(0, c)$.

Vertex Form $f(x) = a(x - h)^2 + k$, with vertex (h, k).

Factored Form $f(x) = a(x - r)(x - s)$, with x-intercepts $(m, 0)$ and $(n, 0)$.

Exercises

Review and Warmup

1. Multiply the polynomials.

$$(x + 6)(x - 2) = \boxed{}$$

2. Multiply the polynomials.

$$(y + 2)(y - 8) = \boxed{}$$

3. Multiply the polynomials.

$$(9y - 4)(y + 9) = \boxed{}$$

4. Multiply the polynomials.

$$(6r - 9)(r + 6) = \boxed{}$$

5. Factor the given polynomial.

$$r^2 + 8r + 12 = \boxed{}$$

6. Factor the given polynomial.

$$r^2 + 19r + 90 = \boxed{}$$

7. Factor the given polynomial.

$$2t^2 + 20t + 18 = \boxed{}$$

8. Factor the given polynomial.

$$5t^2 + 25t + 20 = \boxed{}$$

9. Here is an interval:

Write the interval using set-builder notation.

Write the interval using interval notation.

10. Here is an interval:

Write the interval using set-builder notation.

Write the interval using interval notation.

11. Here is an interval:

Write the interval using set-builder notation.

Write the interval using interval notation.

12. Here is an interval:

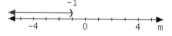

Write the interval using set-builder notation.

Write the interval using interval notation.

Technology and Tables

13. Let $F(x) = x^2 + 3x - 2$. Use technology to make a table of values F.

x	$F(x)$

14. Let $G(x) = x^2 + 2x - 2$. Use technology to make a table of values G.

x	$G(x)$

15. Let $H(x) = -x^2 + 2x + 1$. Use technology to make a table of values H.

x	$H(x)$

16. Let $H(x) = -x^2 - 3x + 1$. Use technology to make a table of values H.

x	$H(x)$

17. Let $K(x) = 3x^2 + 8x - 4$. Use technology to make a table of values K.

x	$K(x)$

18. Let $f(x) = 3x^2 + 3x - 4$. Use technology to make a table of values f.

x	$f(x)$

19. Let $g(x) = 3x^2 - 8x + 35$. Use technology to make a table of values g.

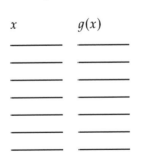

x	$g(x)$

20. Let $g(x) = 3x^2 - 7x + 46$. Use technology to make a table of values g.

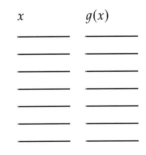

x	$g(x)$

Technology and Graphs

21. Use technology to make a graph of f where $f(x) = x^2 + 3x - 2$.

22. Use technology to make a graph of f where $f(x) = x^2 - 2x - 1$.

23. Use technology to make a graph of f where $f(x) = -x^2 + 3x + 2$.

24. Use technology to make a graph of f where $f(x) = -x^2 + x + 2$.

25. Use technology to make a graph of f where $f(x) = 3x^2 - 6x - 5$.

26. Use technology to make a graph of f where $f(x) = -3x^2 - 8x + 3$.

27. Use technology to make a graph of f where $f(x) = -3x^2 + 4x + 49$.

28. Use technology to make a graph of f where $f(x) = 2x^2 - 2x + 41$.

Technology and Features of Quadratic Function Graphs Use technology to find features of a quadratic function and its graph.

29. Let $h(x) = -2x^2 + 3x - 4$. Use technology to find the following.

    a. The vertex is _____.

    b. The y-intercept is _____.

    c. The x-intercept(s) is/are _____.

    d. The domain of h is _____.

    e. The range of h is _____.

    f. Calculate $h(1)$. _____.

    g. Solve $h(x) = -9$. _____.

    h. Solve $h(x) \leq -9$. _____.

30. Let $F(x) = -x^2 + 3x - 1$. Use technology to find the following.

    a. The vertex is _____.

    b. The y-intercept is _____.

    c. The x-intercept(s) is/are _____.

    d. The domain of F is _____.

    e. The range of F is _____.

    f. Calculate $F(-1)$. _____.

    g. Solve $F(x) = -4$. _____.

    h. Solve $F(x) \leq -4$. _____.

31. Let $G(x) = -1.4x^2 + 1.1x + 0.3$. Use technology to find the following.

a. The vertex is [ ].

b. The y-intercept is [ ].

c. The x-intercept(s) is/are [ ].

d. The domain of G is [ ].

e. The range of G is [ ].

f. Calculate $G(2)$. [ ].

g. Solve $G(x) = -10$. [ ]

h. Solve $G(x) \geq -10$. [ ]

32. Let $H(x) = 0.3x^2 - 0.7x - 5$. Use technology to find the following.

a. The vertex is [ ].

b. The y-intercept is [ ].

c. The x-intercept(s) is/are [ ].

d. The domain of H is [ ].

e. The range of H is [ ].

f. Calculate $H(-7)$. [ ].

g. Solve $H(x) = 5$. [ ]

h. Solve $H(x) \geq 5$. [ ]

33. Let $H(x) = \frac{x^2}{2} + 4.3x + 2.3$. Use technology to find the following.

a. The vertex is [ ].

b. The y-intercept is [ ].

c. The x-intercept(s) is/are [ ].

d. The domain of H is [ ].

e. The range of H is [ ].

f. Calculate $H(-7)$. [ ].

g. Solve $H(x) = -1$. [ ]

h. Solve $H(x) \geq -1$. [ ]

34. Let $K(x) = \frac{x^2}{4} - 3.8x + 4.6$. Use technology to find the following.

a. The vertex is [ ].

b. The y-intercept is [ ].

c. The x-intercept(s) is/are [ ].

d. The domain of K is [ ].

e. The range of K is [ ].

f. Calculate $K(-1)$. [ ].

g. Solve $K(x) = -4$. [ ]

h. Solve $K(x) > -4$. [ ]

Applications

35. An object was launched from the top of a hill with an upward vertical velocity of 50 feet per second. The height of the object can be modeled by the function $h(t) = -16t^2 + 50t + 200$, where t represents the number of seconds after the launch. Assume the object landed on the ground at sea level. Find the answer using graphing technology.

The object's height was [ ] feet when it was launched.

36. An object was launched from the top of a hill with an upward vertical velocity of 70 feet per second. The height of the object can be modeled by the function $h(t) = -16t^2 + 70t + 100$, where t represents the number of seconds after the launch. Assume the object landed on the ground at sea level. Find the answer using graphing technology.

Use a table to list the object's height within the first second after it was launched, at an increment of 0.1 second. Fill in the blanks. Round your answers to two decimal places when needed.

Time in Seconds	Height in Feet
0.1	_____
0.2	_____
0.3	_____

37. An object was launched from the top of a hill with an upward vertical velocity of 90 feet per second. The height of the object can be modeled by the function $h(t) = -16t^2 + 90t + 300$, where t represents the number of seconds after the launch. Assume the object landed on the ground at sea level. Use technology to find the answer.

The object was [   ] feet in the air 4 seconds after it was launched.

38. An object was launched from the top of a hill with an upward vertical velocity of 110 feet per second. The height of the object can be modeled by the function $h(t) = -16t^2 + 110t + 200$, where t represents the number of seconds after the launch. Assume the object landed on the ground at sea level. Find the answer using technology.

[   ] seconds after its launch, the object reached its maximum height of [   ] feet.

39. An object was launched from the top of a hill with an upward vertical velocity of 120 feet per second. The height of the object can be modeled by the function $h(t) = -16t^2 + 120t + 100$, where t represents the number of seconds after the launch. Assume the object landed on the ground at sea level. Find the answer using technology.

[   ] seconds after its launch, the object fell to the ground at sea level.

40. An object was launched from the top of a hill with an upward vertical velocity of 140 feet per second. The height of the object can be modeled by the function $h(t) = -16t^2 + 140t + 250$, where t represents the number of seconds after the launch. Assume the object landed on the ground at sea level. Find the answer using technology. Round your answers to two decimal places. If there is more than one answer, use a comma to separate them.

The object was 483 feet high at the following number of seconds after it was launched: [   ].

41. In a race, a car drove through the starting line at the speed of 6 meters per second. It was accelerating at 3.9 meters per second squared. Its distance from the starting position can be modeled by the function $d(t) = 1.95t^2 + 6t$. Find the answer using technology.

After [   ] seconds, the car was 172.8 meters away from the starting position.

42. In a race, a car drove through the starting line at the speed of 3 meters per second. It was accelerating at 4.4 meters per second squared. Its distance from the starting position can be modeled by the function $d(t) = 2.2t^2 + 3t$. Find the answer using technology.

After [   ] seconds, the car was 473.2 meters away from the starting position.

43. A farmer purchased 800 meters of fencing, and will build a rectangular pen with it. To enclose the largest possible area, what should the pen's length and width be? Model the pen's area with a function, and then find its maximum value.

Use a comma to separate your answers.

To enclose the largest possible area, the pen's length and width should be [   ] meters.

44. A farmer purchased 210 meters of fencing, and will build a rectangular pen along a river. This implies the pen has only 3 fenced sides. To enclose the largest possible area, what should the pen's length and width be? Model the pen's area with a function, and then find its maximum value.

To enclose the largest possible area, the pen's length and width should be [   ] meters.

Quadratic Functions in Vertex Form

45. Find the vertex of the graph of

$$y = -7(x + 10)^2 + 8$$

46. Find the vertex of the graph of

$$y = -5(x - 4)^2 - 5$$

47. Find the vertex of the graph of

$$y = -2(x + 4)^2 + 4$$

48. Find the vertex of the graph of

$$y = 10(x + 8)^2 + 1$$

49. Find the vertex of the graph of

$$y = 2.2(x - 2.7)^2 + 0.6$$

50. Find the vertex of the graph of

$$y = 4.4(x + 4.3)^2 + 9$$

51. A graph of a function f is given. Use the graph to write a formula for f in vertex form. You will need to identify the vertex and also one more point on the graph to find the leading coefficient a.

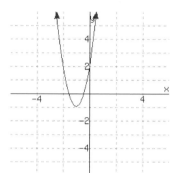

$f(x) =$

52. A graph of a function f is given. Use the graph to write a formula for f in vertex form. You will need to identify the vertex and also one more point on the graph to find the leading coefficient a.

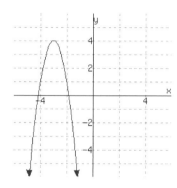

$f(x) =$

53. A graph of a function f is given. Use the graph to write a formula for f in vertex form. You will need to identify the vertex and also one more point on the graph to find the leading coefficient a.

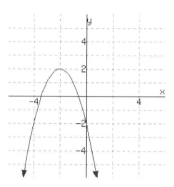

$f(x) =$

54. A graph of a function f is given. Use the graph to write a formula for f in vertex form. You will need to identify the vertex and also one more point on the graph to find the leading coefficient a.

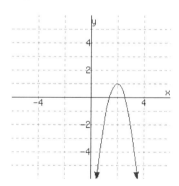

$f(x) =$

55. A graph of a function f is given. Use the graph to write a formula for f in vertex form. You will need to identify the vertex and also one more point on the graph to find the leading coefficient a.

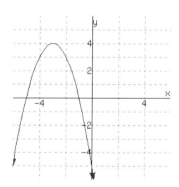

$f(x) =$

56. A graph of a function f is given. Use the graph to write a formula for f in vertex form. You will need to identify the vertex and also one more point on the graph to find the leading coefficient a.

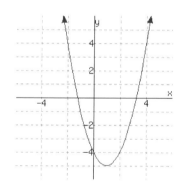

$f(x) =$

57. Write the vertex form for the quadratic function f, whose vertex is $(7, -8)$ and has leading coefficient $a = 7$.

$f(x) =$ [   ]

58. Write the vertex form for the quadratic function f, whose vertex is $(-4, 6)$ and has leading coefficient $a = 2$.

$f(x) =$ [   ]

59. Write the vertex form for the quadratic function f, whose vertex is $(-6, 4)$ and has leading coefficient $a = 4$.

$f(x) =$ [   ]

60. Write the vertex form for the quadratic function f, whose vertex is $(6, -7)$ and has leading coefficient $a = 6$.

$f(x) =$ [   ]

61. Let K be defined by $K(x) = (x + 1)^2 + 1$.

   a. What is the domain of K?

   b. What is the range of K?

62. Let f be defined by $f(x) = (x + 7)^2 + 9$.

   a. What is the domain of f?

   b. What is the range of f?

63. Let g be defined by $g(x) = 7.7(x + 3)^2 - 2$.

   a. What is the domain of g?

   b. What is the range of g?

64. Let g be defined by $g(x) = 4.3(x - 5)^2 - 7$.

   a. What is the domain of g?

   b. What is the range of g?

65. Let h be defined by $h(x) = -8(x + 6)^2 + 7$.

   a. What is the domain of h?

   b. What is the range of h?

66. Let F be defined by $F(x) = 4(x - 2)^2 + 2$.

   a. What is the domain of F?

   b. What is the range of F?

67. Let G be defined by $G(x) = -5\left(x + \frac{4}{3}\right)^2 - \frac{1}{4}$.

   a. What is the domain of G?

   b. What is the range of G?

68. Let G be defined by $G(x) = -2(x - (-3))^2 - \frac{7}{9}$.

   a. What is the domain of G?

   b. What is the range of G?

69. Let H be defined by $H(x) = 7\left(x - \frac{5}{3}\right)^2 + \frac{2}{7}$.

   a. What is the domain of H?

   b. What is the range of H?

70. Let K be defined by $K(x) = -4\left(x - \frac{2}{7}\right)^2 + \frac{5}{7}$.

   a. What is the domain of K?

   b. What is the range of K?

71. Consider the graph of the equation $y = (x - 4)^2 - 6$.

Compared to the graph of $y = x^2$, the vertex has been shifted [＿＿＿] units (□ left □ right) and [＿＿＿] units (□ down □ up) .

72. Consider the graph of the equation $y = (x + 7)^2 + 3$.

Compared to the graph of $y = x^2$, the vertex has been shifted [＿＿＿] units (□ left □ right) and [＿＿＿] units (□ down □ up) .

73. Consider the graph of the equation $y = (x + 46.6)^2 - 41.2$.

Compared to the graph of $y = x^2$, the vertex has been shifted [＿＿＿] units (□ left □ right) and [＿＿＿] units (□ down □ up) .

74. Consider the graph of the equation $y = (x + 24.7)^2 + 88.1$.

Compared to the graph of $y = x^2$, the vertex has been shifted [＿＿＿] units (□ left □ right) and [＿＿＿] units (□ down □ up) .

75. Consider the graph of the equation $y = \left(x + \frac{1}{4}\right)^2 + \frac{1}{5}$.

Compared to the graph of $y = x^2$, the vertex has been shifted [＿＿＿] units (□ left □ right) and [＿＿＿] units (□ down □ up) .

76. Consider the graph of the equation $y = \left(x + \frac{2}{5}\right)^2 + \frac{5}{6}$.

Compared to the graph of $y = x^2$, the vertex has been shifted [＿＿＿] units (□ left □ right) and [＿＿＿] units (□ down □ up) .

Three Forms of Quadratic Functions

77. The quadratic expression $(x - 2)^2 - 1$ is written in vertex form.

    a. Write the expression in standard form.

    b. Write the expression in factored form.

78. The quadratic expression $(x - 3)^2 - 25$ is written in vertex form.

    a. Write the expression in standard form.

    b. Write the expression in factored form.

79. The quadratic expression $(x - 4)^2 - 81$ is written in vertex form.

    a. Write the expression in standard form.

    b. Write the expression in factored form.

80. The quadratic expression $(x - 1)^2 - 36$ is written in vertex form.

    a. Write the expression in standard form.

    b. Write the expression in factored form.

Factored Form and Intercepts

81. The formula for a quadratic function h is $h(x) = (x + 7)(x - 7)$.

    a. The y-intercept is [   ].

    b. The x-intercept(s) is/are [   ].

82. The formula for a quadratic function h is $h(x) = (x + 5)(x + 6)$.

    a. The y-intercept is [   ].

    b. The x-intercept(s) is/are [   ].

83. The formula for a quadratic function K is $K(x) = -2(x - 6)(x - 4)$.

    a. The y-intercept is [   ].

    b. The x-intercept(s) is/are [   ].

84. The formula for a quadratic function F is $F(x) = -(x + 7)(x - 2)$.

    a. The y-intercept is [   ].

    b. The x-intercept(s) is/are [   ].

85. The formula for a quadratic function F is $F(x) = 2x(x + 7)$.

    a. The y-intercept is [   ].

    b. The x-intercept(s) is/are [   ].

86. The formula for a quadratic function K is $K(x) = 4(x - 5)\,x$.

    a. The y-intercept is [   ].

    b. The x-intercept(s) is/are [   ].

87. The formula for a quadratic function h is $h(x) = 6(x + 1)(x + 1)$.

    a. The y-intercept is [   ].

    b. The x-intercept(s) is/are [   ].

88. The formula for a quadratic function H is $H(x) = 8(x + 8)(x + 8)$.

    a. The y-intercept is [   ].

    b. The x-intercept(s) is/are [   ].

89. The formula for a quadratic function h is $h(x) = -9(7x + 6)(3x + 2)$.

    a. The y-intercept is [   ].

    b. The x-intercept(s) is/are [   ].

90. The formula for a quadratic function g is $g(x) = -7(5x - 2)(8x - 9)$.

    a. The y-intercept is [   ].

    b. The x-intercept(s) is/are [   ].

12.2 Completing the Square

In this section, we will learn how to "complete the square" with a quadratic expression. This topic is very useful for solving quadratic equations and putting quadratic functions in vertex form.

12.2.1 Solving Quadratic Equations by Completing the Square

When we have an equation like $(x + 5)^2 = 4$, we can solve it quickly using the square root property:

$$(x + 5)^2 = 4$$

$x + 5 = -2$	or	$x + 5 = 2$
$x = -7$	or	$x = -3$

The method of **completing the square** allows us to solve *any* quadratic equation using the square root property. The challenge is that most quadratic equations don't come with a perfect square already on one side. Let's explore how to do this by looking at some perfect square trinomials to see the pattern.

$$(x + 1)^2 = x^2 + 2x + 1$$
$$(x + 2)^2 = x^2 + 4x + 4$$
$$(x + 3)^2 = x^2 + 6x + 9$$
$$(x + 4)^2 = x^2 + 8x + 16$$
$$(x + 5)^2 = x^2 + 10x + 25$$
$$\vdots$$

There is an important pattern here. Notice that with each middle coefficient on the right, you may cut it in half to get the constant term in the binomial on the left side. And then you may square that number to get the constant term back on the right side. Mathematically, this says:

$$\left(x + \frac{b}{2}\right)^2 = x^2 + bx + \left(\frac{b}{2}\right)^2$$

We will use this fact to make perfect square trinomials.

Fact 12.2.2 The Term that Completes the Square. *For a polynomial $x^2 + bx$, the constant term needed to make a perfect square trinomial is $\left(\frac{b}{2}\right)^2$.*

Example 12.2.3 Solve the quadratic equation $x^2 + 6x = 16$ by completing the square.

Explanation. To solve the quadratic equation $x^2 + 6x = 16$, on the left side we can complete the square by adding $\left(\frac{b}{2}\right)^2$; note that $b = 6$ in this case, which makes $\left(\frac{b}{2}\right)^2 = \left(\frac{6}{2}\right)^2 = 3^2 = 9$. We add it to both sides to maintain equality.

$$x^2 + 6x + 9 = 16 + 9$$
$$x^2 + 6x + 9 = 25$$
$$(x + 3)^2 = 25$$

Now that we have completed the square, we can solve the equation using the square root property.

$$x + 3 = -5 \qquad \text{or} \qquad x + 3 = 5$$
$$x = -8 \qquad \text{or} \qquad x = 2$$

The solution set is $\{-8, 2\}$.

Now let's see the process for completing the square when the quadratic equation is given in standard form.

Example 12.2.4 Solve $x^2 - 14x + 11 = 0$ by completing the square.

Explanation. We will solve $x^2 - 14x + 11 = 0$. We see that the polynomial on the left side is not a perfect square trinomial, so we need to complete the square. We subtract 11 from both sides so we can add the missing term on the left.

$$x^2 - 14x + 11 = 0$$
$$x^2 - 14x = -11$$

Next comes the completing-the-square step. We need to add the correct number to both sides of the equation to make the left side a perfect square. Remember that Fact 12.2.2 states that we need to use $\left(\frac{b}{2}\right)^2$ for this. In our case, $b = -14$, so $\left(\frac{b}{2}\right)^2 = \left(\frac{-14}{2}\right)^2 = 49$

$$x^2 - 14x + 49 = -11 + 49$$
$$(x - 7)^2 = 38$$

$$x - 7 = -\sqrt{38} \qquad \text{or} \qquad x - 7 = \sqrt{38}$$
$$x = 7 - \sqrt{38} \qquad \text{or} \qquad x = 7 + \sqrt{38}$$

The solution set is $\{7 - \sqrt{38}, 7 + \sqrt{38}\}$.

Here are some more examples.

Example 12.2.5 Complete the square to solve for y in $y^2 - 20y - 21 = 0$.

Explanation. To complete the square, we will first move the constant term to the right side of the equation. Then we will use Fact 12.2.2 to find $\left(\frac{b}{2}\right)^2$ to add to both sides.

$$y^2 - 20y - 21 = 0$$
$$y^2 - 20y = 21$$

In our case, $b = -20$, so $\left(\frac{b}{2}\right)^2 = \left(\frac{-20}{2}\right)^2 = 100$

$$y^2 - 20y + 100 = 21 + 100$$
$$(y - 10)^2 = 121$$

$$y - 10 = -11 \qquad \text{or} \qquad y - 10 = 11$$
$$y = -1 \qquad \text{or} \qquad y = 21$$

The solution set is $\{-1, 21\}$.

So far, the value of b has been even each time, which makes $\frac{b}{2}$ a whole number. When b is odd, we will end up adding a fraction to both sides. Here is an example.

Example 12.2.6 Complete the square to solve for z in $z^2 - 3z - 10 = 0$.

Explanation. We will first move the constant term to the right side of the equation:

$$z^2 - 3z - 10 = 0$$
$$z^2 - 3z = 10$$

Next, to complete the square, we will need to find the right number to add to both sides. According to Fact 12.2.2, we need to divide the value of b by 2 and then square the result to find the right number. First, divide by 2:

$$\frac{b}{2} = \frac{-3}{2} = -\frac{3}{2} \tag{12.2.1}$$

and then we square that result:

$$\left(-\frac{3}{2}\right)^2 = \frac{9}{4} \tag{12.2.2}$$

Now we can add the $\frac{9}{4}$ from Equation (12.2.2) to both sides of the equation to complete the square.

$$z^2 - 3z + \frac{9}{4} = 10 + \frac{9}{4}$$

Now, to factor the seemingly complicated expression on the left, just know that it should always factor using the number from the first step in the completing the square process, Equation (12.2.1).

$$\left(z - \frac{3}{2}\right)^2 = \frac{49}{4}$$

$$z - \frac{3}{2} = -\frac{7}{2} \qquad \text{or} \qquad z - \frac{3}{2} = \frac{7}{2}$$
$$z = \frac{3}{2} - \frac{7}{2} \qquad \text{or} \qquad z = \frac{3}{2} + \frac{7}{2}$$
$$z = -\frac{4}{2} \qquad \text{or} \qquad z = \frac{10}{2}$$
$$z = -2 \qquad \text{or} \qquad z = 5$$

The solution set is $\{-2, 5\}$.

In each of the previous examples, the value of a was equal to 1. This is necessary for our missing term formula to work. When a is not equal to 1 we will divide both sides by a. Let's look at an example of that.

Example 12.2.7 Solve for r in $2r^2 + 2r = 3$ by completing the square.

Explanation. Because there is a leading coefficient of 2, we will divide both sides by 2.

$$2r^2 + 2r = 3$$
$$\frac{2r^2}{2} + \frac{2r}{2} = \frac{3}{2}$$
$$r^2 + r = \frac{3}{2}$$

Next, we will complete the square. Since $b = 1$, first,

$$\frac{b}{2} = \frac{1}{2} \tag{12.2.3}$$

and next, squaring that, we have

$$\left(\frac{1}{2}\right)^2 = \frac{1}{4}. \tag{12.2.4}$$

So we will add $\frac{1}{4}$ from Equation (12.2.4) to both sides of the equation:

$$r^2 + r + \frac{1}{4} = \frac{3}{2} + \frac{1}{4}$$
$$r^2 + r + \frac{1}{4} = \frac{6}{4} + \frac{1}{4}$$

Here, remember that we always factor with the number found in the first step of completing the square, Equation (12.2.3).

$$\left(r + \frac{1}{2}\right)^2 = \frac{7}{4}$$

$$r + \frac{1}{2} = -\frac{\sqrt{7}}{2} \qquad \text{or} \qquad r + \frac{1}{2} = \frac{\sqrt{7}}{2}$$

$$r = -\frac{1}{2} - \frac{\sqrt{7}}{2} \qquad \text{or} \qquad r = -\frac{1}{2} + \frac{\sqrt{7}}{2}$$

$$r = \frac{-1 - \sqrt{7}}{2} \qquad \text{or} \qquad r = \frac{-1 + \sqrt{7}}{2}$$

The solution set is $\left\{\frac{-1-\sqrt{7}}{2}, \frac{-1+\sqrt{7}}{2}\right\}$.

12.2.2 Deriving the Vertex Formula and the Quadratic Formula by Completing the Square

In Section 9.2, we learned a formula to find the vertex. In Section 8.4, we learned the Quadratic Formula. You may have wondered where they came from, and now that we know how to complete the square, we can derive them. We will solve the standard form equation $ax^2 + bx + c = 0$ for x.

First, we subtract c from both sides and divide both sides by a.

$$ax^2 + bx + c = 0$$
$$ax^2 + bx = -c$$

$$\frac{ax^2}{a} + \frac{bx}{a} = -\frac{c}{a}$$

$$x^2 + \frac{b}{a}x = -\frac{c}{a}$$

Next, we will complete the square by taking half of the middle coefficient and squaring it. First,

$$\frac{\frac{b}{a}}{2} = \frac{b}{2a} \tag{12.2.5}$$

and then squaring that we have

$$\left(\frac{b}{2a}\right)^2 = \frac{b^2}{4a^2} \tag{12.2.6}$$

We add the $\frac{b^2}{4a^2}$ from Equation (12.2.6) to both sides of the equation:

$$x^2 + \frac{b}{a}x + \frac{b^2}{4a^2} = +\frac{b^2}{4a^2} - \frac{c}{a}$$

Remember that the left side always factors with the value we found in the first step of the completing the square process from Equation (12.2.5). So we have:

$$\left(x + \frac{b}{2a}\right)^2 = \frac{b^2}{4a^2} - \frac{c}{a}$$

To find a common denominator on the right, we multiply by $4a$ in the numerator and denominator on the second term.

$$\left(x + \frac{b}{2a}\right)^2 = \frac{b^2}{4a^2} - \frac{c}{a} \cdot \frac{4a}{4a}$$

$$\left(x + \frac{b}{2a}\right)^2 = \frac{b^2}{4a^2} - \frac{4ac}{4a^2}$$

$$\left(x + \frac{b}{2a}\right)^2 = \frac{b^2 - 4ac}{4a^2}$$

Now that we have completed the square, we can see that the x-value of the vertex is $-\frac{b}{2a}$. That is the vertex formula. Next, we will solve the equation using the square root property to find the Quadratic Formula.

Note on the ± Form. Because of the complexity of the formula we choose to use the ± symbol rather than write out each solution separately. An expression of the form $x = A \pm B$ really means "either $x = A - B$ or $x = A + B$."

$$x + \frac{b}{2a} = \pm\sqrt{\frac{b^2 - 4ac}{4a^2}}$$

$$x + \frac{b}{2a} = \pm\frac{\sqrt{b^2 - 4ac}}{2a}$$

$$x = -\frac{b}{2a} \pm \frac{\sqrt{b^2 - 4ac}}{2a}$$

$$x = \frac{-b \pm \sqrt{b^2 - 4ac}}{2a}$$

This shows us that the solutions to the equation $ax^2 + bx + c = 0$ are $\frac{-b \pm \sqrt{b^2 - 4ac}}{2a}$.

12.2.3 Putting Quadratic Functions in Vertex Form

In Section 12.1, we learned about the vertex form of a parabola, which allows us to quickly read the coordinates of the vertex. We can now use the method of completing the square to put a quadratic function in vertex form. Completing the square with a function is a little different than with an equation so we will start with an example.

Example 12.2.8 Write a formula in vertex form for the function q defined by $q(x) = x^2 + 8x$

Explanation. The formula is in the form $x^2 + bx$, so we will need to add $\left(\frac{b}{2}\right)^2$ to complete the square by Fact 12.2.2. When we had an equation, we could add the same quantity to both sides. But now we do not wish to change the left side, since we are trying to end up with a formula that still says $q(x) = \ldots$. Instead, we will add *and subtract* the term from the right side in order to maintain equality. In this case,

$$\left(\frac{b}{2}\right)^2 = \left(\frac{8}{2}\right)^2$$
$$= 4^2$$
$$= 16$$

To maintain equality, we will both add *and* subtract 16 on the same side of the equation. It is functionally the same as adding 0 on the right, but the 16 makes it possible to factor the expression in a particular way:

$$q(x) = x^2 + 8x + 16 - 16$$
$$= \left(x^2 + 8x + 16\right) - 16$$
$$= (x + 4)^2 - 16$$

Now that we have completed the square, our function is in vertex form. The vertex is $(-4, -16)$. One way to verify that our work is correct is to graph the original version of the function and check that the vertex is where it should be.

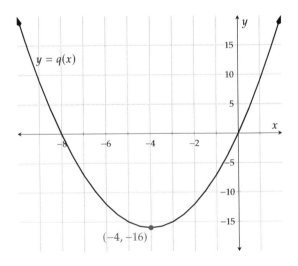

Figure 12.2.9: Graph of $y = x^2 + 8x$

Let's look at a function that has a constant term and see how to complete the square.

Example 12.2.10 Write a formula in vertex form for the function f defined by $f(x) = x^2 - 12x + 3$

Explanation. To complete the square, we need to add and subtract $\left(-\frac{12}{2}\right)^2 = (-6)^2 = 36$ on the right side.

$$f(x) = x^2 - 12x + 36 - 36 + 3$$
$$= \left(x^2 - 12x + 36\right) - 36 + 3$$
$$= (x - 6)^2 - 33$$

The vertex is $(6, -33)$.

In the first two examples, a was equal to 1. When a is not equal to one, we have an additional step. Since we are working with an expression where we intend to preserve the left side as $f(x) = \ldots$, we cannot divide both sides by a. Instead we will factor a out of the first two terms. Let's look at an example of that.

Example 12.2.11 Write a formula in vertex form for the function g defined by $g(x) = 5x^2 + 20x + 25$

Explanation. Before we can complete the square, we will factor the 5 out of the first two terms.

$$g(x) = 5\left(x^2 + 4x\right) + 25$$

Now we will complete the square inside the parentheses by adding and subtracting $\left(\frac{4}{2}\right)^2 = 2^2 = 4$.

$$g(x) = 5\left(x^2 + 4x + 4 - 4\right) + 25$$

Notice that the constant that we subtracted is inside the parentheses, but it will not be part of our perfect square trinomial. In order to bring it outside, we need to multiply it by 5. We are distributing the 5 to that term so we can combine it with the outside term.

$$g(x) = 5\left(\left(x^2 + 4x + 4\right) - 4\right) + 25$$
$$= 5\left(x^2 + 4x + 4\right) - 5 \cdot 4 + 25$$
$$= 5\left(x + 2\right)^2 - 20 + 25$$
$$= 5\left(x + 2\right)^2 + 5$$

The vertex is $(-2, 5)$.

Here is an example that includes fractions.

Example 12.2.12 Write a formula in vertex form for the function h defined by $h(x) = -3x^2 - 4x - \frac{7}{4}$

Explanation. First, we will factor the leading coefficient out of the first two terms.

$$h(x) = -3x^2 - 4x - \frac{7}{4}$$
$$= -3\left(x^2 + \frac{4}{3}x\right) - \frac{7}{4}$$

Next, we will complete the square for $x^2 + \frac{4}{3}x$ inside the grouping symbols by adding and subtracting the right number. To find that number, we divide the value of b by two and square the result. That looks like:

$$\frac{b}{2} = \frac{\frac{4}{3}}{2} = \frac{4}{3} \cdot \frac{1}{2} = \frac{2}{3} \tag{12.2.7}$$

and then,

$$\left(\frac{2}{3}\right)^2 = \frac{2^2}{3^2} = \frac{4}{9} \tag{12.2.8}$$

Adding and subtracting the value from Equation (12.2.8), we have:

$$h(x) = -3\left(x^2 + \frac{4}{3}x + \frac{4}{9} - \frac{4}{9}\right) - \frac{7}{4}$$

$$= -3\left(\left(x^2 + \frac{4}{3}x + \frac{4}{9}\right) - \frac{4}{9}\right) - \frac{7}{4}$$

$$= -3\left(x^2 + \frac{4}{3}x + \frac{4}{9}\right) - \left(3 \cdot -\frac{4}{9}\right) - \frac{7}{4}$$

Remember that when completing the square, the expression should always factor with the number found in the first step of the completing-the-square process, Equation (12.2.7).

$$= -3\left(x + \frac{2}{3}\right)^2 + \frac{4}{3} - \frac{7}{4}$$

$$= -3\left(x + \frac{2}{3}\right)^2 + \frac{16}{12} - \frac{21}{12}$$

$$= -3\left(x + \frac{2}{3}\right)^2 - \frac{5}{12}$$

The vertex is $\left(-\frac{2}{3}, -\frac{5}{12}\right)$.

Completing the square can also be used to find a minimum or maximum in an application.

Example 12.2.13 In Example 6.4.19, we learned that artist Tyrone's annual income from paintings can be modeled by $I(x) = -100x^2 + 1000x + 20000$, where x is the number of times he will raise the price per painting by \$20.00. To maximize his income, how should Tyrone set his price per painting? Find the maximum by completing the square.

Explanation. To find the maximum is essentially the same as finding the vertex, which we can find by completing the square. To complete the square for $I(x) = -100x^2 + 1000x + 20000$, we start by factoring out the -100 from the first two terms:

$$I(x) = -100x^2 + 1000x + 20000$$

$$= -100\left(x^2 - 10x\right) + 20000$$

Next, we will complete the square for $x^2 - 10x$ by adding and subtracting $\left(-\frac{10}{2}\right)^2 = (-5)^2 = 25$.

$$I(x) = -100\left(x^2 - 10x + 25 - 25\right) + 20000$$

$$= -100\left(\left(x^2 - 10x + 25\right) - 25\right) + 20000$$

$$= -100\left(x^2 - 10x + 25\right) - (100 \cdot -25) + 20000$$

$$= -100(x - 5)^2 + 2500 + 20000$$

$$= -100(x - 5)^2 + 22500$$

The vertex is the point $(5, 22500)$. This implies Tyrone should raise the price per painting 5 times, which is $5 \cdot 20 = 100$ dollars. He would sell $100 - 5(5) = 75$ paintings. This would make the price per painting $200 + 100 = 300$ dollars, and his annual income from paintings would become $22500 by this model.

12.2.4 Graphing Quadratic Functions by Hand

Now that we know how to put a quadratic function in vertex form, let's review how to graph a parabola by hand.

Example 12.2.14 Graph the function h defined by $h(x) = 2x^2 + 4x - 6$ by determining its key features algebraically.

Explanation. To start, we'll note that this function opens upward because the leading coefficient, 2, is positive.

Now we will complete the square to find the vertex. We will factor the 2 out of the first two terms, and then add and subtract $\left(\frac{2}{2}\right)^2 = 1^2 = 1$ on the right side.

$$\begin{aligned}
h(x) &= 2\left(x^2 + 2x\right) - 6 \\
&= 2\left[x^2 + 2x + 1 - 1\right] - 6 \\
&= 2\left[(x^2 + 2x + 1) - 1\right] - 6 \\
&= 2(x^2 + 2x + 1) - (2 \cdot 1) - 6 \\
&= 2(x + 1)^2 - 2 - 6 \\
&= 2(x + 1)^2 - 8
\end{aligned}$$

The vertex is $(-1, -8)$ so the axis of symmetry is the line $x = -1$.

To find the y-intercept, we'll replace x with 0 or read the value of c from the function in standard form:

$$\begin{aligned}
h(0) &= 2(0)^2 + 2(0) - 6 \\
&= -6
\end{aligned}$$

The y-intercept is $(0, -6)$ and we will find its symmetric point on the graph, which is $(-2, -6)$.

Next, we'll find the horizontal intercepts. We see this function factors so we will write the factored form to get the horizontal intercepts.

$$\begin{aligned}
h(x) &= 2x^2 + 4x - 6 \\
&= 2\left(x^2 + 2x - 3\right) \\
&= 2(x - 1)(x + 3)
\end{aligned}$$

The x-intercepts are $(1, 0)$ and $(-3, 0)$.

Now we will plot all of the key points and draw the parabola.

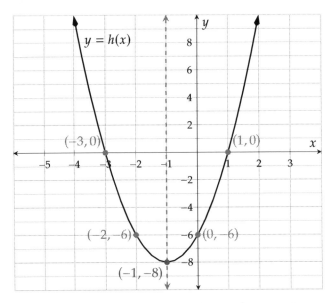

Figure 12.2.15: The graph of $y = 2x^2 + 4x - 6$.

Example 12.2.16 Write a formula in vertex form for the function p defined by $p(x) = -x^2 - 4x - 1$, and find the graph's key features algebraically. Then sketch the graph.

Explanation.   In this function, the leading coefficient is negative so it will open downward. To complete the square we first factor -1 out of the first two terms.

$$p(x) = -x^2 - 4x - 1$$
$$= -\left(x^2 + 4x\right) - 1$$

Now, we add and subtract the correct number on the right side of the function: $\left(\frac{b}{2}\right)^2 = \left(\frac{4}{2}\right)^2 = 2^2 = 4.$

$$p(x) = -\left(x^2 + 4x + 4 - 4\right) - 1$$
$$= -\left(\left(x^2 + 4x + 4\right) - 4\right) - 1$$
$$= -\left(x^2 + 4x + 4\right) - (-4) - 1$$
$$= -(x + 2)^2 + 4 - 1$$
$$= -(x + 2)^2 + 3$$

The vertex is $(-2, 3)$ so the axis of symmetry is the line $x = -2$.

We find the y-intercept by looking at the value of c, which is -1. So, the y-intercept is $(0, -1)$ and we will find its symmetric point on the graph, $(-4, -1)$.

The original expression, $-x^2 - 4x - 1$, does not factor so to find the x-intercepts we need to set $p(x) = 0$ and complete the square or use the quadratic formula. Since we just went through the process of completing the square above, we can use that result to save several repetitive steps.

$$p(x) = 0$$
$$-(x + 2)^2 + 3 = 0$$

$$-(x+2)^2 = -3$$
$$(x+2)^2 = 3$$

$x + 2 = -\sqrt{3}$	or	$x + 2 = \sqrt{3}$
$x = -2 - \sqrt{3}$	or	$x = -2 + \sqrt{3}$
$x \approx -3.73$	or	$x \approx -0.268$

The x-intercepts are approximately $(-3.7, 0)$ and $(-0.3, 0)$. Now we can plot all of the points and draw the parabola.

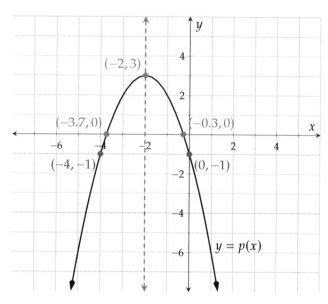

Figure 12.2.17: The graph of $y = -x^2 - 4x - 1$.

Exercises

Review and Warmup

1. Use a square root to solve $\left(y - 9\right)^2 = 4$.

2. Use a square root to solve $\left(y + 4\right)^2 = 25$.

3. Use a square root to solve $\left(4y + 7\right)^2 = 4$.

4. Use a square root to solve $(9r - 4)^2 = 64$.

5. Use a square root to solve $(r + 3)^2 = 14$.

6. Use a square root to solve $(t - 4)^2 = 3$.

7. Use a square root to solve $t^2 + 18t + 81 = 36$.

8. Use a square root to solve $x^2 + 4x + 4 = 81$.

9. Use a square root to solve $16x^2 - 16x + 4 = 25$.

10. Use a square root to solve $81y^2 + 108y + 36 = 9$.

11. Use a square root to solve $36y^2 - 72y + 36 = 17$.

12. Use a square root to solve $9y^2 + 12y + 4 = 13$.

Completing the Square to Solve Equations Solve the equation by completing the square.

13. $r^2 + 2r = 63$

14. $r^2 + 10r = -9$

15. $t^2 - t = 42$

16. $t^2 - 3t = 10$

17. $x^2 + 4x = 6$

18. $x^2 - 8x = 8$

19. $y^2 - 6y + 5 = 0$

20. $y^2 - 8y - 9 = 0$

21. $y^2 + 15y + 56 = 0$

22. $r^2 - r - 42 = 0$

23. $r^2 - 2r - 6 = 0$

24. $t^2 - 10t + 3 = 0$

25. $12t^2 + 28t + 15 = 0$

26. $12x^2 + 20x + 7 = 0$

27. $2x^2 - x - 5 = 0$

28. $2x^2 + 5x + 1 = 0$

Converting to Vertex Form

29. Consider $h(y) = y^2 + 4y + 4$.

    a. Give the formula for h in vertex form.

    b. What is the vertex of the parabola graph of h?

30. Consider $F(t) = t^2 - 6t + 1$.

    a. Give the formula for F in vertex form.

    b. What is the vertex of the parabola graph of F?

31. Consider $G(r) = r^2 + r - 2$.

    a. Give the formula for G in vertex form.

    b. What is the vertex of the parabola graph of G?

32. Consider $G(y) = y^2 + 9y - 5$.

    a. Give the formula for G in vertex form.

    b. What is the vertex of the parabola graph of G?

33. Consider $H(t) = 5t^2 + 25t + 5$.

    a. Give the formula for H in vertex form.

    b. What is the vertex of the parabola graph of H?

34. Consider $K(r) = 8r^2 + 64r + 2$.

    a. Give the formula for K in vertex form.

    b. What is the vertex of the parabola graph of K?

Domain and Range   Complete the square to convert the quadratic function from standard form to vertex form, and use the result to find the function's domain and range.

35. $f(x) = x^2 + 20x + 94$

The domain of f is ☐

The range of f is ☐

36. $f(x) = x^2 + 16x + 72$

The domain of f is ☐

The range of f is ☐

37. $f(x) = -x^2 - 10x - 34$

The domain of f is ☐

The range of f is ☐

38. $f(x) = -x^2 - 6x - 16$

The domain of f is ☐

The range of f is ☐

39. $f(x) = 5x^2 + 10x + 12$

The domain of f is ☐

The range of f is ☐

40. $f(x) = 3x^2 - 12x + 11$

The domain of f is ☐

The range of f is ☐

41. $f(x) = -4x^2 + 32x - 72$

The domain of f is ☐

The range of f is ☐

42. $f(x) = -2x^2 + 24x - 66$

The domain of f is ☐

The range of f is ☐

Sketching Graphs of Quadratic Functions   Graph each function by algebraically determining its key features. Then state the domain and range of the function.

43. $f(x) = x^2 - 7x + 12$

44. $f(x) = x^2 + 5x - 14$

45. $f(x) = -x^2 - x + 20$

46. $f(x) = -x^2 + 4x + 21$

47. $f(x) = x^2 - 8x + 16$

48. $f(x) = x^2 + 6x + 9$

49. $f(x) = x^2 - 4$

50. $f(x) = x^2 - 9$

51. $f(x) = x^2 + 6x$

52. $f(x) = x^2 - 8x$

53. $f(x) = -x^2 + 5x$

54. $f(x) = -x^2 + 16$

55. $f(x) = x^2 + 4x + 7$

56. $f(x) = x^2 - 2x + 6$

57. $f(x) = x^2 + 2x - 5$

58. $f(x) = x^2 - 6x + 2$

59. $f(x) = -x^2 + 4x - 1$

60. $f(x) = -x^2 - x + 3$

61. $f(x) = 2x^2 - 4x - 30$ **62.** $f(x) = 3x^2 + 21x + 36$

Information from Vertex Form

63. Find the minimum value of the function

$$f(x) = 10x^2 - x + 1$$

64. Find the minimum value of the function

$$f(x) = x^2 - 9x + 10$$

65. Find the maximum value of the function

$$f(x) = 5x - 2x^2 - 2$$

66. Find the maximum value of the function

$$f(x) = 6 - \left(3x^2 + 2x\right)$$

67. Find the range of the function

$$f(x) = -\left(4x^2 + 10x + 6\right)$$

68. Find the range of the function

$$f(x) = 4x - 5x^2 + 3$$

69. Find the range of the function

$$f(x) = 6x^2 - 3x - 9$$

70. Find the range of the function

$$f(x) = 7x^2 + 10x - 1$$

71. If a ball is throw straight up with a speed of $66\ \frac{ft}{s}$, its height at time t (in seconds) is given by

$$h(t) = -8t^2 + 66t + 2$$

Find the maximum height the ball reaches.

72. If a ball is throw straight up with a speed of $68\ \frac{ft}{s}$, its height at time t (in seconds) is given by

$$h(t) = -8t^2 + 68t + 2$$

Find the maximum height the ball reaches.

Challenge

73. Let $f(x) = x^2 + bx + c$. Let b and c be real numbers. Complete the square to find the vertex of $f(x) = x^2 + bx + c$. Write $f(x)$ in vertex form and then state the vertex.

12.3 More on Complex Solutions to Quadratic Equations

When we solve a quadratic equation, sometimes there are no real solutions. In this section we will explore when that happens and what it means on a graph. We will also learn how to handle complex solutions algebraically.

12.3.1 Applications with Real or Complex Solutions

Let's look at an application where we will determine whether the solutions are real or complex. Iman is a pilot and in a stunt plane performance, she plans to dive the plane toward the ground and then back up. The plane's height can be modeled by a quadratic function. If one possible function is h, where $h(t) = \frac{1}{2}t^2 - 5t + 12$, with t standing for time in seconds after the stunt begins, determine whether the plane would hit the ground during the stunt.

To check whether the plane on that flight path would hit the ground, we will solve the equation $h(t) = 0$. We will solve this equation with the quadratic formula. First, we identify that $a = \frac{1}{2}$, $b = -5$ and $c = 12$.

$$t = \frac{-b \pm \sqrt{b^2 - 4ac}}{2a}$$

$$= \frac{-(-5) \pm \sqrt{(-5)^2 - 4(1/2)(12)}}{2(1/2)}$$

$$= \frac{5 \pm \sqrt{25 - 24}}{1}$$

$$= 5 \pm \sqrt{1}$$

$$= 5 \pm 1$$

So, either:

$$t = 6 \qquad \text{or} \qquad t = 4$$

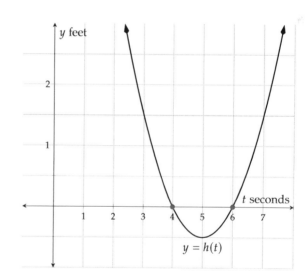

Figure 12.3.2: Graph of $y = h(t)$

This equation has two real solutions and we can see from the graph that the real solutions are the zeros of h. The solution 4 shows that the plane would hit the ground 4 seconds into the stunt, so this is not a good flight path.

To avoid hitting the ground, Iman adjusted the function to p, where $p(t) = \frac{1}{2}t^2 - 5t + 12.5$. To see whether the plane on this flight path would hit the ground, we will solve the equation $p(t) = 0$. We will again use the quadratic formula to solve this equation. We identify that $a = \frac{1}{2}$, $b = -5$ and $c = 12.5$.

$$t = \frac{-b \pm \sqrt{b^2 - 4ac}}{2a}$$

$$= \frac{-(-5) \pm \sqrt{(-5)^2 - 4(1/2)(12.5)}}{2(1/2)}$$

$$= \frac{5 \pm \sqrt{25 - 25}}{1}$$

$$= 5 \pm \sqrt{0}$$

$$= 5 \pm 0$$

$$= 5$$

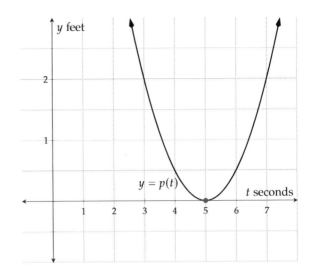

Figure 12.3.3: Graph of $y = p(t)$

This equation has one real solution because p has one zero. This time the plane would hit the ground 5 seconds into the stunt. This is also not a good flight path.

Iman again adjusted the flight path to q, where $q(t) = \frac{1}{2}t^2 - 5t + 13$. We will solve the equation $q(t) = 0$ using the quadratic formula. Identify that $a = \frac{1}{2}$, $b = -5$ and $c = 13$.

$$t = \frac{-b \pm \sqrt{b^2 - 4ac}}{2a}$$

$$= \frac{-(-5) \pm \sqrt{(-5)^2 - 4(1/2)(13)}}{2(1/2)}$$

$$= \frac{5 \pm \sqrt{25 - 26}}{1}$$

$$= 5 \pm \sqrt{-1}$$

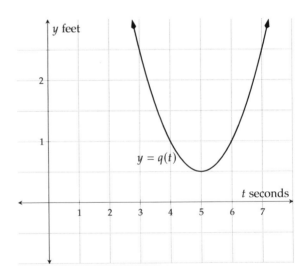

Figure 12.3.4: Graph of $y = q(t)$

Because the radicand is negative, there are no real solutions and the function has no horizontal intercepts. This means the plane will not touch the ground and Iman can complete her stunt using this path.

In general, the radicand of the quadratic formula, $b^2 - 4ac$ is called the **discriminant**. The sign of the discriminant will tell us how many horizontal intercepts a quadratic function will have

- When a quadratic function h has two horizontal intercepts, the equation $h(t) = 0$ has two real solutions. The discriminant will be a positive number so that the $\pm$ from the quadratic formula will provide two solutions.

- When a quadratic function p has one horizontal intercept, the equation $p(t) = 0$ has one real solution. The discriminant will be zero so that the $\pm$ from the quadratic formula will provide only one solution.

- When a quadratic function q has no horizontal intercepts, the equation $q(t) = 0$ has no real solutions, but it has two complex solutions. The discriminant will be a negative number so that the $\sqrt{}$ from the quadratic formula will provide imaginary numbers, and then the $\pm$ will provide two complex solutions.

Example 12.3.5 Futsal[a] is a form of what is usually called soccer in the United States. The game is played on a hard court surface and is usually indoors. The ceiling is out of bounds, so if the ball hits the ceiling it goes to the opposing team.

Borna kicks the ball from the ground with an upward velocity of 8 meters per second. The ball's height in meters can be modeled by the quadratic function h, where $h(t) = -4.9t^2 + 8t$, with t standing for time in seconds after the ball was kicked. If the ceiling height is 4 meters, the minimum height allowed by regulation, determine whether the ball will hit the ceiling.

Explanation. To see whether their ball will hit the ceiling, we will solve the equation $h(t) = 4$. We could complete the square or use the quadratic formula. Because this equation has decimal coefficients we will use the quadratic formula. We put the equation in standard form and identify that $a = -4.9$, $b = 8$ and $c = -4$.

$$-4.9t^2 + 8t = 4$$
$$-4.9t^2 + 8t - 4 = 0$$

$$
\begin{aligned}
t &= \frac{-b \pm \sqrt{b^2 - 4ac}}{2a} \\
&= \frac{-8 \pm \sqrt{8^2 - 4(-4.9)(-4)}}{2(-4.9)} \\
&= \frac{-8 \pm \sqrt{64 - 78.4}}{-9.8} \\
&= \frac{-8 \pm \sqrt{-14.4}}{-9.8}
\end{aligned}
$$

The radicand is negative so we can conclude that there are no real solutions to the equation $h(t) = 4$. That means the parabola will not cross the line $y = 4$ and the ball will not hit the ceiling.

---

[a]en.wikipedia.org/wiki/Futsal

Example 12.3.6 Emma kicks the ball from the ground with an upward velocity of 10 meters per second. This gives us the quadratic function for the height of the ball $h(t) = -4.9t^2 + 10t$, with t standing for time in seconds after the ball was kicked. If the ceiling height is 4.5 meters, determine whether the ball will hit the ceiling.

Explanation. To see whether her ball will hit the ceiling, we will solve the equation $h(t) = 4.5$. We will use the quadratic formula because this equation has decimal coefficients. We put the equation in standard form and identify that $a = -4.9$, $b = 10$ and $c = -4.5$.

$$-4.9t^2 + 10t = 4.5$$

$$-4.9t^2 + 10t - 4.5 = 0$$

$$t = \frac{-b \pm \sqrt{b^2 - 4ac}}{2a}$$

$$= \frac{-10 \pm \sqrt{10^2 - 4(-4.9)(-4.5)}}{2(-4.9)}$$

$$= \frac{-10 \pm \sqrt{100 - 88.2}}{-9.8}$$

$$= \frac{-10 \pm \sqrt{11.8}}{-9.8}$$

The radicand is positive so there are two real solutions to the equation $h(t) = 4.5$. That means the parabola will cross the line $y = 4.5$ and the ball will hit the ceiling.

12.3.2 Solving Equations with Complex Solutions

In a physical context we may only want to know whether solutions are real or complex. Or we may want to find the solutions. When the radicand is negative, we need to go into the complex number system. First we will revisit the definition of complex numbers. Recall that i is defined as $\sqrt{-1}$.

Definition 12.3.7 Complex Number. A **complex number**[1] is a number that can be expressed in the form $a + bi$, where a and b are real numbers and i is the imaginary unit. In this expression, a is the **real part** and b (not bi) is the **imaginary part**.

Here are some examples of solving equations that have complex solutions.

Example 12.3.8 Solve for s in $s^2 - 10s = -34$.

Explanation. We will use the method of completing the square. To do so, we need to add $\left(\frac{b}{2}\right)^2 = (-5)^2 = 25$ to both sides to complete the square.

$$s^2 - 10s = -34$$

$$s^2 - 10s + 25 = -34 + 25$$

$$(s - 5)^2 = -9$$

$s - 5 = -\sqrt{-9}$	or	$s - 5 = \sqrt{-9}$
$s - 5 = -\sqrt{9} \cdot \sqrt{-1}$	or	$s - 5 = \sqrt{9} \cdot \sqrt{-1}$
$s - 5 = -3i$	or	$s - 5 = 3i$
$s = 5 - 3i$	or	$s = 5 + 3i$

---

[1]en.wikipedia.org/wiki/Complex_number

The solution set is $\{5 - 3i, 5 + 3i\}$.

Checkpoint 12.3.9. Solve for x in $2x^2 + 12x + 26 = 0$.

Explanation. We will use the completing-the-square method again. To do so, we first need to divide both sides by the leading coefficient, 2.

$$2x^2 + 12x = -26$$
$$\frac{2x^2}{2} + \frac{12x}{2} = \frac{-26}{2}$$
$$x^2 + 6x = -13$$

Now we can add $\left(\frac{b}{2}\right)^2 = (3)^2 = 9$ to both sides to complete the square.

$$x^2 + 6x + 9 = -13 + 9$$
$$(x + 3)^2 = -4$$

$$x + 3 = -\sqrt{-4} \qquad \text{or} \quad x + 3 = \sqrt{-4}$$
$$x + 3 = -\sqrt{4} \cdot \sqrt{-1} \quad \text{or} \quad x + 3 = \sqrt{4} \cdot \sqrt{-1}$$
$$x + 3 = -2i \qquad \text{or} \quad x + 3 = 2i$$
$$x = -3 - 2i \qquad \text{or} \qquad x = -3 + 2i$$

The solution set is $\{-3 - 2i, -3 + 2i\}$.

The quadratic formula can also be used to solve for complex solutions. Here is an example where it makes more sense to use the quadratic formula.

Example 12.3.10 Solve for x in $5x^2 - 2x = -3$.

Explanation. If we were to complete the square, we would divide both sides by 5 and have lots of fractions in our equation. Instead, we will put the equation in standard form and use the quadratic formula.

$$5x^2 - 2x = -3$$
$$5x^2 - 2x + 3 = 0$$

We identify that $a = 5$, $b = -2$ and $c = 3$ and substitute them into the Quadratic Formula:

$$x = \frac{-b \pm \sqrt{b^2 - 4ac}}{2a}$$
$$= \frac{-(-2) \pm \sqrt{(-2)^2 - 4(5)(3)}}{2(5)}$$
$$= \frac{2 \pm \sqrt{4 - 60}}{10}$$
$$= \frac{2 \pm \sqrt{-56}}{10}$$
$$= \frac{2 \pm \sqrt{-1 \cdot 4 \cdot 14}}{10}$$

$$= \frac{2 \pm \sqrt{-1} \cdot \sqrt{4} \cdot \sqrt{14}}{10}$$

$$= \frac{2 \pm i \cdot 2 \cdot \sqrt{14}}{10}$$

Now we need to put the solutions in standard form which is $a + bi$.

$$x = \frac{2}{10} \pm \frac{2i\sqrt{14}}{10}$$

$$x = \frac{1}{5} \pm \frac{\sqrt{14}}{5}i$$

The solution set is $\left\{ \frac{1}{5} - \frac{\sqrt{14}}{5}i, \frac{1}{5} + \frac{\sqrt{14}}{5}i \right\}$.

Exercises

Review and Warmup Simplify the radical and write it into a complex number.

1. $\sqrt{-30} = $ [   ]

2. $\sqrt{-105} = $ [   ]

3. $\sqrt{-56} = $ [   ]

4. $\sqrt{-28} = $ [   ]

5. $\sqrt{-252} = $ [   ]

6. $\sqrt{-112} = $ [   ]

Real Versus Complex Solutions Determine the nature of the solutions to this quadratic equation.

7. $-7r^2 - 18r - 10 = 0$

  (□ two real solutions □ two non-real solutions □ one doubled real solution □ none of these)

8. $7x^2 - x + 5 = 0$

  (□ two real solutions □ two non-real solutions □ one doubled real solution □ none of these)

9. $8x^2 - x + 1 = 0$

  (□ two real solutions □ two non-real solutions □ one doubled real solution □ none of these)

10. $-8y^2 - 8y - 6 = 0$

  (□ two real solutions □ two non-real solutions □ one doubled real solution □ none of these)

11. $6y^2 + 2y + 9 = 0$

  (□ two real solutions □ two non-real solutions □ one doubled real solution □ none of these)

12. $-z^2 - 4z + 3 = 0$

  (□ two real solutions □ two non-real solutions □ one doubled real solution □ none of these)

13. $-9z^2 + z - 3 = 0$

(□ two real solutions □ two non-real solutions □ one doubled real solution □ none of these)

14. $5t^2 + 5t - 9 = 0$

(□ two real solutions □ two non-real solutions □ one doubled real solution □ none of these)

Solving Equations with Complex Solutions Solve the quadratic equation. Solutions could be complex numbers.

15. $t^2 = -25$

16. $t^2 = -4$

17. $-4x^2 - 1 = 255$

18. $10x^2 - 2 = -252$

19. $-x^2 - 8 = 9$

20. $-2y^2 + 3 = 7$

21. $3(y - 8)^2 - 2 = -50$

22. $-9(r + 2)^2 - 3 = 897$

23. $5r^2 - 8 = -233$

24. $-7t^2 + 9 = 93$

25. $t^2 - 6t + 10 = 0$

26. $x^2 + 2x + 5 = 0$

27. $x^2 - 4x + 7 = 0$

28. $x^2 + 6x + 16 = 0$

Applications

29. A remote control aircraft will perform a stunt by flying toward the ground and then up. Its height can be modeled by the function $h(t) = 1.7t^2 - 17t + 42.5$. The plane (□ will □ will not) hit the ground during this stunt.

30. A remote control aircraft will perform a stunt by flying toward the ground and then up. Its height can be modeled by the function $h(t) = 1.2t^2 - 14.4t + 39.2$. The plane (□ will □ will not) hit the ground during this stunt.

31. A submarine is traveling in the sea. Its depth can be modeled by $d(t) = -0.2t^2 + 2.8t - 5.8$, where t stands for time in seconds. The submarine (□ will □ will not) hit the sea surface along this route.

32. A submarine is traveling in the sea. Its depth can be modeled by $d(t) = -0.9t^2 + 14.4t - 57.6$, where t stands for time in seconds. The submarine (□ will □ will not) hit the sea surface along this route.

12.4 Complex Number Operations

Complex numbers[1] are used in many math, science and engineering applications. In this section, we will learn the basics of complex number operations.

12.4.1 Adding and Subtracting Complex Numbers

Adding and subtracting complex numbers is just like combining like terms. We combine the terms that are real and the terms that are imaginary. Here are some examples

> **Example 12.4.2** Simplify the expression $(1 - 7i) + (5 + 4i)$.
>
> $$(1 - 7i) + (5 + 4i) = 1 + 5 - 7i + 4i$$
> $$= 6 - 3i$$

> **Example 12.4.3** Simplify the expression $(3 - 10i) - (4 - 6i)$.
>
> $$(3 - 10i) - (4 - 6i) = 3 - 10i - 4 + 6i$$
> $$= -1 - 4i$$

Checkpoint 12.4.4. Simplify the expression $(8 + 2i) - (5 + 3i)$.

Explanation.

$$(8 + 2i) - (5 + 3i) = 8 + 2i - 5 - 3i$$
$$= 3 - i$$

12.4.2 Multiplying Complex Numbers

Now let's learn how to multiply complex numbers. It is very similar to multiplying polynomials.

> **Example 12.4.5** Simplify the expression $2i(3 - 2i)$.
>
> We distribute the $2i$ to both terms, then we simplify any powers of i.
>
> $$2i(3 - 2i) = 2i \cdot 3 - 2i \cdot 2i$$
> $$= 6i - 4i^2$$
> $$= 6i - 4(-1)$$
> $$= 6i + 4$$
> $$= 4 + 6i$$
>
> Note that we always write a complex number in standard form, which is $a + bi$.

When we multiply two complex numbers we can use the distributive method, FOIL method, or generic rectangles. Here is an example of each method.

---

[1]en.wikipedia.org/wiki/Complex_number#Applications

Example 12.4.6 Multiply $(1 + 5i)(2 - 7i)$.

We will use the distributive method to multiply the two binomials.

$$
\begin{aligned}
(1 + 5i)(2 - 7i) &= 2(1 + 5i) - 7i(1 + 5i) \\
&= 2 + 10i - 7i - 35i^2 \\
&= 2 + 10i - 7i - 35(-1) \\
&= 2 + 3i + 35 \\
&= 37 + 3i
\end{aligned}
$$

Example 12.4.7 Expand and simplify the expression $(3 - 4i)^2$.

Explanation. We will use the FOIL method to expand this perfect square.

$$
\begin{aligned}
(3 - 4i)^2 &= (3 - 4i)(3 - 4i) \\
&= 9 - 12i - 12i + 16i^2 \\
&= 9 - 24i + 16(-1) \\
&= 9 - 24i - 16 \\
&= -7 - 24i
\end{aligned}
$$

Figure 12.4.8: Using the FOIL method to expand $(3 - 4i)^2$.

Example 12.4.9 Multiply $(3 + 4i)(3 - 4i)$.

Explanation. We will use the Generic Rectangle Method to multiply those two binomials.

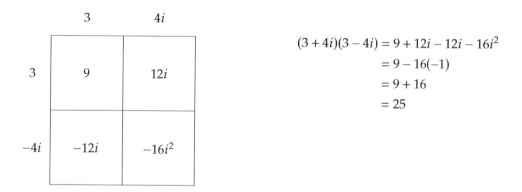

$$
\begin{aligned}
(3 + 4i)(3 - 4i) &= 9 + 12i - 12i - 16i^2 \\
&= 9 - 16(-1) \\
&= 9 + 16 \\
&= 25
\end{aligned}
$$

Figure 12.4.10: Using the Generic Rectangle Method to multiply $(3 + 4i)(3 - 4i)$.

As the last example shows, it is possible to multiply two complex numbers and get a real number result. Notice that the middle terms, $12i$ and $-12i$, are opposites, which makes the result a real number. This happens when we multiply a sum and difference of the same real and imaginary parts, called **complex conjugates**. This pair of factors results in the difference of squares:

$$(a + b)(a - b) = a^2 - b^2.$$

Example 12.4.11 Here is an example using the sum and difference formula to multiply $(5 + 2i)(5 - 2i)$:

$$
\begin{aligned}
(5 + 2i)(5 - 2i) &= 5^2 - (2i)^2 \\
&= 25 - 4i^2 \\
&= 25 - 4(-1) \\
&= 25 + 4 \\
&= 29
\end{aligned}
$$

Checkpoint 12.4.12. Multiply $(7 - 9i)(7 + 9i)$.

Explanation.

$$
\begin{aligned}
(7 - 9i)(7 + 9i) &= (7)^2 - (9i)^2 \\
&= 49 - 81i^2 \\
&= 49 - 81(-1) \\
&= 49 + 81 \\
&= 130
\end{aligned}
$$

12.4.3 Dividing Complex Numbers

When we divide by i we use a process that is similar to rationalizing the denominator. We use the property $\sqrt{x} \cdot \sqrt{x} = x$ when we rationalize the denominator, and we use the property $i \cdot i = -1$ when we have complex numbers. Let's compare these two problems $\frac{2}{\sqrt{2}}$ and $\frac{2}{i}$:

$$
\begin{aligned}
\frac{2}{\sqrt{2}} &= \frac{2 \cdot \sqrt{2}}{\sqrt{2} \cdot \sqrt{2}} \\
&= \frac{2\sqrt{2}}{2} \\
&= \sqrt{2}
\end{aligned}
$$

$$
\begin{aligned}
\frac{2}{i} &= \frac{2 \cdot i}{i \cdot i} \\
&= \frac{2i}{-1} \\
&= -2i
\end{aligned}
$$

Example 12.4.13 Rationalize the denominator in the expression $-\frac{7}{4i}$.

$$
\begin{aligned}
-\frac{7}{4i} &= -\frac{7 \cdot i}{4i \cdot i} \\
&= -\frac{7i}{4(-1)} \\
&= \frac{7i}{4}
\end{aligned}
$$

$$= \frac{7}{4}i$$

Checkpoint 12.4.14. Rationalize the denominator in the expression $\frac{5}{3i}$.

Explanation.

$$\frac{5}{3i} = \frac{5 \cdot i}{3i \cdot i}$$
$$= \frac{5i}{3(-1)}$$
$$= -\frac{5i}{3}$$
$$= -\frac{5}{3}i$$

When the denominator is in the form $a + bi$, we need to use the complex conjugate to remove the imaginary terms from the denominator. Here is an example.

Example 12.4.15 Simplify the expression $\frac{1}{4+3i}$.

Explanation. To get a real result in the denominator we multiply the numerator and denominator by $4 - 3i$, and we have:

$$\frac{1}{4+3i} = \frac{1}{4+3i} \cdot \frac{(4-3i)}{(4-3i)}$$
$$= \frac{4-3i}{16 - 12i + 12i - 9i^2}$$
$$= \frac{4-3i}{16 - 9(-1)}$$
$$= \frac{4-3i}{16+9}$$
$$= \frac{4-3i}{25}$$
$$= \frac{4}{25} - \frac{3}{25}i$$

Note that we always write complex numbers in standard form which is $a + bi$.

Now we can divide two complex numbers as in the next example.

Example 12.4.16 Simplify the expression $\frac{1+2i}{2-4i}$.

Explanation. To divide complex numbers, we rationalize the denominator using the conjugate $2 + 4i$:

$$\frac{1+2i}{2-4i} = \frac{(1+2i)}{(2-4i)} \cdot \frac{(2+4i)}{(2+4i)}$$
$$= \frac{2 + 4i + 4i + 8i^2}{4 + 8i - 8i - 16i^2}$$
$$= \frac{2 + 8i + 8(-1)}{4 - 16(-1)}$$
$$= \frac{2 + 8i - 8}{4 + 16}$$

$$= \frac{-6 + 8i}{20}$$

$$= \frac{-6}{20} + \frac{8i}{20}$$

$$= -\frac{3}{10} + \frac{2}{5}i$$

Checkpoint 12.4.17. Simplify the expression $\frac{4-7i}{5+i}$.

Explanation. To divide, we rationalize the denominator using the conjugate $5 - i$:

$$\frac{4 - 7i}{5 + i} = \frac{(4 - 7i)}{(5 + i)} \cdot \frac{(5 - i)}{(5 - i)}$$

$$= \frac{20 - 4i - 35i + 7i^2}{25 - 5i + 5i - i^2}$$

$$= \frac{20 - 39i + 7(-1)}{25 - 1(-1)}$$

$$= \frac{20 - 39i - 7}{25 + 1}$$

$$= \frac{13 - 39i}{26}$$

$$= \frac{13}{26} - \frac{39i}{26}$$

$$= \frac{1}{2} - \frac{3}{2}i$$

Exercises

Adding and Subtracting Complex Numbers

1. Add up the following complex numbers:

$(-7 + 6i) + (2 + 5i) = \boxed{}$

2. Add up the following complex numbers:

$(-4 - 3i) + (12 - 2i) = \boxed{}$

3. Subtract the following complex numbers:

$(-1 - 11i) - (-3 - 8i) = \boxed{}$

4. Subtract the following complex numbers:

$(1 + 5i) - (8 + 10i) = \boxed{}$

5. Write the complex number in standard form.

$$(3 - 3i) + (-6 + 2i)$$

6. Write the complex number in standard form.

$$(6 - 10i) + (3 - 4i)$$

7. Write the complex number in standard form.

$$(8 + 3i) + (-9 - 9i)$$

8. Write the complex number in standard form.

$$(10 - 4i) + (6i)$$

9. Write the complex number in standard form.

$$(-8 + 10i) - (8)$$

10. Write the complex number in standard form.

$$(-6 + 2i) - (-4 - 6i)$$

11. Write the complex number in standard form.

$$(-4 - 5i) - (5 + 9i)$$

12. Write the complex number in standard form.

$$(-1 + 9i) - (-7 + 4i)$$

Multiplying Complex Numbers

13. Multiply the following complex numbers:

$$i(1 + 2i) = \boxed{}$$

14. Multiply the following complex numbers:

$$i(4 - 7i) = \boxed{}$$

15. Multiply the following complex numbers:

$$(7 + 9i)(-3 + 8i) = \boxed{}$$

16. Multiply the following complex numbers:

$$(10 + i)(8 + i) = \boxed{}$$

17. Multiply the following complex numbers:

$$(12 - 8i)^2 = \boxed{}$$

18. Multiply the following complex numbers:

$$(-10 + 8i)^2 = \boxed{}$$

19. Multiply the following complex numbers:

$$(-7 - 11i)(-7 + 11i) = \boxed{}$$

20. Multiply the following complex numbers:

$$(-4 - 9i)(-4 + 9i) = \boxed{}$$

21. Write the complex number in standard form.

$$(-1 + 6i)(8 - 7i)$$

22. Write the complex number in standard form.

$$(1 - i)(-4 + 8i)$$

23. Write the complex number in standard form.

$$(3 - 8i)(5 + 3i)$$

24. Write the complex number in standard form.

$$(6 + 5i)(-7 - 3i)$$

Dividing Complex Numbers

25. Rewrite the following expression into the form of a+bi:

$$\frac{6}{i} = \boxed{}$$

26. Rewrite the following expression into the form of a+bi:

$$\frac{2}{i} = \boxed{}$$

27. Rewrite the following expression into the form of a+bi:

$$\frac{-8 + 4i}{-2 + 6i} = \boxed{}$$

28. Rewrite the following expression into the form of a+bi:

$$\frac{-2 + 4i}{-4 - 2i} = \boxed{}$$

29. Rewrite the following expression into the form of a+bi:

$$\frac{-3 - 8i}{-5 + 8i} = \boxed{}$$

30. Rewrite the following expression into the form of a+bi:

$$\frac{3 - 8i}{2 + 3i} = \boxed{}$$

31. Write the complex number in standard form.

$$\frac{1 - 4i}{-9 - 2i}$$

32. Write the complex number in standard form.

$$\frac{3 + 10i}{-8i}$$

33. Write the complex number in standard form.

$$\frac{6 + 3i}{8 + 7i}$$

34. Write the complex number in standard form.

$$\frac{8 - 5i}{-4 + i}$$

12.5 More on Quadratic Functions Chapter Review

12.5.1 Graphs and Vertex Form

In Section 12.1 we covered the use of technology in analyzing quadratic functions, the vertex form of a quadratic function and how it affects horizontal and vertical shifts of the graph of a parabola, and the factored form of a quadratic function.

> **Example 12.5.1 Exploring Quadratic Functions with Graphing Technology.** Use technology to graph and make a table of the quadratic function g defined by $g(x) = -x^2 + 5x - 6$ and find each of the key points or features.
>
> | | |
> |---|---|
> | a. Find the vertex. | e. Solve $g(x) = -6$ using the graph. |
> | b. Find the vertical intercept. | f. Solve $g(x) \le -6$ using the graph. |
> | c. Find the horizontal intercept(s). | g. State the domain and range of the function. |
> | d. Find $g(-1)$. | |
>
> **Explanation.**
>
> The specifics of how to use any one particular technology tool vary. Whether you use an app, a physical calculator, or something else, a table and graph should look like:
>
> | x | $g(x)$ |
> |-----|--------|
> | -1 | -12 |
> | 0 | -6 |
> | 1 | -2 |
> | 2 | 0 |
> | 2 | 0 |
> | 3 | 0 |
> | 4 | -2 |
>
>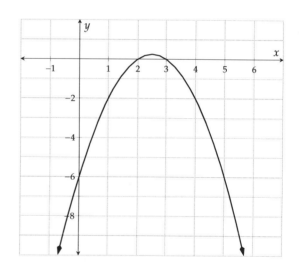

Additional features of your technology tool can enhance the graph to help answer these questions. You may be able to make the graph appear like:

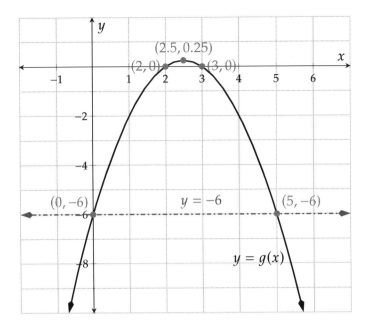

a. The vertex is $(2.5, 0.25)$.

b. The vertical intercept is $(0, -6)$.

c. The horizontal intercepts are $(2, 0)$ and $(3, 0)$.

d. $g(-1) = -2$.

e. The solutions to $g(x) = -6$ are the x-values where $y = 6$. We graph the horizontal line $y = -6$ and find the x-values where the graphs intersect. The solution set is $\{0, 5\}$.

f. The solutions are all x-values where the function below (or touching) the line $y = -6$. The solution set is $(-\infty, 0] \cup [5, \infty)$.

g. The domain is $(-\infty, \infty)$ and the range is $(-\infty, 0.25]$.

Example 12.5.2 The Vertex Form of a Parabola. Recall that the vertex form of a quadratic function tells us the location of the vertex of a parabola.

a. State the vertex of the quadratic function $r(x) = -8(x + 1)^2 + 7$.

b. State the vertex of the quadratic function $u(x) = 5(x - 7)^2 - 3$.

c. Write the formula for a parabola with vertex $(-5, 3)$ and $a = 2$.

d. Write the formula for a parabola with vertex $(1, -17)$ and $a = -4$.

Explanation.

a. The vertex of the quadratic function $r(x) = -8(x+1)^2 + 7$ is $(-1, 7)$.

b. The vertex of the quadratic function $u(x) = 5(x-7)^2 - 3$ is $(7, -3)$.

c. The formula for a parabola with vertex $(-5, 3)$ and $a = 2$ is $y = 2(x+5)^2 + 3$.

d. The formula for a parabola with vertex $(1, -17)$ and $a = -4$ is $y = 4(x-1)^2 - 17$.

Example 12.5.3 Horizontal and Vertical Shifts. Identify the horizontal and vertical shifts compared with $f(x) = x^2$.

a. $s(x) = (x+1)^2 + 7$.

b. $v(x) = (x-7)^2 - 3$.

Explanation.

a. The graph of the quadratic function $s(x) = -8(x+1)^2 + 7$ is the same as the graph of $f(x) = x^2$ shifted to the left 1 unit and up 7 units.

b. The graph of the quadratic function $v(x) = 5(x-7)^2 - 3$ is the same as the graph of $f(x) = x^2$ shifted to the right 7 units and down 3 units.

Example 12.5.4 The Factored Form of a Parabola. Recall that the factored form of a quadratic function tells us the horizontal intercepts very quickly.

a. $n(x) = 13(x-1)(x+6)$.

b. $p(x) = -6(x - \frac{2}{3})(x + \frac{1}{2})$.

Explanation.

a. The horizontal intercepts of n are $(1, 0)$ and $(-6, 0)$.

b. The horizontal intercepts of p are $(\frac{2}{3}, 0)$ and $(-\frac{1}{2}, 0)$.

12.5.2 Completing the Square

In Section 12.2 we covered how to complete the square to both solve quadratic equations in one variable and to put quadratic functions into vertex form.

Example 12.5.5 Solving Quadratic Equations by Completing the Square. Solve the equations by completing the square.

a. $k^2 - 18k + 1 = 0$

b. $4p^2 - 3p = 2$

Explanation.

a. To complete the square in the equation $k^2 - 18k + 1 = 0$, we first we will first move the constant term to the right side of the equation. Then we will use Fact 12.2.2 to find $\left(\frac{b}{2}\right)^2$ to add to both sides.

$$k^2 - 18k + 1 = 0$$

$$k^2 - 18k = -1$$

In our case, $b = -18$, so $\left(\frac{b}{2}\right)^2 = \left(\frac{-18}{2}\right)^2 = 81$

$$k^2 - 18k + 81 = -1 + 81$$
$$(k - 9)^2 = 80$$

$k - 9 = -\sqrt{80}$	or	$k - 9 = \sqrt{80}$
$k - 9 = -4\sqrt{5}$	or	$k - 9 = 4\sqrt{5}$
$k = 9 - 4\sqrt{5}$	or	$k = 9 + 4\sqrt{5}$

The solution set is $\{9 + 4\sqrt{5}, 9 - 4\sqrt{5}\}$.

b. To complete the square in the equation $4p^2 - 3p = 2$, we first divide both sides by 4 since the leading coefficient is 4.

$$\frac{4p^2}{4} - \frac{3p}{4} = \frac{2}{4}$$
$$p^2 - \frac{3}{4}p = \frac{1}{2}$$
$$p^2 - \frac{3}{4}p = \frac{1}{2}$$

Next, we will complete the square. Since $b = -\frac{3}{4}$, first,

$$\frac{b}{2} = \frac{-\frac{3}{4}}{2} = -\frac{3}{8} \tag{12.5.1}$$

and next, squaring that, we have

$$\left(-\frac{3}{8}\right)^2 = \frac{9}{64}. \tag{12.5.2}$$

So we will add $\frac{9}{64}$ from Equation (12.5.2) to both sides of the equation:

$$p^2 - \frac{3}{4}p + \frac{9}{64} = \frac{1}{2} + \frac{9}{64}$$
$$p^2 - \frac{3}{4}p + \frac{9}{64} = \frac{32}{64} + \frac{9}{64}$$
$$p^2 - \frac{3}{4}p + \frac{9}{64} = \frac{41}{64}$$

Here, remember that we always factor with the number found in the first step of completing the square, Equation (12.5.1).

$$\left(p - \frac{3}{8}\right)^2 = \frac{41}{64}$$

$p - \frac{3}{8} = -\frac{\sqrt{41}}{8}$	or	$p - \frac{3}{8} = \frac{\sqrt{41}}{8}$

$$p = \frac{3}{8} - \frac{\sqrt{41}}{8} \qquad \text{or} \qquad p = \frac{3}{8} + \frac{\sqrt{41}}{8}$$

$$p = \frac{3 - \sqrt{41}}{8} \qquad \text{or} \qquad p = \frac{3 + \sqrt{41}}{8}$$

The solution set is $\left\{ \frac{3-\sqrt{41}}{8}, \frac{3+\sqrt{41}}{8} \right\}$.

Example 12.5.6 Putting Quadratic Functions in Vertex Form. Write a formula in vertex form for the function T defined by $T(x) = 4x^2 + 20x + 24$.

Explanation. Before we can complete the square, we will factor the 4 out of the first two terms. Don't be tempted to factor the 4 out of the constant term.

$$T(x) = 4\left(x^2 + 5x\right) + 24$$

Now we will complete the square inside the parentheses by adding and subtracting $\left(\frac{5}{2}\right)^2 = \frac{25}{4}$.

$$T(x) = 4\left(x^2 + 5x + \frac{25}{4} - \frac{25}{4}\right) + 24$$

Notice that the constant that we subtracted is inside the parentheses, but it will not be part of our perfect square trinomial. In order to bring it outside, we need to multiply it by 4. We are distributing the 4 to that term so we can combine it with the outside term.

$$
\begin{aligned}
T(x) &= 4\left(\left(x^2 + 5x + \frac{25}{4}\right) - \frac{25}{4}\right) + 24 \\
&= 4\left(x^2 + 5x + \frac{25}{4}\right) - 4 \cdot \frac{25}{4} + 24 \\
&= 4\left(x + \frac{5}{2}\right)^2 - 25 + 24 \\
&= 4\left(x + \frac{5}{2}\right)^2 - 1
\end{aligned}
$$

Note that The vertex is $\left(-\frac{5}{2}, -1\right)$.

Example 12.5.7 Graphing Quadratic Functions by Hand. Graph the function H defined by $H(x) = -x^2 - 8x - 15$ by determining its key features algebraically.

Explanation. To start, we'll note that this function opens downward because the leading coefficient, -1, is negative.

Now we will complete the square to find the vertex. We will factor the -1 out of the first two terms, and then add and subtract $\left(\frac{8}{2}\right)^2 = 4^2 = 16$ on the right side.

$$
\begin{aligned}
H(x) &= -\left[x^2 + 8x\right] - 15 \\
&= -\left[x^2 + 8x + 16 - 16\right] - 15 \\
&= -\left[\left(x^2 + 8x + 16\right) - 16\right] - 15
\end{aligned}
$$

$$= -\left(x^2 + 8x + 16\right) - (-1 \cdot 16) - 15$$
$$= -(x + 4)^2 + 16 - 15$$
$$= -(x + 4)^2 + 1$$

The vertex is $(-4, 1)$ so the axis of symmetry is the line $x = -4$.

To find the y-intercept, we'll replace x with 0 or read the value of c from the function in standard form:

$$H(0) = -(0)^2 - 8(0) - 15$$
$$= -15$$

The y-intercept is $(0, -15)$ and we will find its symmetric point on the graph, which is $(-8, -15)$.

Next, we'll find the horizontal intercepts. We see this function factors so we will write the factored form to get the horizontal intercepts.

$$H(x) = -x^2 - 8x - 15$$
$$= -\left(x^2 + 8x + 15\right)$$
$$= -(x + 3)(x + 5)$$

The x-intercepts are $(-3, 0)$ and $(-5, 0)$.

Now we will plot all of the key points and draw the parabola.

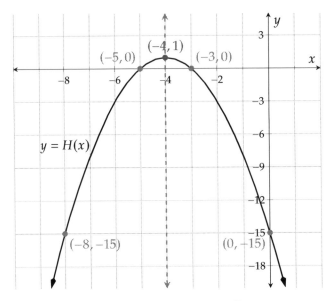

Figure 12.5.8: The graph of $y = -x^2 - 8x - 15$.

12.5.3 More on Complex Solutions to Quadratic Equations

In Section 12.3 we covered the definition of a complex number, and discussed both quadratic applications and equations where complex numbers appear as solutions.

Example 12.5.9 Applications with Real or Complex Solutions. One day, Samar was bouncing a ball inside the house. The trajectory of his bounce followed the quadratic function $H(t) = -16t^2 + 24t$, where $H(t)$ describes the height of the ball, in feet, at time t seconds after it bounced off the ground. If the ceilings in Samar's house were 10 feet tall, find out if the ball will hit the ceiling.

Explanation. To find out if the ball will hit the ceiling, we need to set the formula for the function equal to 10 and solve.

$$H(t) = -16t^2 + 24t$$
$$10 = -16t^2 + 24t$$
$$0 = -16t^2 + 24t - 10$$

This is a quadratic equation where verything is divisible by 2. We will divide every term by 2 which can simplify the process.

$$\frac{0}{2} = \frac{-16t^2}{2} + \frac{24t}{2} - \frac{10}{2}$$
$$0 = -8t^2 + 12t - 5$$

Since the equation doesn't seem to factor easily, we will use the quadratic formula to solve it. Note that $a = -8$, $b = 12$, and $c = -5$.

$$t = \frac{-b \pm \sqrt{b^2 - 4ac}}{2a}$$
$$t = \frac{-(12) \pm \sqrt{(12)^2 - 4(-8)(-5)}}{2(-8)}$$
$$t = \frac{-12 \pm \sqrt{-16}}{-16}$$

Note that the discriminant is negative, which means that the equation has no real solutions. Just for practice, we will finish the simplification process, but we are ready to make our conclusion here.

$$t = \frac{-12 \pm \sqrt{16 \cdot -1}}{-16}$$
$$t = \frac{-12 \pm 4i}{-16}$$
$$t = \frac{-12}{-16} \pm \frac{4i}{-16}$$
$$t = \frac{3}{4} \mp \frac{i}{4}$$

Since the solutions to the equation are complex numbers, the reality of the situation must be that the ball never does hit the ceiling. Samar's ceiling lights are safe for now.

Example 12.5.10 Solving Equations with Complex Solutions. Solve for x in $3x^2 - 12x + 36 = 0$.

Explanation. We will use the completing-the-square method. To do so, we first need to divide both sides by the leading coefficient, 3.

$$3x^2 - 12x + 36 = 0$$
$$3x^2 - 12x = -36$$

$$\frac{3x^2}{3} - \frac{12x}{3} = \frac{-36}{3}$$
$$x^2 - 4x = -12$$

Now we can add $\left(\frac{b}{2}\right)^2 = (-2)^2 = 4$ to both sides to complete the square.

$$x^2 - 4x + 4 = -12 + 4$$
$$(x - 2)^2 = -8$$

$x - 2 = -\sqrt{-8}$	or	$x - 2 = \sqrt{-8}$
$x - 2 = -\sqrt{4 \cdot -1 \cdot 2}$	or	$x - 2 = \sqrt{4 \cdot -1 \cdot 2}$
$x - 2 = -2i\sqrt{2}$	or	$x - 2 = 2i\sqrt{2}$
$x = 2 - 2i\sqrt{2}$	or	$x = 2 + 2i\sqrt{2}$

The solution set is $\{2 - 2i\sqrt{2}, 2 + 2i\sqrt{2}\}$.

12.5.4 Complex Number Operations

In Section 12.4 we covered the essential algebra of complex numbers.

Example 12.5.11 Adding and Subtracting Complex Numbers. Simplify the expression $(5 - 3i) - (1 - 7i)$.

Explanation.

$$(5 - 3i) - (1 - 7i) = 5 - 3i - 1 + 7i$$
$$= 4 + 4i$$

Example 12.5.12 Multiplying Complex Numbers. Multiply $(3 + 2i)(5 - 6i)$.

Explanation. We will use the FOIL method to multiply the two binomials.

$$(1 + 5i)(2 - 7i) = 15 - 18i + 10i - 12i^2$$
$$= 15 - 8i - 12(-1)$$
$$= 15 - 8i + 12$$
$$= 27 - 8i$$

Example 12.5.13 Dividing Complex Numbers. Simplify the expression $\frac{3+5i}{5-6i}$.

Explanation. To divide complex numbers, we rationalize the denominator using the conjugate $2 + 4i$:

$$\frac{3 + 5i}{5 - 6i} = \frac{(3 + 5i)}{(5 - 6i)} \cdot \frac{(5 + 6i)}{(5 + 6i)}$$
$$= \frac{15 + 18i + 25i + 30i^2}{25 + 30i - 30i - 36i^2}$$

$$= \frac{15 + 43i + 30(-1)}{25 - 36(-1)}$$

$$= \frac{15 + 43i - 30}{25 + 36}$$

$$= \frac{-15 + 43i}{61}$$

$$= -\frac{15}{61} + \frac{43i}{61}$$

Exercises

Graphs and Vertex Form

1. Use technology to make a table of values for the function K defined by $K(x) = 3x^2 - 2x - 3$.

x	$K(x)$

2. Use technology to make a table of values for the function f defined by $f(x) = -3x^2 - 8x + 37$.

x	$f(x)$

3. Use technology to make a graph of f where $f(x) = 3x^2 - 6x - 5$.

4. Use technology to make a graph of f where $f(x) = -3x^2 - 8x + 3$.

5. Let $g(x) = -x^2 + x + 3$. Use technology to find the following.

   a. The vertex is _____.

   b. The y-intercept is _____.

   c. The x-intercept(s) is/are _____.

   d. The domain of g is _____.

   e. The range of g is _____.

   f. Calculate $g(3)$. _____.

   g. Solve $g(x) = 2$. _____.

   h. Solve $g(x) > 2$. _____.

6. Let $h(x) = -x^2 - 4x - 2$. Use technology to find the following.

   a. The vertex is _____.

   b. The y-intercept is _____.

   c. The x-intercept(s) is/are _____.

   d. The domain of h is _____.

   e. The range of h is _____.

   f. Calculate $h(-1)$. _____.

   g. Solve $h(x) = -5$. _____.

   h. Solve $h(x) \geq -5$. _____.

7. An object was launched from the top of a hill with an upward vertical velocity of 110 feet per second. The height of the object can be modeled by the function $h(t) = -16t^2 + 110t + 200$, where t represents the number of seconds after the launch. Assume the object landed on the ground at sea level. Find the answer using technology.

[ ] seconds after its launch, the object reached its maximum height of [ ] feet.

8. An object was launched from the top of a hill with an upward vertical velocity of 130 feet per second. The height of the object can be modeled by the function $h(t) = -16t^2 + 130t + 100$, where t represents the number of seconds after the launch. Assume the object landed on the ground at sea level. Find the answer using technology.

[ ] seconds after its launch, the object fell to the ground at sea level.

9. Find the vertex of the graph of

$$y = 3(x - 7)^2 - 6$$

10. Find the vertex of the graph of

$$y = 6(x - 3)^2 - 3$$

11. Write the vertex form for the quadratic function f, whose vertex is $(-7, -8)$ and has leading coefficient $a = 7$.

$f(x) =$ [ ]

12. Write the vertex form for the quadratic function f, whose vertex is $(6, 6)$ and has leading coefficient $a = 9$.

$f(x) =$ [ ]

13. A graph of a function f is given. Use the graph to write a formula for f in vertex form. You will need to identify the vertex and also one more point on the graph to find the leading coefficient a.

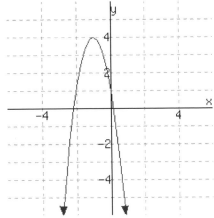

$f(x) =$ [ ]

14. A graph of a function f is given. Use the graph to write a formula for f in vertex form. You will need to identify the vertex and also one more point on the graph to find the leading coefficient a.

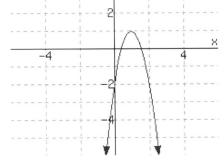

$f(x) =$ [ ]

15. Let h be defined by $h(x) = (x - 5)^2 + 5$.

    a. What is the domain of h?

    b. What is the range of h?

16. Let h be defined by $h(x) = (x + 2)^2 - 7$.

    a. What is the domain of h?

    b. What is the range of h?

17. Consider the graph of the equation $y = (x - 1)^2 - 8$.

    Compared to the graph of $y = x^2$, the vertex has been shifted ☐☐☐ units (☐ left ☐ right) and ☐☐☐ units (☐ down ☐ up) .

18. Consider the graph of the equation $y = (x - 3)^2 + 4$.

    Compared to the graph of $y = x^2$, the vertex has been shifted ☐☐☐ units (☐ left ☐ right) and ☐☐☐ units (☐ down ☐ up) .

19. The quadratic expression $(x - 4)^2 - 4$ is written in vertex form.

    a. Write the expression in standard form.

    b. Write the expression in factored form.

20. The quadratic expression $(x - 4)^2 - 1$ is written in vertex form.

    a. Write the expression in standard form.

    b. Write the expression in factored form.

21. The formula for a quadratic function g is $g(x) = (x - 9)(x - 3)$.

    a. The y-intercept is ☐☐☐ .

    b. The x-intercept(s) is/are ☐☐☐ .

22. The formula for a quadratic function G is $G(x) = (x + 8)(x + 3)$.

    a. The y-intercept is ☐☐☐ .

    b. The x-intercept(s) is/are ☐☐☐ .

Completing the Square

23. Solve $x^2 - 4x = 5$ by completing the square.

24. Solve $y^2 + 8y = -12$ by completing the square.

25. Solve $y^2 + 7y + 6 = 0$ by completing the square.

26. Solve $r^2 - 5r - 24 = 0$ by completing the square.

27. Solve $12r^2 - 4r - 5 = 0$ by completing the square.

28. Solve $3t^2 + 8t + 5 = 0$ by completing the square.

29. Complete the square to convert the quadratic function from standard form to vertex form, and use the result to find the function's domain and range.

$$f(x) = -4x^2 + 64x - 248$$

The domain of f is ⬚

The range of f is ⬚

30. Complete the square to convert the quadratic function from standard form to vertex form, and use the result to find the function's domain and range.

$$f(x) = -5x^2 + 100x - 499$$

The domain of f is ⬚

The range of f is ⬚

31. Graph $f(x) = x^2 - 7x + 12$ by algebraically determining its key features. Then state the domain and range of the function.

32. Graph $f(x) = -x^2 + 4x + 21$ by algebraically determining its key features. Then state the domain and range of the function.

33. Graph $f(x) = x^2 - 8x + 16$ by algebraically determining its key features. Then state the domain and range of the function.

34. Graph $f(x) = x^2 + 6x + 9$ by algebraically determining its key features. Then state the domain and range of the function.

35. Graph $f(x) = x^2 + 4x + 7$ by algebraically determining its key features. Then state the domain and range of the function.

36. Graph $f(x) = x^2 - 2x + 6$ by algebraically determining its key features. Then state the domain and range of the function.

37. Graph $f(x) = 2x^2 - 4x - 30$ by algebraically determining its key features. Then state the domain and range of the function.

38. Graph $f(x) = 3x^2 + 21x + 36$ by algebraically determining its key features. Then state the domain and range of the function.

39. Find the minimum value of the function

$$f(x) = x^2 - 7x - 2$$

40. Find the minimum value of the function

$$f(x) = 2x^2 + 7x + 7$$

More on Complex Solutions to Quadratic Equations

41. Solve the quadratic equation. Solutions could be complex numbers.

$$-10(y + 8)^2 - 3 = 357$$

42. Solve the quadratic equation. Solutions could be complex numbers.

$$8(y - 3)^2 - 4 = -76$$

43. Solve the quadratic equation. Solutions could be complex numbers.

$$r^2 - 10r + 32 = 0$$

44. Solve the quadratic equation. Solutions could be complex numbers.

$$r^2 - 8r + 19 = 0$$

45. A remote control aircraft will perform a stunt by flying toward the ground and then up. Its height can be modeled by the function $h(t) = 1.7t^2 - 30.6t + 133.7$. The plane (□ will □ will not) hit the ground during this stunt.

46. A remote control aircraft will perform a stunt by flying toward the ground and then up. Its height can be modeled by the function $h(t) = 0.6t^2 - 10.8t + 51.6$. The plane (□ will □ will not) hit the ground during this stunt.

Complex Number Operations

47. Write the complex number in standard form.

$$(10 - 2i) - (5 - 4i)$$

48. Write the complex number in standard form.

$$(-9 - 9i) - (-7 - 10i)$$

49. Write the complex number in standard form.

$$(-6 + 4i)(2 + 6i)$$

50. Write the complex number in standard form.

$$(-4 - 3i)(10)$$

51. Rewrite the following expression into the form of a+bi:

$$\frac{-1 - 8i}{-2 - 5i} = \boxed{}$$

52. Rewrite the following expression into the form of a+bi:

$$\frac{3 + 6i}{7 - 5i} = \boxed{}$$

CHAPTER 13

Rational Functions and Equations

13.1 Introduction to Rational Functions

In this chapter we will learn about rational functions, which are ratios of two polynomial functions. Creating this ratio inherently requires division, and we'll explore the effect this has on the graphs of rational functions and their domain and range.

13.1.1 Graphs of Rational Functions

When a drug is injected into a patient, the drug's concentration in the patient's bloodstream can be modeled by the function C, with formula

$$C(t) = \frac{3t}{t^2 + 8}$$

where $C(t)$ gives the drug's concentration, in milligrams per liter, t hours since the injection. A new injection is needed when the concentration falls to 0.35 milligrams per liter. Let's use graphing technology to explore this situation.

a. What is the concentration after 10 hours?

b. After how many hours since the first injection is the drug concentration greatest?

c. After how many hours since the first injection should the next injection be given?

d. What happens to the drug concentration if no further injections are given?

Using graphing technology, we will graph $y = \frac{3t}{t^2+8}$ and $y = 0.35$.

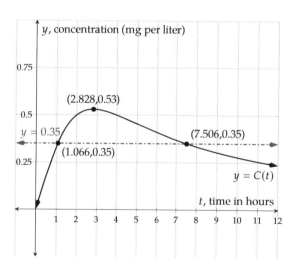

Figure 13.1.2: Graph of $C(t) = \frac{3t}{t^2+8}$

To determine the concentration after 10 hours, we will evaluate C at $t = 10$. After 10 hours, the concentration will be about $0.2777 \frac{mg}{L}$.

$$C(10) = \frac{3(10)}{10^2 + 8}$$
$$= \frac{30}{108}$$
$$= \frac{5}{18} \approx 0.2777$$

Using the graph, we can see that the maximum concentration of the drug will be $0.53 \frac{mg}{L}$ and will occur after about 2.828 hours.

The approximate points of intersection $(1.066, 0.35)$ and $(7.506, 0.35)$ tell us that the concentration of the drug will reach $0.35 \frac{mg}{L}$ after about 1.066 hours and again after about 7.506 hours. Given the rising, then falling shape of the graph, this means that another dose will need to be administered after about 7.506 hours.

From the initial graph, it appears that the concentration of the drug will diminish to zero with enough time passing. Exploring further, we can see both numerically and graphically that for larger and larger values of t, the function values get closer and closer to zero. This is shown in Table 13.1.3 and Figure 13.1.4.

t	$C(t)$
24	0.123...
48	0.062...
72	0.041...
96	0.031...
120	0.020...

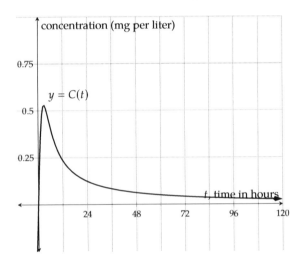

Table 13.1.3: Numerical Values for $C(t) = \frac{3t}{t^2+8}$

Figure 13.1.4: Graph of $C(t) = \frac{3t}{t^2+8}$

In Section 13.5, we'll explore how to algebraically solve $C(t) = 0.35$. For now, we just relied on technology to make the graph and determine intersection points.

The function C, where $C(t) = \frac{3t}{t^2+8}$, is a *rational function*, which is a type of function defined as follows.

Definition 13.1.5 Rational Function. A **rational function** f is a function in the form

$$f(x) = \frac{P(x)}{Q(x)}$$

where P and Q are polynomial functions, but Q is not the constant zero function.

Checkpoint 13.1.6. Identify which of the following are rational functions and which are not.

a. f defined by $f(x) = \frac{25x^2+3}{25x^2+3}$ (□ is   □ is not)   a rational function.

b. Q defined by $Q(x) = \frac{5x^2+3\sqrt{x}}{2x}$ ($\square$ is $\square$ is not)   a rational function.

c. g defined by $g(t) = \frac{t\sqrt{5}-t^3}{2t+1}$ ($\square$ is $\square$ is not)   a rational function.

d. P defined by $P(x) = \frac{5x+3}{|2x+1|}$ ($\square$ is $\square$ is not)   a rational function.

e. h defined by $h(x) = \frac{3^x+1}{x^2+1}$ ($\square$ is $\square$ is not)   a rational function.

Explanation.

a. f defined by $f(x) = \frac{25x^2+3}{25x^2+3}$ is a rational function as its formula is a polynomial divided by another polynomial.

b. Q defined by $Q(x) = \frac{5x^2+3\sqrt{x}}{2x}$ is not a rational function because the numerator contains $\sqrt{x}$ and is therefore not a polynomial.

c. g defined by $g(t) = \frac{t\sqrt{5}-t^3}{2t+1}$ is a rational function as its formula is a polynomial divided by another polynomial.

d. P defined by $P(x) = \frac{5x+3}{|2x+1|}$ is not a rational function because the denominator contains the absolute value of an expression with variables in it.

e. h defined by $h(x) = \frac{3^x+1}{x^2+1}$ is not a rational function because the numerator contains 3^x, which has a variable in the exponent.

A rational function's graph is not always smooth like the one shown in Example 13.1.2. It could have breaks, as we'll see now.

Example 13.1.7 Build a table and sketch the graph of the function f where $f(x) = \frac{1}{x-2}$. Find the function's domain and range.

Since $x = 2$ makes the denominator 0, the function will be undefined for $x = 2$. We'll start by choosing various x-values and plotting the associated points.

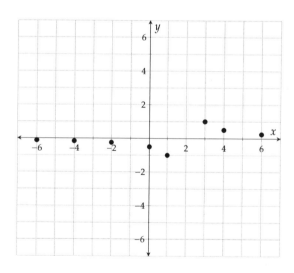

x	$f(x)$	Point
-6	$\frac{1}{-6-2} = -0.125$	$(-6, -0.125)$
-4	$\frac{1}{-4-2} \approx -0.167$	$\left(-4, -\frac{1}{6}\right)$
-2	$\frac{1}{-2-2} \approx -0.25$	$(-2, -0.25)$
0	$\frac{1}{0-2} = -0.5$	$(0, -0.5)$
1	$\frac{1}{1-2} = -1$	$(1, -1)$
2	undefined	
3	$\frac{1}{3-2} = 1$	$(3, 1)$
4	$\frac{1}{4-2} = 0.5$	$(4, 0.5)$

Table 13.1.8: Initial Values of $f(x) = \frac{1}{x-2}$

Figure 13.1.9: Initial Points for $f(x) = \frac{1}{x-2}$

Note that extra points were chosen near $x = 2$ in the Table 13.1.8, but it's still not clear on the graph what happens really close to $x = 2$. It will be essential that we include at least one x-value between 1 and 2 and also between 2 and 3.

Further, we'll note that dividing one number by a number that is close to 0 yields a large number. For example, $\frac{1}{0.0005} = 2000$. In fact, the smaller the number is that we divide by, the larger our result becomes. So when x gets closer and closer to 2, then $x - 2$ gets closer and closer to 0. And then $\frac{1}{x-2}$ takes very large values.

When we plot additional points closer and closer to 2, we get larger and larger results. To the left of 2, the results are negative, so the connected curve has an arrow pointing downward there. The opposite happens to the right of $x = 2$, and an arrow points upward. We'll also draw the vertical line $x = 2$ as a dashed line to indicate that the graph never actually touches it.

x	$f(x)$	Point
-6	$\frac{1}{-6-2} = -0.125$	$(-6, -0.125)$
-4	$\frac{1}{-4-2} \approx -0.167$	$\left(-4, -\frac{1}{6}\right)$
-2	$\frac{1}{-2-2} \approx -0.25$	$(-2, -0.25)$
0	$\frac{1}{0-2} = -0.5$	$(0, -0.5)$
1	$\frac{1}{1-2} = -1$	$(1, -1)$
1.5	$\frac{1}{1.5-2} = -2$	$(1.5, -2)$
1.9	$\frac{1}{1.9-2} = -10$	$(1, -10)$
2	undefined	
2.1	$\frac{1}{2.1-2} = 10$	$(2.1, 10)$
2.5	$\frac{1}{2.5-2} = 2$	$(2.5, 2)$
3	$\frac{1}{3-2} = 1$	$(3, 1)$
4	$\frac{1}{4-2} = 0.5$	$(4, 0.5)$

Table 13.1.10: Values of $f(x) = \frac{1}{x-2}$

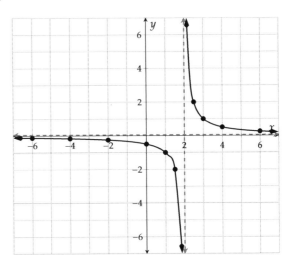

Figure 13.1.11: Full Graph of $f(x) = \frac{1}{x-2}$

Note that in Figure 13.1.11, the line $y = 0$ was also drawn as a dashed line. This is because the values of $y = f(x)$ will get closer and closer to zero as the inputs become more and more positive (or negative).

We know that the domain of this function is $(-\infty, 2) \cup (2, \infty)$ as the function is undefined at 2. We can determine this algebraically, and it is also evident in the graph.

We can see from the graph that the range of the function is $(-\infty, 0) \cup (0, \infty)$. See Checkpoint 10.2.27 for a discussion of how to see the range using a graph like this one.

Remark 13.1.12. The line $x = 2$ in Example 13.1.7 is referred to as a **vertical asymptote**. The line $y = 0$ is referred to as a **horizontal asymptote**. We'll use this vocabulary when referencing such lines, but the classification of vertical asymptotes and horizontal asymptotes is beyond the scope of this book.

Example 13.1.13 Algebraically find the domain of $g(x) = \frac{3x^2}{x^2-2x-24}$. Use technology to sketch a graph of this function.

Explanation. To find a rational function's domain, we set the denominator equal to 0 and solve:

$$x^2 - 2x - 24 = 0$$
$$(x - 6)(x + 4) = 0$$

$$x - 6 = 0 \qquad \text{or} \qquad x + 4 = 0$$
$$x = 6 \qquad \text{or} \qquad x = -4$$

Since $x = 6$ and $x = -4$ will cause the denominator to be 0, they are excluded from the domain. The function's domain is $\{x \mid x \neq 6, x \neq -4\}$. In interval notation, the domain is $(-\infty, -4) \cup (-4, 6) \cup (6, \infty)$.

To begin creating this graph, we'll use technology to create a table of function values, making sure to include values near both -4 and 6. We'll sketch an initial plot of these.

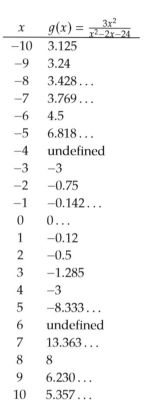

x	$g(x) = \frac{3x^2}{x^2-2x-24}$
-10	3.125
-9	3.24
-8	$3.428\ldots$
-7	$3.769\ldots$
-6	4.5
-5	$6.818\ldots$
-4	undefined
-3	-3
-2	-0.75
-1	$-0.142\ldots$
0	$0\ldots$
1	-0.12
2	-0.5
3	-1.285
4	-3
5	$-8.333\ldots$
6	undefined
7	$13.363\ldots$
8	8
9	$6.230\ldots$
10	$5.357\ldots$

Table 13.1.14: Numerical Values for g

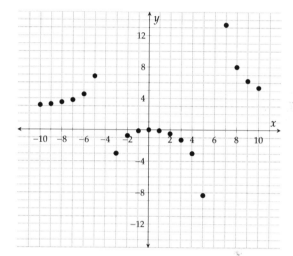

Figure 13.1.15: Initial Set-Up to Graph g

We can now begin to see what happens near $x = -4$ and $x = 6$. These are referred to as vertical asymptotes and will be graphed as dashed vertical lines as they are features of the graph but do not include function values.

The last thing we need to consider is what happens for large positive values of x and large negative values of x. Choosing a few values, we find:

x	$g(x)$
1000	3.0060...
2000	3.0030...
3000	3.0020...
4000	3.0015...

x	$g(x)$
−1000	2.9940...
−2000	2.9970...
−3000	2.9980...
−4000	2.9985...

Table 13.1.16: Values for Large Positive x **Table 13.1.17:** Values for Large Negative x

Thus for really large positive x and for really large negative x, we see that the function values get closer and closer to $y = 3$. This is referred to as the horizontal asymptote, and will be graphed as a dashed horizontal line on the graph.

Putting all of this together, we can sketch a graph of this function.

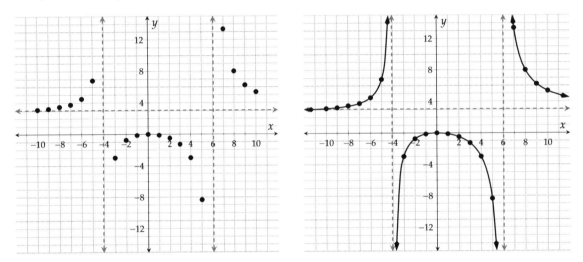

Figure 13.1.18: Asymptotes Added for Graphing $g(x) = \frac{3x^2}{x^2-2x-24}$

Figure 13.1.19: Full Graph of $g(x) = \frac{3x^2}{x^2-2x-24}$

Let's look at another example where a rational function is used to model real life data.

Example 13.1.20 The monthly operation cost of Saqui's shoe company is approximately $300,000.00. The cost of producing each pair of shoes is $30.00. As a result, the cost of producing x pairs of shoes is $30x + 300000$ dollars, and the average cost of producing each pair of shoes can be modeled by

$$\bar{C}(x) = \frac{30x + 300000}{x}.$$

Answer the following questions with technology.

a. What's the average cost of producing 100 pairs of shoes? Of 1,000 pairs? Of 10,000 pairs? What's the pattern?

b. To make the average cost of producing each pair of shoes cheaper than $50.00, at least how many pairs of shoes must Saqui's company produce?

c. Assume that her company's shoes are very popular. What happens to the average cost of producing shoes if more and more people keep buying them?

Explanation. We will graph the function with technology. After adjusting window settings, we have:

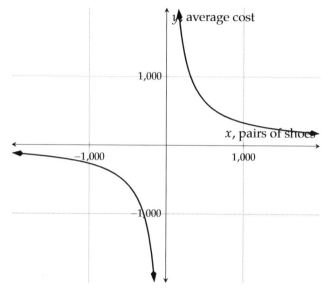

Figure 13.1.21: Graph of $\bar{C}(x) = \frac{30x+300000}{x}$

a. What's the average cost of producing 100 pairs of shoes? 1,000 pairs? 10,000 pairs? What's the pattern?

To answer this question, we locate the points where x values are 100, 1,000 and 10,000. They are $(100, 3030)$, $(1000, 330)$ and $(10000, 60)$. They imply:

- If the company produces 100 pairs of shoes, the average cost of producing one pair is $3,030.00.

- If the company produces 1,000 pairs of shoes, the average cost of producing one pair is $330.00.

- If the company produces 10,000 pairs of shoes, the average cost of producing one pair is $60.00.

We can see the more shoes her company produces, the lower the average cost.

b. To make the average cost of producing each pair of shoes cheaper than $50.00, at least how many pairs of shoes must Saqui's company produce?

To answer this question, we locate the point where its y-value is 50. With technology, we graph both $y = \bar{C}(x)$ and $y = 50$, and locate their intersection.

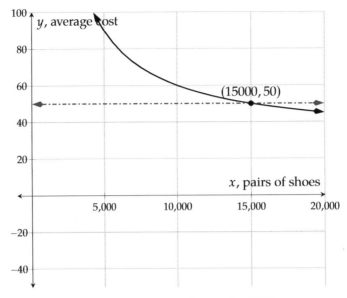

Figure 13.1.22: Intersection of $\bar{C}(x) = \frac{30x+300000}{x}$ and $y = 50$

The intersection $(15000, 50)$ implies the average cost of producing one pair is $50.00 if her company produces 15,000 pairs of shoes.

c. Assume her company's shoes are very popular. What happens to the average cost of producing shoes if more and more people keep buying them?

To answer this question, we substitute x with some large numbers, and use technology to create a table of values:

x	$g(x)$
100000	33
1000000	31
10000000	30.03
100000000	30.003

Table 13.1.23: Values for Large Positive x

We can estimate that the average cost of producing one pair is getting closer and closer to $30.00 as her company produces more and more pairs of shoes.

Note that the cost of producing each pair is $30.00. This implies, for big companies whose products are very popular, the cost of operations can be ignored when calculating the average cost of producing each unit of product.

Exercises

Rational Functions in Context

1. The population of deer in a forest can be modeled by

$$P(x) = \frac{3920x + 1540}{8x + 7}$$

where x is the number of years in the future. Answer the following questions.

   a. How many deer live in this forest this year?

   b. How many deer will live in this forest 9 years later? Round your answer to an integer.

   c. After how many years, the deer population will be 472? Round your answer to an integer.

   d. Use a calculator to answer this question: As time goes on, the population levels off at about how many deer?

2. The population of deer in a forest can be modeled by

$$P(x) = \frac{2970x + 180}{9x + 2}$$

where x is the number of years in the future. Answer the following questions.

   a. How many deer live in this forest this year?

   b. How many deer will live in this forest 9 years later? Round your answer to an integer.

   c. After how many years, the deer population will be 327? Round your answer to an integer.

   d. Use a calculator to answer this question: As time goes on, the population levels off at about how many deer?

3. In a certain store, cashiers can serve 60 customers per hour on average. If x customers arrive at the store in a given hour, then the average number of customers C waiting in line can be modeled by the function

$$C(x) = \frac{x^2}{3600 - 60x}$$

where $x < 60$.

Answer the following questions with a graphing calculator. Round your answers to integers.

   a. If 44 customers arrived in the store in the past hour, there are approximately [   ] customers waiting in line.

   b. If there are 7 customers waiting in line, approximately [   ] customers arrived in the past hour.

4. In a certain store, cashiers can serve 55 customers per hour on average. If x customers arrive at the store in a given hour, then the average number of customers C waiting in line can be modeled by the function

$$C(x) = \frac{x^2}{3025 - 55x}$$

where $x < 55$.

Answer the following questions with a graphing calculator. Round your answers to integers.

   a. If 48 customers arrived in the store in the past hour, there are approximately [   ] customers waiting in line.

   b. If there are 2 customers waiting in line, approximately [   ] customers arrived in the past hour.

Identify Rational Functions   Select all rational functions. There are several correct answers.

5.

□ $n(x) = \frac{3x^2+6\sqrt{x}-4}{3-2x^3}$ □ $m(x) = \frac{3x+6}{3x+6}$ □ $r(x) = \frac{3x^2+6x-4}{3-2x^{-3}}$ □ $t(x) = \frac{3-2x^3}{3x^{0.7}+6x-4}$

□ $b(x) = \frac{3x^2+6x-4}{3}$ □ $s(x) = \frac{\sqrt{3x^2+6x-4}}{3-2x^3}$ □ $c(x) = \frac{3x^2+6x-4}{3+|x|}$ □ $a(x) = \frac{3x^2+6x-4}{3-2x^3}$

□ $h(x) = \frac{3}{3x^2+6x-4}$

6.

□ $t(x) = \frac{8-2x^3}{3x^{0.7}+3x-7}$ □ $s(x) = \frac{\sqrt{3x^2+3x-7}}{8-2x^3}$ □ $m(x) = \frac{3x+3}{3x+3}$ □ $r(x) = \frac{3x^2+3x-7}{8-2x^{-3}}$

□ $a(x) = \frac{3x^2+3x-7}{8-2x^3}$ □ $c(x) = \frac{3x^2+3x-7}{8+|x|}$ □ $h(x) = \frac{8}{3x^2+3x-7}$ □ $b(x) = \frac{3x^2+3x-7}{8}$

□ $n(x) = \frac{3x^2+3\sqrt{x}-7}{8-2x^3}$

Domain

7. Find the domain of h where
$$h(x) = \frac{2x}{x-8}.$$

8. Find the domain of F where
$$F(x) = \frac{5x}{x-1}.$$

9. Find the domain of G where
$$G(x) = \frac{10-6x}{x^2-16x+63}.$$

10. Find the domain of H where
$$H(x) = \frac{7x-2}{x^2-x-72}.$$

11. Find the domain of H where
$$H(x) = \frac{7x+2}{x^2-10x}.$$

12. Find the domain of K where
$$K(x) = -\frac{7x+5}{x^2-4x}.$$

13. Find the domain of f where
$$f(x) = \frac{6x+3}{x^2-100}.$$

14. Find the domain of g where
$$g(x) = -\frac{x+9}{x^2-9}.$$

15. Find the domain of the function c defined by $c(x) = \frac{x-4}{x^2}$

16. Find the domain of the function p defined by $p(x) = \frac{x-2}{x^5}$

17. Find the domain of the function q defined by $q(x) = \frac{x-2}{x^2+16}$

18. Find the domain of the function m defined by $m(x) = \frac{x+3}{x^2+81}$

19. Find the domain of the function m defined by $m(x) = \frac{x+5}{x+5}$

20. Find the domain of the function b defined by $b(x) = \frac{x+7}{x+7}$

A function is graphed.

21.

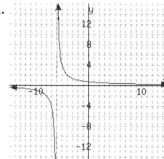

What is its domain?

22.

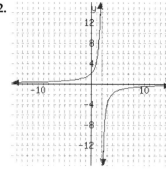

What is its domain?

23.

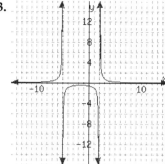

What is its domain?

24.

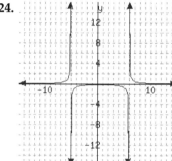

What is its domain?

25.

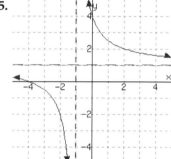

This function has domain ☐
and range ☐.

26.

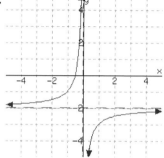

This function has domain ☐
and range ☐.

27.

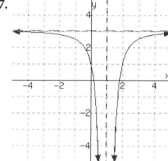

This function has domain ☐
and range ☐.

28.

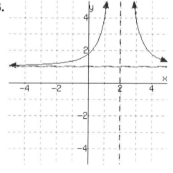

This function has domain ☐
and range ☐.

Graphing Technology

29. In a forest, the number of deer can be modeled by the function $f(x) = \frac{180t+270}{0.6t+3}$, where t stands for the number of years from now. Answer the question with technology. Round your answer to a whole number.

    After 30 years, there would be approximately [   ] deer living in the forest.

30. In a forest, the number of deer can be modeled by the function $f(x) = \frac{180t+900}{0.4t+9}$, where t stands for the number of years from now. Answer the question with technology. Round your answer to one decimal place.

    After [   ] years, there would be approximately 350 deer living in the forest.

31. In a forest, the number of deer can be modeled by the function $f(x) = \frac{25t+180}{0.1t+6}$, where t stands for the number of years from now. Answer the question with technology. Round your answer to one decimal place.

    As time goes on, the population levels off at approximately [   ] deer living in the forest.

32. The concentration of a drug in a patient's blood stream, in milligrams per liter, can be modeled by the function $C(t) = \frac{3t}{t^2+6}$, where t is the number of hours since the drug is injected. Answer the following question with technology. Round your answer to two decimal places if needed.

    The drug's concentration after 8 hours is [   ] milligrams per liter.

33. The concentration of a drug in a patient's blood stream, in milligrams per liter, can be modeled by the function $C(t) = \frac{4t}{t^2+9}$, where t is the number of hours since the drug is injected. Answer the following question with technology. Round your answer to two decimal places if needed. If there are more than one answer, use commas to separate them.

    [   ] hours since injection, the drug's concentration is 0.22 milligrams per liter.

34. The concentration of a drug in a patient's blood stream, in milligrams per liter, can be modeled by the function $C(t) = \frac{5t}{t^2+7}$, where t is the number of hours since the drug is injected. Answer the following question with technology. Round your answer to two decimal places if needed.

    [   ] hours since injection, the drug's concentration is at the maximum value of

    [   ] milligrams per liter.

35. The concentration of a drug in a patient's blood stream, in milligrams per liter, can be modeled by the function $C(t) = \frac{7t}{t^2+4}$, where t is the number of hours since the drug is injected. Answer the following question with technology. Round your answer to two decimal places if needed.

    As time goes on, the drug's concentration in the patient's blood stream levels off at approximately

    [   ] milligrams per liter.

13.2 Multiplication and Division of Rational Expressions

In the last section, we learned some rational function applications. In this section, we will learn how to simplify rational expressions, and how to multiply and divide them.

13.2.1 Simplifying Rational Expressions

Consider the two rational functions below. At first glance, which function *looks* simpler?

$$f(x) = \frac{8x^3 - 12x^2 + 8x - 12}{2x^3 - 3x^2 + 10x - 15} \qquad\qquad g(x) = \frac{4(x^2 + 1)}{x^2 + 5}, \text{ for } x \neq \frac{3}{2}$$

It can be argued that the function g is simpler, at least with regard to the ease with which we can determine its domain, quickly evaluate it, and also determine where its function value is zero. All of these things are considerably more difficult with the function f.

These two functions are actually the *same* function. Using factoring and the same process of canceling that's used with numerical ratios, we will learn how to simplify the function f into the function g. (The full process for simplifying $f(x) = \frac{8x^3 - 12x^2 + 8x - 12}{2x^3 - 3x^2 + 10x - 15}$ will be shown in Example 13.2.8.)

To see a simple example of the process for simplifying a rational function or expression, let's look at simplifying $\frac{14}{21}$ and $\frac{(x+2)(x+7)}{(x+3)(x+7)}$ by canceling common factors:

$$\frac{14}{21} = \frac{2 \cdot \cancel{7}}{3 \cdot \cancel{7}} \qquad\qquad \frac{(x+2)(x+7)}{(x+3)(x+7)} = \frac{(x+2)\cancel{(x+7)}}{(x+3)\cancel{(x+7)}}$$
$$= \frac{2}{3} \qquad\qquad\qquad\qquad = \frac{x+2}{x+3}, \text{ for } x \neq -7$$

The statement "for $x \neq -7$" was added when the factors of $x + 7$ were canceled. This is because $\frac{(x+2)(x+7)}{(x+3)(x+7)}$ was undefined where $x = -7$, so the simplified version must also be undefined for $x = -7$.

Warning 13.2.2 Cancel Factors, not Terms. It may be tempting to want to try to simplify $\frac{x+2}{x+3}$ into $\frac{2}{3}$ by canceling each x that appears. But these x's are *terms* (pieces that are added with other pieces), not *factors*. Canceling (an act of division) is only possible with *factors*.

The process of canceling factors is key to simplifying rational expressions. If the expression is not given in factored form, then this will be our first step. We'll now look at a few more examples.

Example 13.2.3 Simplify the rational function formula $Q(x) = \frac{3x-12}{x^2+x-20}$ and state the domain of Q.

Explanation.

To start, we'll factor the numerator and denominator. We'll then cancel any factors common to both the numerator and denominator.

$$Q(x) = \frac{3x - 12}{x^2 + x - 20}$$
$$Q(x) = \frac{3\cancel{(x-4)}}{(x + 5)\cancel{(x-4)}}$$
$$Q(x) = \frac{3}{x + 5}, \text{ for } x \neq 4$$

The domain of this function will incorporate the *explicit* domain restriction $x \neq 4$ that was stated when the factor of $x-4$ was canceled from both the numerator and denominator. We will also exclude -5 from the domain as this value would make the denominator zero. Thus the domain of Q is $\{x \mid x \neq -5, 4\}$.

Warning 13.2.4. When simplifying the function Q in Example 13.2.3, we cannot simply write $Q(x) = \frac{3}{x+5}$. The reason is that this would result in our simplified version of the function Q having a different domain than the original Q. More specifically, for our original function Q it held that $Q(4)$ was undefined, and this still needs to be true for the simplified form of Q.

Example 13.2.5 Simplify the rational function formula $R(y) = \frac{-y-2y^2}{2y^3-y^2-y}$ and state the domain of R.

Explanation.

$$R(y) = \frac{-y - 2y^2}{2y^3 - y^2 - y}$$

$$R(y) = \frac{-2y^2 - y}{y(2y^2 - y - 1)}$$

$$R(y) = \frac{-\cancel{y}(2y+1)}{\cancel{y}(2y+1)(y - 1)}$$

$$R(y) = -\frac{1}{y - 1}, \text{ for } y \neq 0, y \neq -\frac{1}{2}$$

The domain of this function will incorporate the explicit restrictions $y \neq 0, y \neq -\frac{1}{2}$ that were stated when the factors of y and $2y+1$ were canceled from both the numerator and denominator. Since the factor $y - 1$ is still in the denominator, we also need the restriction that $y \neq 1$. Therefore the domain of R is $\{y \mid y \neq -\frac{1}{2}, 0, 1\}$.

Example 13.2.6 Simplify the expression $\frac{9y+2y^2-5}{y^2-25}$.

Explanation.

To start, we need to recognize that $9y + 2y^2 - 5$ is not written in standard form (where terms are written from highest degree to lowest degree). Before attempting to factor this expression, we'll re-write it as $2y^2 + 9y - 5$.

$$\frac{9y + 2y^2 - 5}{y^2 - 25} = \frac{2y^2 + 9y - 5}{y^2 - 25}$$

$$= \frac{(2y - 1)\cancel{(y+5)}}{\cancel{(y+5)}(y - 5)}$$

$$= \frac{2y - 1}{y - 5}, \text{ for } y \neq -5$$

Example 13.2.7 Simplify the expression $\frac{-48z+24z^2-3z^3}{4-z}$.

Explanation. To begin simplifying this expression, we will rewrite each polynomial in descending order. Then we'll factor out the GCF, including the constant -1 from both the numerator and denominator because their leading terms are negative.

$$\frac{-48z + 24z^2 - 3z^3}{4 - z} = \frac{-3z^3 + 24z^2 - 48z}{-z + 4}$$

$$= \frac{-3z(z^2 - 8z + 16)}{-(z - 4)}$$

$$= \frac{-3z(z - 4)^2}{-(z - 4)}$$

$$= \frac{-3z(z - 4)\cancel{(z - 4)}}{-\cancel{(z - 4)}}$$

$$= \frac{3z(z - 4)}{1}, \text{ for } z \neq 4$$

$$= 3z(z - 4), \text{ for } z \neq 4$$

Example 13.2.8 Simplify the rational function formula $f(x) = \frac{8x^3 - 12x^2 + 8x - 12}{2x^3 - 3x^2 + 10x - 15}$ and state the domain of f.

Explanation.

To simplify this rational function, we'll first note that both the numerator and denominator have four terms. To factor them we'll need to use factoring by grouping. (Note that if this technique didn't work, very few other approaches would be possible.) Once we've used factoring by grouping, we'll cancel any factors common to both the numerator and denominator and state the associated restrictions.

$$f(x) = \frac{8x^3 - 12x^2 + 8x - 12}{2x^3 - 3x^2 + 10x - 15}$$

$$f(x) = \frac{4(2x^3 - 3x^2 + 2x - 3)}{2x^3 - 3x^2 + 10x - 15}$$

$$f(x) = \frac{4(x^2(2x - 3) + (2x - 3))}{x^2(2x - 3) + 5(2x - 3)}$$

$$f(x) = \frac{4(x^2 + 1)\cancel{(2x - 3)}}{(x^2 + 5)\cancel{(2x - 3)}}$$

$$f(x) = \frac{4(x^2 + 1)}{x^2 + 5}, \text{ for } x \neq \frac{3}{2}$$

In determining the domain of this function, we'll need to account for any implicit and explicit restrictions. When the factor $2x - 3$ was canceled, the explicit statement of $x \neq \frac{3}{2}$ was given. The denominator in the final simplified form of this function has $x^2 + 5$. There is no value of x for which $x^2 + 5 = 0$, so the only restriction is that $x \neq \frac{3}{2}$. Therefore the domain is $\left\{ x \mid x \neq \frac{3}{2} \right\}$.

Example 13.2.9 Simplify the expression $\frac{3y - x}{x^2 - xy - 6y^2}$. In this example, there are two variables. It is still possible that in examples like this, there can be domain restrictions when simplifying rational expressions. However since we are not studying *functions* of more than one variable, this textbook ignores domain restrictions with examples like this one.

Explanation.

$$\frac{3y - x}{x^2 - xy - 6y^2} = \frac{-\cancel{(x - 3y)}}{\cancel{(x - 3y)}(x + 2y)}$$

$$= \frac{-1}{x + 2y}$$

13.2.2 Multiplication and Division of Rational Functions and Expressions

Recall the property for multiplying fractions 1.2.16, which states that the product of two fractions is equal to the product of their numerators divided by the product of their denominators. We will use this same property for multiplying rational expressions.

When multiplying fractions, one approach is to multiply the numerator and denominator, and then simplify the fraction that results by determining the greatest common factor in both the numerator and denominator, like this:

$$\frac{14}{9} \cdot \frac{3}{10} = \frac{14 \cdot 3}{9 \cdot 10}$$
$$= \frac{42}{90}$$
$$= \frac{7 \cdot \cancel{6}}{15 \cdot \cancel{6}}$$
$$= \frac{7}{15}$$

This approach works great when we can easily identify that 6 is the greatest common factor in both 42 and 90. But in more complicated instances, it isn't always an easy approach. It also won't work particularly well when we have $(x + 2)$ instead of 2 as a factor, as we'll see shortly.

Another approach to multiplying and simplifying fractions involves utilizing the prime factorization of each the numerator and denominator, like this:

$$\frac{14}{9} \cdot \frac{3}{10} = \frac{2 \cdot 7}{3^2} \cdot \frac{3}{2 \cdot 5}$$
$$= \frac{\cancel{2} \cdot 7 \cdot \cancel{3}}{\cancel{3} \cdot 3 \cdot \cancel{2} \cdot 5}$$
$$= \frac{7}{15}$$

The method for multiplying and simplifying rational expressions is nearly identical, as shown here:

$$\frac{x^2 + 9x + 14}{x^2 + 6x + 9} \cdot \frac{x + 3}{x^2 + 7x + 10} = \frac{(x + 2)(x + 7)}{(x + 3)^2} \cdot \frac{x + 3}{(x + 2)(x + 5)}$$
$$= \frac{\cancel{(x + 2)}(x + 7)\cancel{(x + 3)}}{\cancel{(x + 3)}(x + 3)\cancel{(x + 2)}(x + 5)}$$
$$= \frac{(x + 7)}{(x + 3)(x + 5)}, \text{ for } x \neq -2$$

This process will be used for both multiplying and dividing rational expressions. The main distinctions in various examples will be in the factoring methods required.

> **Example 13.2.10** Multiply the rational expressions: $\frac{x^2 - 4x}{x^2 - 4} \cdot \frac{4 - 4x + x^2}{20 - x - x^2}$.
>
> **Explanation.** Note that to factor the second rational expression, we'll want to re-write the terms in descending order for both the numerator and denominator. In the denominator, we'll first factor out -1 as the leading term is $-x^2$.
>
> $$\frac{x^2 - 4x}{x^2 - 4} \cdot \frac{4 - 4x + x^2}{20 - x - x^2} = \frac{x^2 - 4x}{x^2 - 4} \cdot \frac{x^2 - 4x + 4}{-x^2 - x + 20}$$

$$= \frac{x^2 - 4x}{x^2 - 4} \cdot \frac{x^2 - 4x + 4}{-(x^2 + x - 20)}$$

$$= \frac{x(x-4)}{(x+2)(x-2)} \cdot \frac{(x-2)(x-2)}{-(x+5)(x-4)}$$

$$= -\frac{x(x-2)}{(x+2)(x+5)}, \text{ for } x \neq 2, x \neq 4$$

Example 13.2.11 Multiply the rational expressions: $\frac{p^2q^4}{3r} \cdot \frac{9r^2}{pq^2}$. Note this book ignores domain restrictions on multivariable expressions.

Explanation. We won't need to factor anything in this example, and can simply multiply across and then simplify.

$$\frac{p^2q^4}{3r} \cdot \frac{9r^2}{pq^2} = \frac{p^2q^2 \cdot 9r^2}{3r \cdot pq^2}$$

$$= \frac{pq^2 \cdot 3r}{1}$$

$$= 3pq^2r$$

We can divide rational expressions using the property for dividing fractions 1.2.18, which simply requires that we change dividing by an expression to multiplying by its reciprocal. Let's look at a few examples.

Example 13.2.12 Divide the rational expressions: $\frac{x+2}{x+5} \div \frac{x+2}{x-3}$.

Explanation.

$$\frac{x+2}{x+5} \div \frac{x+2}{x-3} = \frac{x+2}{x+5} \cdot \frac{x-3}{x+2}, \text{ for } x \neq 3$$

$$= \frac{x-3}{x+5}, \text{ for } x \neq -2, x \neq 3$$

Example 13.2.13 Simplify the rational expression using division: $\frac{\frac{3x-6}{2x+10}}{\frac{x^2-4}{3x+15}}$.

Explanation. To begin, we'll note that the larger fraction bar is denoting division, so we will use multiplication by the reciprocal. After that, we'll factor each expression and cancel any common factors.

$$\frac{\frac{3x-6}{2x+10}}{\frac{x^2-4}{3x+15}} = \frac{3x-6}{2x+10} \div \frac{x^2-4}{3x+15}$$

$$= \frac{3x-6}{2x+10} \cdot \frac{3x+15}{x^2-4}$$

$$= \frac{3(x-2)}{2(x+5)} \cdot \frac{3(x+5)}{(x+2)(x-2)}$$

$$= \frac{3 \cdot 3}{2(x+2)}, \text{ for } x \neq -5, x \neq 2$$

$$= \frac{9}{2x+4}, \text{ for } x \neq -5, x \neq 2$$

Example 13.2.14 Divide the rational expressions: $\frac{x^2-5x-14}{x^2+7x+10} \div \frac{x-7}{x+4}$.

Explanation.

$$\frac{x^2-5x-14}{x^2+7x+10} \div \frac{x-7}{x+4} = \frac{x^2-5x-14}{x^2+7x+10} \cdot \frac{x+4}{x-7}, \text{ for } x \neq -4$$
$$= \frac{(x-7)(x+2)}{(x+5)(x+2)} \cdot \frac{x+4}{x-7}, \text{ for } x \neq -4$$
$$= \frac{x+4}{x+5}, \text{ for } x \neq -4, x \neq -2, x \neq 7$$

Example 13.2.15 Divide the rational expressions: $(p^4 - 16) \div \frac{p^4-2p^3}{2p}$.

Explanation.

$$(p^4 - 16) \div \frac{p^4-2p^3}{2p} = \frac{p^4-16}{1} \cdot \frac{2p}{p^4-2p^3}$$
$$= \frac{(p^2+4)(p+2)(p-2)}{1} \cdot \frac{2p}{p^3(p-2)}$$
$$= \frac{2(p^2+4)(p+2)}{p^2}, \text{ for } p \neq 2$$

Example 13.2.16 Divide the rational expressions: $\frac{3x^2}{x^2-9y^2} \div \frac{6x^3}{x^2-2xy-15y^2}$. Note this book ignores domain restrictions on multivariable expressions.

Explanation.

$$\frac{3x^2}{x^2-9y^2} \div \frac{6x^3}{x^2-2xy-15y^2} = \frac{3x^2}{x^2-9y^2} \cdot \frac{x^2-2xy-15y^2}{6x^3}$$
$$= \frac{3x^2}{(x+3y)(x-3y)} \cdot \frac{(x+3y)(x-5y)}{6x^3}$$
$$= \frac{1}{x-3y} \cdot \frac{x-5y}{2x}$$
$$= \frac{x-5y}{2x(x-3y)}$$

Example 13.2.17 Divide the rational expressions: $\frac{m^2n^2-3mn-4}{2mn} \div (m^2n^2 - 16)$. Note this book ignores domain restrictions on multivariable expressions.

Explanation.

$$\frac{m^2n^2-3mn-4}{2mn} \div (m^2n^2 - 16) = \frac{m^2n^2-3mn-4}{2mn} \cdot \frac{1}{m^2n^2-16}$$
$$= \frac{(mn-4)(mn+1)}{2mn} \cdot \frac{1}{(mn+4)(mn-4)}$$

$$= \frac{mn+1}{2mn} \cdot \frac{1}{mn+4}$$

$$= \frac{mn+1}{2mn(mn+4)}$$

Exercises

Review and Warmup

1. Multiply: $-\frac{6}{11} \cdot \frac{11}{21}$

2. Multiply: $-\frac{15}{11} \cdot \frac{11}{15}$

3. Multiply: $-\frac{20}{17} \cdot \left(-\frac{13}{18}\right)$

4. Multiply: $-\frac{20}{11} \cdot \left(-\frac{7}{24}\right)$

5. Divide: $\frac{1}{5} \div \frac{9}{4}$

6. Divide: $\frac{1}{2} \div \frac{8}{3}$

7. Divide: $\frac{8}{15} \div \left(-\frac{7}{20}\right)$

8. Divide: $\frac{5}{12} \div \left(-\frac{8}{15}\right)$

Factor the given polynomial.

9. $r^2 - 1 = \boxed{}$

10. $r^2 - 81 = \boxed{}$

11. $r^2 + 9r + 20 = \boxed{}$

12. $t^2 + 19t + 90 = \boxed{}$

13. $t^2 - 10t + 21 = \boxed{}$

14. $x^2 - 10x + 21 = \boxed{}$

15. $2x^2 - 14x + 20 = \boxed{}$

16. $2y^2 - 18y + 28 = \boxed{}$

17. $2y^4 + 16y^3 + 30y^2 = \boxed{}$

18. $3r^7 + 12r^6 + 9r^5 = \boxed{}$

19. $64r^2 - 16r + 1 = \boxed{}$

20. $16r^2 - 8r + 1 = \boxed{}$

Simplifying Rational Expressions with One Variable

21. Select all correct simplifications, ignoring possible domain restrictions.

☐ $\frac{7x+9}{x+9} = 7$ ☐ $\frac{x}{7x} = \frac{1}{7}$ ☐ $\frac{7x+9}{7} = x+9$ ☐ $\frac{9}{x+9} = \frac{1}{x+1}$ ☐ $\frac{x+9}{x} = 9$ ☐ $\frac{9x}{x} = 9$

☐ $\frac{x+9}{x+7} = \frac{9}{7}$ ☐ $\frac{x+9}{9} = x$ ☐ $\frac{x+9}{x+9} = 1$ ☐ $\frac{7(x-9)}{x-9} = 7$ ☐ $\frac{9}{x+9} = \frac{1}{x}$

22. Select all correct simplifications, ignoring possible domain restrictions.

☐ $\frac{9(x-10)}{x-10} = 9$ ☐ $\frac{x+10}{x+9} = \frac{10}{9}$ ☐ $\frac{x+10}{x} = 10$ ☐ $\frac{x+10}{10} = x$ ☐ $\frac{9x+10}{9} = x+10$

☐ $\frac{x+10}{x+10} = 1$ ☐ $\frac{10x}{x} = 10$ ☐ $\frac{9x+10}{x+10} = 9$ ☐ $\frac{10}{x+10} = \frac{1}{x}$ ☐ $\frac{10}{x+10} = \frac{1}{x+1}$ ☐ $\frac{x}{9x} = \frac{1}{9}$

23. Simplify the following expressions, and if applicable, write the restricted domain on the simplified expression.

a. $\dfrac{x+3}{x+3} = $ ▭

b. $\dfrac{x+3}{3+x} = $ ▭

c. $\dfrac{x-3}{x-3} = $ ▭

d. $\dfrac{x-3}{3-x} = $ ▭

24. Simplify the following expressions, and if applicable, write the restricted domain on the simplified expression.

a. $\dfrac{x+9}{x+9} = $ ▭

b. $\dfrac{x+9}{9+x} = $ ▭

c. $\dfrac{x-9}{x-9} = $ ▭

d. $\dfrac{x-9}{9-x} = $ ▭

Simplify the following expression, and if applicable, write the restricted domain on the simplified expression.

25. $\dfrac{y-6}{(y-2)(y-6)} = $ ▭

26. $\dfrac{y+3}{(y-8)(y+3)} = $ ▭

27. $\dfrac{-8(r-8)}{(r-6)(r-8)} = $ ▭

28. $\dfrac{-3(r+5)}{(r-3)(r+5)} = $ ▭

29. $\dfrac{(r+2)(r-10)}{10-r} = $ ▭

30. $\dfrac{(t-8)(t-7)}{7-t} = $ ▭

31. $\dfrac{5t-25}{t-5} = $ ▭

32. $\dfrac{-2x+4}{x-2} = $ ▭

33. $\dfrac{-9x}{x^2+8x} = $ ▭

34. $\dfrac{-6y}{y^2-5y} = $ ▭

35. $\dfrac{6y-y^2}{y^2-9y+18} = $ ▭

36. $\dfrac{4y-y^2}{y^2+2y-24} = $ ▭

37. $\dfrac{r^2+2r}{4-r^2} = $ ▭

38. $\dfrac{r^2-6r}{36-r^2} = $ ▭

39. $\dfrac{-t^2+4t}{-4+5t-t^2} = $ ▭

40. $\dfrac{-t^2-2t}{12+4t-t^2} = $ ▭

41. $\dfrac{5x^2 + 11x + 6}{-7x - 5 - 2x^2} =$ [ ]   **42.** $\dfrac{3x^2 + x - 2}{-2x + 3 - 5x^2} =$ [ ]

43. $\dfrac{y^2 + y - 2}{2y - y^2 - 1} =$ [ ]   **44.** $\dfrac{y^2 - 11y + 30}{10y - y^2 - 25} =$ [ ]

45. $\dfrac{-y^2 - y + 12}{y^2 - 9} =$ [ ]   **46.** $\dfrac{-r^2 + 3r - 2}{r^2 - 4} =$ [ ]

47. $\dfrac{5r^2 + 9r + 4}{-7r - 5 - 2r^2} =$ [ ]   **48.** $\dfrac{3t^2 + t - 2}{-11t - 5 - 6t^2} =$ [ ]

49. $\dfrac{3t^2 - t^3}{t^2 - 5t + 6} =$ [ ]   **50.** $\dfrac{-2x^4 - x^5}{x^2 - 3x - 10} =$ [ ]

51. $\dfrac{x^5 - x^4 - 20x^3}{x^5 - 11x^4 + 30x^3} =$ [ ]   **52.** $\dfrac{y^4 - 5y^3 - 6y^2}{y^4 - 4y^3 - 5y^2} =$ [ ]

53. $\dfrac{y^3 + 64}{y^2 - 16} =$ [ ]   **54.** $\dfrac{y^3 - 27}{y^2 - 9} =$ [ ]

Simplifying Rational Expressions with More Than One Variable   Simplify this expression.

55. $\dfrac{5rx - r^2x^2}{r^2x^2 - 10rx + 25} =$ [ ]   **56.** $\dfrac{3rx - r^2x^2}{r^2x^2 - 9rx + 18} =$ [ ]

57. $\dfrac{4t + 8y}{t^2 - 2ty - 8y^2} =$ [ ]   **58.** $\dfrac{6t - 36x}{t^2 - 5tx - 6x^2} =$ [ ]

59. $\dfrac{-x^2 - 9xr - 18r^2}{x^2 - 36r^2} =$ [ ]   **60.** $\dfrac{-x^2 + 4xy + 12y^2}{x^2 - 4y^2} =$ [ ]

61. $\dfrac{2y^2t^2 - yt - 3}{-11yt - 6 - 5y^2t^2} =$ [ ]   **62.** $\dfrac{5y^2r^2 + 3yr - 2}{-7yr - 3 - 4y^2r^2} =$ [ ]

Simplifying Rational Functions   Simplify the function formula, and if applicable, write the restricted domain.

63. $g(y) = \dfrac{y + 2}{y^2 + 12y + 20}$

Reduced $g(y) = $ ⬚

64. $K(r) = \dfrac{r - 6}{r^2 - 15r + 54}$

Reduced $K(r) = $ ⬚

65. $F(r) = \dfrac{r^3 - 100r}{r^3 + 16r^2 + 60r}$

Reduced $F(r) = $ ⬚

66. $g(t) = \dfrac{t^3 - 16t}{t^3 + 11t^2 + 28t}$

Reduced $g(t) = $ ⬚

67. $K(t) = \dfrac{t^4 + 6t^3 + 9t^2}{3t^4 + 11t^3 + 6t^2}$

Reduced $K(t) = $ ⬚

68. $F(x) = \dfrac{x^4 - 6x^3 + 9x^2}{2x^4 - 9x^3 + 9x^2}$

Reduced $F(x) = $ ⬚

69. $g(x) = \dfrac{3x^3 + 5x^2}{3x^3 + 8x^2 + 5x}$

Reduced $g(x) = $ ⬚

70. $H(y) = \dfrac{2y^3 + 3y^2}{2y^3 - 7y^2 - 15y}$

Reduced $H(y) = $ ⬚

Multiplying and Dividing Rational Expressions with One Variable

71. Select all correct equations:

☐ $5 \cdot \dfrac{x}{y} = \dfrac{5x}{5y}$ ☐ $5 \cdot \dfrac{x}{y} = \dfrac{x}{5y}$ ☐ $5 \cdot \dfrac{x}{y} = \dfrac{5x}{y}$ ☐ $-\dfrac{x}{y} = \dfrac{-x}{-y}$ ☐ $-\dfrac{x}{y} = \dfrac{-x}{y}$ ☐ $-\dfrac{x}{y} = \dfrac{x}{-y}$

72. Select all correct equations:

☐ $6 \cdot \dfrac{x}{y} = \dfrac{6x}{y}$ ☐ $6 \cdot \dfrac{x}{y} = \dfrac{6x}{6y}$ ☐ $6 \cdot \dfrac{x}{y} = \dfrac{x}{6y}$ ☐ $-\dfrac{x}{y} = \dfrac{-x}{y}$ ☐ $-\dfrac{x}{y} = \dfrac{x}{-y}$ ☐ $-\dfrac{x}{y} = \dfrac{-x}{-y}$

73. Simplify the following expressions, and if applicable, write the restricted domain.

$-\dfrac{r^5}{r + 5} \cdot r^2 = $ ⬚

$-\dfrac{r^5}{r + 5} \cdot \dfrac{1}{r^2} = $ ⬚

74. Simplify the following expressions, and if applicable, write the restricted domain.

$-\dfrac{y^4}{y + 4} \cdot y^2 = $ ⬚

$-\dfrac{y^4}{y + 4} \cdot \dfrac{1}{y^2} = $ ⬚

Simplify this expression, and if applicable, write the restricted domain.

75. $\dfrac{t^2 + 3t + 2}{t + 5} \cdot \dfrac{3t + 15}{t + 1} = $ ▭

76. $\dfrac{t^2 + 2t - 24}{t - 5} \cdot \dfrac{4t - 20}{t - 4} = $ ▭

77. $\dfrac{x^2 - 16x}{x^2 - 16} \cdot \dfrac{x^2 - 4x}{x^2 - 15x - 16} = $ ▭

78. $\dfrac{x^2 - 9x}{x^2 - 9} \cdot \dfrac{x^2 - 3x}{x^2 - 5x - 36} = $ ▭

79. $\dfrac{25y + 25}{24 - 2y - 2y^2} \cdot \dfrac{y^2 - 6y + 9}{5y^2 + 5y} = $ ▭

80. $\dfrac{20y + 20}{-135 - 72y - 9y^2} \cdot \dfrac{y^2 + 6y + 9}{5y^2 + 5y} = $ ▭

81. $\dfrac{6y^2 - 13y + 7}{36y^6 - 24y^5} \cdot \dfrac{4y^5 - 6y^6}{36y^2 - 49} = $ ▭

82. $\dfrac{3r^2 + 4r - 7}{80r^5 - 120r^4} \cdot \dfrac{15r^4 - 10r^5}{9r^2 - 49} = $ ▭

83. $\dfrac{r}{r + 15} \div 3r^4 = $ ▭

84. $\dfrac{t}{t - 4} \div 4t^3 = $ ▭

85. $9t \div \dfrac{3}{t^4} = $ ▭

86. $8x \div \dfrac{4}{x^4} = $ ▭

87. $(2x + 2) \div (12x + 12) = $ ▭

88. $(4y + 4) \div (16y + 16) = $ ▭

89. $\dfrac{25y^2 - 16}{5y^2 + 9y + 4} \div (4 - 5y) = $ ▭

90. $\dfrac{9y^2 - 4}{3y^2 + 5y + 2} \div (2 - 3y) = $ ▭

91. $\dfrac{r^5}{r^2 - 5r} \div \dfrac{1}{r^2 + r + (-30)} = $ ▭

92. $\dfrac{r^5}{r^2 + 2r} \div \dfrac{1}{r^2 + (-1)r + (-6)} = $ ▭

93. $\dfrac{\frac{9m+2}{m}}{\frac{m+7}{m}} = $ ▭

94. $\dfrac{\frac{6m-10}{m}}{\frac{m-9}{m}} = $ ▭

95. $\dfrac{\frac{z}{(z-2)^2}}{\frac{5z}{z^2-4}} = $ ▭

96. $\dfrac{\frac{z}{(z-8)^2}}{\frac{9z}{z^2-64}} = $ ▭

97. $\dfrac{x^2 - 3x}{x^2 - 1} \div \dfrac{x^2 - 9}{x^2 - 3x + 2} = $ ▭

98. $\dfrac{x^2 - 3x}{x^2 - 25} \div \dfrac{x^2 - 9}{x^2 - 3x - 10} = $ ▭

Multiplying and Dividing Rational Expressions with More Than One Variable   Simplify this expression.

99. $\dfrac{6(y + r)}{y - r} \cdot \dfrac{y - r}{2(3y + r)} =$ ⬜

100. $\dfrac{8(r + y)}{r - y} \cdot \dfrac{r - y}{4(3r + y)} =$ ⬜

101. $\dfrac{5r^3 x}{2r} \cdot \dfrac{8r^3 x^3}{15x^5} =$ ⬜

102. $\dfrac{2tr}{5t^3} \cdot \dfrac{5t^4 r^3}{4r^5} =$ ⬜

103. $\dfrac{t^2 - 8ty + 15y^2}{t - y} \cdot \dfrac{6t - 6y}{t - 5y} =$ ⬜

104. $\dfrac{x^2 + xt - 2t^2}{x + 3t} \cdot \dfrac{2x + 6t}{x + 2t} =$ ⬜

105. $\dfrac{x^2 r^5}{5} \div \dfrac{x^2 r^3}{30} =$ ⬜

106. $\dfrac{y^5 x^4}{4} \div \dfrac{y^5 x^2}{12} =$ ⬜

107. $(y^3 - 2y^2 t + yt^2) \div (y^5 - y^4 t) =$ ⬜

108. $(y^3 + 10y^2 r + 25yr^2) \div (y^5 + 5y^4 r) =$ ⬜

109. $\dfrac{1}{r^2 - 4ry + 3y^2} \div \dfrac{r^4}{r^2 - 3ry} =$ ⬜

110. $\dfrac{1}{r^2 + 9rt + 18t^2} \div \dfrac{r^4}{r^2 + 3rt} =$ ⬜

111. $\dfrac{t^3}{t^2 r - 5t} \div \dfrac{1}{t^2 r^2 - 9tr + 20} =$ ⬜

112. $\dfrac{t^5}{t^2 x + 5t} \div \dfrac{1}{t^2 x^2 + 2tx - 15} =$ ⬜

113. $\dfrac{30x^3 t^3}{x - 5t} \div \dfrac{6x^4 t}{x^2 - 25t^2} =$ ⬜

114. $\dfrac{9x^5 r^5}{x - 5r} \div \dfrac{3x^6 r}{x^2 - 25r^2} =$ ⬜

115. $\dfrac{\frac{a}{b}}{\frac{3a}{2b^2}} =$ ⬜

116. $\dfrac{\frac{a}{b}}{\frac{6a}{5b^2}} =$ ⬜

117. $\dfrac{\frac{st^2}{6u}}{\frac{s}{7tu}} =$ ⬜

118. $\dfrac{\frac{st^2}{4u}}{\frac{s}{2tu}} =$ ⬜

Challenge

119. Simplify the following: $\frac{1}{x+1} \div \frac{x+2}{x+1} \div \frac{x+3}{x+2} \div \frac{x+4}{x+3} \div \cdots \div \frac{x+75}{x+74}$. For this exercise, you do not have to write the restricted domain of the simplified expression.

13.3 Addition and Subtraction of Rational Expressions

In the last section, we learned how to multiply and divide rational expressions. In this section, we will learn how to add and subtract rational expressions.

13.3.1 Introduction

Example 13.3.2 Julia is taking her family on a boat trip 12 miles down the river and back. The river flows at a speed of 2 miles per hour and she wants to drive the boat at a constant speed, v miles per hour downstream and back upstream. Due to the current of the river, the actual speed of travel is $v + 2$ miles per hour going downstream, and $v - 2$ miles per hour going upstream. If Julia plans to spend 8 hours for the whole trip, how fast should she drive the boat?

We need to review three forms of the formula for movement at a constant rate:

$$d = vt \qquad\qquad v = \frac{d}{t} \qquad\qquad t = \frac{d}{v}$$

where d stands for distance, v represents speed, and t stands for time. According to the third form, the time it takes the boat to travel downstream is $\frac{12}{v+2}$, and the time it takes to get back upstream is $\frac{12}{v-2}$.

The function to model the time of the whole trip is

$$t(v) = \frac{12}{v-2} + \frac{12}{v+2}$$

where t stands for time in hours, and v is the boat's speed in miles per hour. Let's look at the graph of this function in Figure 13.3.3. Note that since the speed v and the time $t(v)$ should be positive in context, it's only the first quadrant of Figure 13.3.3 that matters.

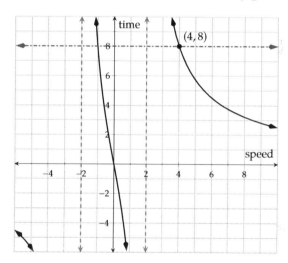

Figure 13.3.3: Graph of $t(v) = \frac{12}{v-2} + \frac{12}{v+2}$ and $t = 8$

To find the speed that Julia should drive the boat to make the round trip last 8 hours we can use graphing technology to solve the equation

$$\frac{12}{v-2} + \frac{12}{v+2} = 8$$

graphically and we see that $v = 4$. This tells us that a speed of 4 miles per hour will give a total time of 8 hours to complete the trip. To go downstream it would take $\frac{12}{v+2} = \frac{12}{4+2} = 2$ hours; and to go upstream it would take $\frac{12}{v-2} = \frac{12}{4-2} = 6$ hours.

The point of this section is to work with expressions like $\frac{12}{v-2} + \frac{12}{v+2}$, where two rational expressions are added (or subtracted). There are times when it is useful to combine them into a single fraction. We will learn that the expression $\frac{12}{v-2} + \frac{12}{v+2}$ is equal to the expression $\frac{24v}{v^2-4}$, and we will learn how to make that simplification.

13.3.2 Addition and Subtraction of Rational Expressions with the Same Denominator

The process of adding and subtracting rational expressions will be very similar to the process of adding and subtracting purely numerical fractions.

If the two expressions have the same denominator, then we can rely on the property of adding and subtracting fractions and simplify that result.

Let's review how to add fractions with the same denominator:

$$\begin{aligned} \frac{1}{10} + \frac{3}{10} &= \frac{1+3}{10} \\ &= \frac{4}{10} \\ &= \frac{2}{5} \end{aligned}$$

We can add and subtract rational expressions in the same way:

$$\begin{aligned} \frac{2}{3x} - \frac{5}{3x} &= \frac{2-5}{3x} \\ &= \frac{-3}{3x} \\ &= -\frac{1}{x} \end{aligned}$$

Identify the LCD Determine the least common denominator of all of the denominators.

Build If necessary, build each expression so that the denominators are the same.

Add/Subtract Combine the numerators using the properties of adding and subtracting fractions.

Simplify Simplify the resulting rational expression as much as possible. This may require factoring the numerator.

List 13.3.4: Steps to Adding/Subtracting Rational Expressions

Example 13.3.5 Add the rational expressions: $\dfrac{2x}{x+y} + \dfrac{2y}{x+y}$.

Explanation. These expressions already have a common denominator:

$$\begin{aligned} \frac{2x}{x+y} + \frac{2y}{x+y} &= \frac{2x+2y}{x+y} \\ &= \frac{2\cancel{(x+y)}}{\cancel{x+y}} \\ &= \frac{2}{1} \\ &= 2 \end{aligned}$$

Note that we didn't stop at $\frac{2x+2y}{x+y}$. If possible, we must simplify the numerator and denominator. Since this is a multivariable expression, this textbook ignores domain restrictions while canceling.

13.3.3 Addition and Subtraction of Rational Expressions with Different Denominators

To add rational expressions with different denominators, we'll need to build each fraction to the least common denominator, in the same way we do with numerical fractions. Let's briefly review this process by adding $\frac{3}{5}$ and $\frac{1}{6}$:

$$\frac{3}{5} + \frac{1}{6} = \frac{3}{5} \cdot \frac{6}{6} + \frac{1}{6} \cdot \frac{5}{5}$$
$$= \frac{18}{30} + \frac{5}{30}$$
$$= \frac{18+5}{30}$$
$$= \frac{23}{30}$$

This exact method can be used when adding rational expressions containing variables. The key is that the expressions *must* have the same denominator before they can be added or subtracted. If they don't have this initially, then we'll identify the least common denominator and build each expression so that it has that denominator.

Let's apply this to adding the two expressions with denominators that are $v - 2$ and $v + 2$ from Example 13.3.2.

Example 13.3.6 Add the rational expressions and fully simplify the function given by $t(v) = \frac{12}{v-2} + \frac{12}{v+2}$.

Explanation.

$$t(v) = \frac{12}{v-2} + \frac{12}{v+2}$$
$$t(v) = \frac{12}{v-2} \cdot \frac{v+2}{v+2} + \frac{12}{v+2} \cdot \frac{v-2}{v-2}$$
$$t(v) = \frac{12v+24}{(v-2)(v+2)} + \frac{12v-24}{(v+2)(v-2)}$$
$$t(v) = \frac{(12v+24)+(12v-24)}{(v+2)(v-2)}$$
$$t(v) = \frac{24v}{(v+2)(v-2)}$$

Example 13.3.7 Add the rational expressions: $\dfrac{2}{5x^2y} + \dfrac{3}{20xy^2}$

Explanation. The least common denominator of $5x^2y$ and $20xy^2$ must include two x's and two y's, as well as 20. Thus it is $20x^2y^2$. We will build both denominators to $20x^2y^2$ before doing addition.

$$\frac{2}{5x^2y} + \frac{3}{20xy^2} = \frac{2}{5x^2y} \cdot \frac{4y}{4y} + \frac{3}{20xy^2} \cdot \frac{x}{x}$$

$$= \frac{8y}{20x^2y^2} + \frac{3x}{20x^2y^2}$$

$$= \frac{8y + 3x}{20x^2y^2}$$

Let's look at a few more complicated examples.

Example 13.3.8 Subtract the rational expressions: $\dfrac{y}{y-2} - \dfrac{8y-8}{y^2-4}$

Explanation. To start, we'll make sure each denominator is factored. Then we'll find the least common denominator and build each expression to that denominator. Then we will be able to combine the numerators and simplify the expression.

$$\frac{y}{y-2} - \frac{8y-8}{y^2-4} = \frac{y}{y-2} - \frac{8y-8}{(y+2)(y-2)}$$

$$= \frac{y}{y-2} \cdot \frac{y+2}{y+2} - \frac{8y-8}{(y+2)(y-2)}$$

$$= \frac{y^2+2y}{(y+2)(y-2)} - \frac{8y-8}{(y+2)(y-2)}$$

$$= \frac{y^2+2y - \overset{\downarrow}{(}8y \overset{\downarrow}{-} 8)}{(y+2)(y-2)}$$

$$= \frac{y^2+2y-8y+8}{(y+2)(y-2)}$$

$$= \frac{y^2-6y+8}{(y+2)(y-2)}$$

$$= \frac{(y-2)(y-4)}{(y+2)(y-2)}$$

$$= \frac{y-4}{y+2}, \text{ for } y \neq 2$$

Note that we must factor the numerator in $\frac{y^2-6y+8}{(y+2)(y-2)}$ and try to reduce the fraction (which we did).

Warning 13.3.9. In Example 13.3.8, be careful to subtract the entire numerator of $8y-8$. When this expression is in the numerator of $\frac{8y-8}{(y+2)(y-2)}$, it's implicitly grouped and doesn't need parentheses. But once $8y-8$ is subtracted from y^2+2y, we need to add parentheses so the entire expression is subtracted.

In the next example, we'll look at adding a rational expression to a polynomial. Much like adding a fraction and an integer, we'll rely on writing that expression as itself over one in order to build its denominator.

Example 13.3.10 Add the expressions: $-\dfrac{2}{r-1} + r$

Explanation.

$$-\frac{2}{r-1} + r = -\frac{2}{r-1} + \frac{r}{1}$$

$$= -\frac{2}{r-1} + \frac{r}{1} \cdot \frac{r-1}{r-1}$$

$$= \frac{-2}{r-1} + \frac{r^2-r}{r-1}$$

$$= \frac{-2+r^2-r}{r-1}$$

$$= \frac{r^2-r-2}{r-1}$$

$$= \frac{(r-2)(r+1)}{r-1}$$

Note that we factored the numerator to reduce the fraction if possible. Even though it was not possible in this case, leaving it in factored form makes it easier to see that it is reduced.

Example 13.3.11 Subtract the expressions: $\dfrac{6}{x^2-2x-8} - \dfrac{1}{x^2+3x+2}$

Explanation. To start, we'll need to factor each of the denominators. After that, we'll identify the LCD and build each denominator accordingly. Then we can combine the numerators and simplify the resulting expression.

$$\frac{6}{x^2-2x-8} - \frac{1}{x^2+3x+2} = \frac{6}{(x-4)(x+2)} - \frac{1}{(x+2)(x+1)}$$

$$= \frac{6}{(x-4)(x+2)} \cdot \frac{x+1}{x+1} - \frac{1}{(x+2)(x+1)} \cdot \frac{x-4}{x-4}$$

$$= \frac{6x+6}{(x-4)(x+2)(x+1)} - \frac{x-4}{(x+2)(x+1)(x-4)}$$

$$= \frac{6x+6-(x-4)}{(x-4)(x+2)(x+1)}$$

$$= \frac{6x+6-x+4}{(x-4)(x+2)(x+1)}$$

$$= \frac{5x+10}{(x-4)(x+2)(x+1)}$$

$$= \frac{5\cancel{(x+2)}}{(x-4)\cancel{(x+2)}(x+1)}$$

$$= \frac{5}{(x-4)(x+1)}, \text{ for } x \neq -2$$

Exercises

Review and Warmup

1. Add: $\dfrac{17}{24} + \dfrac{11}{24}$

2. Add: $\dfrac{7}{12} + \dfrac{7}{12}$

3. Add: $\dfrac{7}{10} + \dfrac{5}{6}$

4. Add: $\dfrac{9}{10} + \dfrac{5}{6}$

5. Subtract: $\dfrac{23}{21} - \dfrac{20}{21}$

6. Subtract: $\dfrac{11}{14} - \dfrac{3}{14}$

7. Subtract: $\dfrac{3}{7} - \dfrac{1}{21}$

8. Subtract: $\dfrac{3}{7} - \dfrac{19}{21}$

Factor the given polynomial.

9. $r^2 - 64 =$ [   ]

10. $t^2 - 16 =$ [   ]

11. $t^2 + 11t + 10 =$ [   ]

12. $x^2 + 12x + 35 =$ [   ]

13. $x^2 - 12x + 27 =$ [   ]

14. $x^2 - 13x + 30 =$ [   ]

15. $9y^2 - 27y + 18 =$ [   ]

16. $8y^2 - 24y + 16 =$ [   ]

Addition and Subtraction of Rational Expressions with One Variable Add or subtract the rational expressions to a single rational expression and then simplify. If applicable, state the restricted domain.

17. $\dfrac{4r}{r+2} + \dfrac{8}{r+2} =$ [   ]

18. $\dfrac{6r}{r+6} + \dfrac{36}{r+6} =$ [   ]

19. $\dfrac{3t}{t+5} + \dfrac{15}{t+5} =$ [   ]

20. $\dfrac{6t}{t+3} + \dfrac{18}{t+3} =$ [   ]

21. $\dfrac{1}{x^2 - x - 30} - \dfrac{x-5}{x^2 - x - 30} =$ [   ]

22. $\dfrac{6}{x^2 - 12x + 20} - \dfrac{x-4}{x^2 - 12x + 20} =$ [   ]

23. $\dfrac{4}{x^2 - 25} - \dfrac{x-1}{x^2 - 25} =$ [   ]

24. $\dfrac{5}{y^2 - 8y + 7} - \dfrac{y-2}{y^2 - 8y + 7} =$ [   ]

25. $\dfrac{4y}{5} + \dfrac{y}{30} =$ [   ]

26. $\dfrac{5r}{6} + \dfrac{r}{18} =$ [   ]

27. $\dfrac{6}{r+2} - \dfrac{5}{r+4} =$ [   ]

28. $\dfrac{5}{t-5} - \dfrac{2}{t-4} =$ [   ]

29. $\dfrac{3}{t-1} - \dfrac{2}{t+5} =$ [   ]

30. $\dfrac{5}{x+6} - \dfrac{2}{x+2} =$ [   ]

31. $\dfrac{1}{x+2} + \dfrac{4}{x^2 - 4} =$ [   ]

32. $\dfrac{1}{x+3} + \dfrac{6}{x^2 - 9} =$ [   ]

33. $\dfrac{1}{y-1} - \dfrac{2}{y^2 - 1} =$ [   ]

34. $\dfrac{1}{y-3} - \dfrac{6}{y^2 - 9} =$ [   ]

35. $\dfrac{5}{r+2} - \dfrac{10r}{r^2-4} = \boxed{}$

36. $\dfrac{4}{r-3} - \dfrac{8r}{r^2-9} = \boxed{}$

37. $\dfrac{5}{t+5} - \dfrac{10t}{t^2-25} = \boxed{}$

38. $\dfrac{4}{t-2} - \dfrac{8t}{t^2-4} = \boxed{}$

39. $\dfrac{x}{x+2} - \dfrac{6x+16}{x^2+2x} = \boxed{}$

40. $\dfrac{x}{x+2} - \dfrac{4x+12}{x^2+2x} = \boxed{}$

41. $\dfrac{x}{x-2} - \dfrac{3x-2}{x^2-2x} = \boxed{}$

42. $\dfrac{y}{y-9} - \dfrac{15y-54}{y^2-9y} = \boxed{}$

43. $\dfrac{6}{y^2-9} + \dfrac{3}{y+3} - \dfrac{1}{y-3} = \boxed{}$

44. $\dfrac{6}{r^2-1} + \dfrac{5}{r+1} - \dfrac{3}{r-1} = \boxed{}$

45. $\dfrac{4r}{r^2-7r+12} + \dfrac{4r}{r-3} = \boxed{}$

46. $-\dfrac{3t}{t^2-5t+6} - \dfrac{3t}{t-2} = \boxed{}$

47. $\dfrac{4t}{t^2-11t+30} + \dfrac{4t}{t-5} = \boxed{}$

48. $-\dfrac{3x}{x^2+5x+6} - \dfrac{3x}{x+3} = \boxed{}$

49. $\dfrac{x^2-5}{x^2+5x} - \dfrac{x-1}{x} = \boxed{}$

50. $\dfrac{x^2+15}{x^2-3x} - \dfrac{x-5}{x} = \boxed{}$

51. $\dfrac{3}{y+5} + 4 = \boxed{}$

52. $-\dfrac{5}{y+1} - 4 = \boxed{}$

53. $\dfrac{4r}{r+2} + \dfrac{r}{r-2} - 5 = \boxed{}$

54. $\dfrac{6r}{r+1} + \dfrac{r}{r-1} - 7 = \boxed{}$

Addition and Subtraction of Rational Expressions with More Than Variable Add or subtract the rational expressions to a single rational expression and then simplify.

55. $\dfrac{25t^2}{5t-8r} - \dfrac{64r^2}{5t-8r} = \boxed{}$

56. $\dfrac{81t^2}{9t+4y} - \dfrac{16y^2}{9t+4y} = \boxed{}$

57. $\dfrac{5x}{18t} + \dfrac{x}{6t} = \boxed{}$

58. $\dfrac{4x}{15r} + \dfrac{4x}{3r} = \boxed{}$

59. $\dfrac{2x}{3y^3} - \dfrac{2}{5xy} = $ [_____]

60. $-\dfrac{3y}{2t^4} + \dfrac{6}{5yt} = $ [_____]

61. $\dfrac{3}{yr - 6} - \dfrac{6yr}{y^2r^2 - 36} = $ [_____]

62. $\dfrac{3}{rx + 2} - \dfrac{6rx}{r^2x^2 - 4} = $ [_____]

63. $-\dfrac{6rt}{r^2 + 4rt + 3t^2} + \dfrac{3r}{r + t} = $ [_____]

64. $\dfrac{6tr}{t^2 - 13tr + 40r^2} - \dfrac{2t}{t - 8r} = $ [_____]

13.4 Complex Fractions

In this section, we will learn how to simplify complex fractions, which have fractions in the numerator and/or denominator of another fraction.

13.4.1 Simplifying Complex Fractions

Consider the rational expression

$$\frac{\frac{6}{x-4}}{\frac{6}{x-4}+3}.$$

It's difficult to quickly evaluate this expression, or determine the important information such as its domain. This type of rational expression, which contains a "fraction within a fraction," is referred to as a **complex fraction**. Our goal is to simplify such a fraction so that it has a *single* numerator and a *single* denominator, neither of which contain any fractions themselves.

A complex fraction may have fractions in its numerator and/or denominator. Here is an example to show how we use division to simplify a complex fraction.

$$
\begin{aligned}
\frac{\frac{1}{2}}{3} &= \frac{1}{2} \div 3 \\
&= \frac{1}{2} \div \frac{3}{1} \\
&= \frac{1}{2} \cdot \frac{1}{3} \\
&= \frac{1}{6}
\end{aligned}
$$

What if the expression had something more complicated in the denominator, like $\frac{\frac{1}{2}}{\frac{1}{3}+\frac{1}{4}}$? We would no longer be able to simply multiply by the reciprocal of the denominator, since we don't immediately know the reciprocal of that denominator. Instead, we could multiply the "main" numerator and denominator by something that eliminates all of the "internal" denominators. (We'll use the LCD to determine this). For example, with $\frac{\frac{1}{2}}{3}$, we can multiply by $\frac{2}{2}$:

$$
\begin{aligned}
\frac{\frac{1}{2}}{3} &= \frac{\frac{1}{2}}{3} \cdot \frac{2}{2} \\
&= \frac{1}{6}
\end{aligned}
$$

Remark 13.4.2. In the last example, it's important to identify which fraction bar is the "main" fraction bar, and which fractions are "internal." Comparing the two expressions below, both of which are "one over two over three", we see that they are not equivalent.

$$
\begin{aligned}
\frac{\frac{1}{2}}{3} &= \frac{\frac{1}{2}}{3} \cdot \frac{2}{2} \\
&= \frac{1}{6}
\end{aligned}
\qquad \text{versus} \qquad
\begin{aligned}
\frac{1}{\frac{2}{3}} &= \frac{1}{\frac{2}{3}} \cdot \frac{3}{3} \\
&= \frac{3}{2}
\end{aligned}
$$

For the first of these, the "main" fraction bar is above the 3, but for the second of these, the "main" fraction bar is above the $\frac{2}{3}$.

To attack multiple fractions in a complex fraction, we need to multiply the numerator and denominator by the LCD of all the internal fractions, as we will show in the next example.

Example 13.4.3 Simplify the complex fraction $\dfrac{\frac{1}{2}}{\frac{1}{3}+\frac{1}{4}}$.

Explanation.

The internal denominators are 2, 3, and 4, so the LCD is 12. We will thus multiply the main numerator and denominator by 12 and simplify the result:

$$\frac{\frac{1}{2}}{\frac{1}{3}+\frac{1}{4}} = \frac{\frac{1}{2}}{\frac{1}{3}+\frac{1}{4}} \cdot \frac{12}{12}$$

$$= \frac{\frac{1}{2}\cdot 12}{\left(\frac{1}{3}+\frac{1}{4}\right)\cdot 12}$$

$$= \frac{6}{4+3}$$

$$= \frac{6}{7}$$

Next we will evaluate a function whose formula is a complex fraction and then simplify the result.

Example 13.4.4 Find each function value for $f(x) = \dfrac{\frac{x+2}{x+3}}{\frac{2}{x+3}-\frac{3}{x-1}}$.

 a. $f(4)$ b. $f(0)$ c. $f(-3)$ d. $f(-11)$

Explanation. We will determine each function value by replacing x with the specified number and then simplify the complex fraction:

a. $f(4) = \dfrac{\frac{4+2}{4+3}}{\frac{2}{4+3}-\frac{3}{4-1}}$

$= \dfrac{\frac{6}{7}}{\frac{2}{7}-\frac{3}{3}}$

$= \dfrac{\frac{6}{7}}{\frac{2}{7}-1}\cdot\dfrac{7}{7}$

$= \dfrac{6}{2-7}$

$= -\dfrac{6}{5}$

b. $f(0) = \dfrac{\frac{0+2}{0+3}}{\frac{2}{0+3}-\frac{3}{0-1}}$

$= \dfrac{\frac{2}{3}}{\frac{2}{3}-\frac{3}{-1}}$

$= \dfrac{\frac{2}{3}}{\frac{2}{3}+3}\cdot\dfrac{3}{3}$

$= \dfrac{2}{2+9}$

$= \dfrac{2}{11}$

c. When evaluating f at -3, we can quickly see that this results in division by zero:

$f(-3) = \dfrac{\frac{-3+2}{-3+3}}{\frac{2}{-3+3}-\frac{3}{-3-1}}$

$= \dfrac{\frac{2}{0}}{\frac{2}{0}-\frac{3}{-4}}$

Thus $f(-3)$ is undefined.

d. $f(-11) = \dfrac{\frac{-11+2}{-11+3}}{\frac{2}{-11+3}-\frac{3}{-11-1}}$

$= \dfrac{\frac{-8}{-9}}{\frac{2}{-8}-\frac{3}{-12}}$

$= \dfrac{\frac{8}{9}}{-\frac{1}{4}+\frac{1}{4}}$

$= \dfrac{\frac{8}{9}}{0}$

Therefore $f(-11)$ is undefined.

We have simplified complex fractions involving numbers and now we will apply the same concept to complex fractions with variables.

Example 13.4.5 Simplify the complex fraction $\dfrac{3}{\frac{1}{y}+\frac{5}{y^2}}$.

Explanation.

To start, we look at the internal denominators and identify the LCD as y^2. We'll multiply the main numerator and denominator by the LCD, and then simplify. Since we are multiplying by $\frac{y^2}{y^2}$, it is important to note that y cannot be 0, since $\frac{0}{0}$ is undefined.

$$\frac{3}{\frac{1}{y} + \frac{5}{y^2}} = \frac{3}{\frac{1}{y} + \frac{5}{y^2}} \cdot \frac{y^2}{y^2}$$

$$= \frac{3 \cdot y^2}{\frac{1}{y} \cdot y^2 + \frac{5}{y^2} \cdot y^2}$$

$$= \frac{3y^2}{y + 5}, \text{ for } y \neq 0$$

Example 13.4.6 Simplify the complex fraction $\dfrac{\frac{5x-6}{2x+1}}{\frac{3x+2}{2x+1}}$.

Explanation.

The internal denominators are both $2x+1$, so this is the LCD and we will multiply the main numerator and denominator by this expression. Since we are multiplying by $\frac{2x+1}{2x+1}$, what x-value would cause $2x+1$ to equal 0? Solving $2x+1 = 0$ leads to $x = -\frac{1}{2}$. So x cannot be $-\frac{1}{2}$, since $\frac{0}{0}$ is undefined.

$$\frac{\frac{5x-6}{2x+1}}{\frac{3x+2}{2x+1}} = \frac{\frac{5x-6}{2x+1}}{\frac{3x+2}{2x+1}} \cdot \frac{2x+1}{2x+1}$$

$$= \frac{5x-6}{3x+2}, \text{ for } x \neq -\frac{1}{2}$$

Example 13.4.7 Completely simplify the function defined by $f(x) = \dfrac{\frac{x+2}{x+3}}{\frac{2}{x+3} - \frac{3}{x-1}}$. Then determine the domain of this function.

Explanation. The LCD of the internal denominators is $(x+3)(x-1)$. We will thus multiply the main numerator and denominator by the expression $(x+3)(x-1)$ and then simplify the resulting expression.

$$f(x) = \frac{\frac{x+2}{x+3}}{\frac{2}{x+3} - \frac{3}{x-1}}$$

$$f(x) = \frac{\frac{x+2}{x+3}}{\frac{2}{x+3} - \frac{3}{x-1}} \cdot \frac{(x+3)(x-1)}{(x+3)(x-1)}$$

$$f(x) = \frac{\frac{x+2}{x+3} \cdot (x+3)(x-1)}{\left(\frac{2}{x+3} - \frac{3}{x-1}\right) \cdot (x+3)(x-1)}$$

$$f(x) = \frac{\frac{x+2}{x+3} \cdot (x+3)(x-1)}{\frac{2}{x+3} \cdot (x+3)(x-1) - \frac{3}{x-1} \cdot (x+3)(x-1)}$$

$$f(x) = \frac{(x+2)(x-1)}{2(x-1) - 3(x+3)}, \text{ for } x \neq -3, x \neq 1$$

$$f(x) = \frac{(x+2)(x-1)}{2x - 2 - 3x - 9}, \text{ for } x \neq -3, x \neq 1$$

$$f(x) = \frac{(x+2)(x-1)}{-x - 11}, \text{ for } x \neq -3, x \neq 1$$

$$f(x) = \frac{(x+2)(x-1)}{-(x+11)}, \text{ for } x \neq -3, x \neq 1$$

In the original (unsimplified) function, we could see that $x \neq -3$ and $x \neq 1$. In the simplified function, we need $x + 11 \neq 0$, so we can also see that $x \neq -11$. Therefore the domain of the function f is $\{x \mid x \neq -11, -3, 1\}$.

Example 13.4.8 Simplify the complex fraction $\dfrac{2\left(\frac{-4x+3}{x-2}\right) + 3}{\frac{-4x+3}{x-2} + 4}$.

Explanation. The only internal denominator is $x - 2$, so we will begin by multiplying the main numerator and denominator by this. Then we'll simplify the resulting expression.

$$\frac{2\left(\frac{-4x+3}{x-2}\right) + 3}{\frac{-4x+3}{x-2} + 4} = \frac{2\left(\frac{-4x+3}{x-2}\right) + 3}{\frac{-4x+3}{x-2} + 4} \cdot \frac{x-2}{x-2}$$

$$= \frac{2\left(\frac{-4x+3}{x-2}\right) \cdot (x-2) + 3 \cdot (x-2)}{\left(\frac{-4x+3}{x-2}\right) \cdot (x-2) + 4 \cdot (x-2)}$$

$$= \frac{2(-4x+3) + 3(x-2)}{(-4x+3) + 4(x-2)}, \text{ for } x \neq 2$$

$$= \frac{-8x + 6 + 3x - 6}{-4x + 3 + 4x - 8}, \text{ for } x \neq 2$$

$$= \frac{-5x}{-5}, \text{ for } x \neq 2$$

$$= x, \text{ for } x \neq 2$$

Example 13.4.9 Simplify the complex fraction $\dfrac{\frac{5}{x} + \frac{4}{y}}{\frac{3}{x} - \frac{2}{y}}$. Recall that with a multivariable expression, this textbook ignores domain restrictions.

Explanation.

We multiply the numerator and denominator by the common denominator of x and y, which is xy:

$$\frac{\frac{5}{x} + \frac{4}{y}}{\frac{3}{x} - \frac{2}{y}} = \frac{\frac{5}{x} + \frac{4}{y}}{\frac{3}{x} - \frac{2}{y}} \cdot \frac{xy}{xy}$$

$$= \frac{\left(\frac{5}{x} + \frac{4}{y}\right) \cdot xy}{\left(\frac{3}{x} - \frac{2}{y}\right) \cdot xy}$$

$$= \frac{\frac{5}{x} \cdot xy + \frac{4}{y} \cdot xy}{\frac{3}{x} \cdot xy - \frac{2}{y} \cdot xy}$$

$$= \frac{5y + 4x}{3y - 2x}$$

Example 13.4.10 Simplify the complex fraction $\dfrac{\frac{t}{t+3} + \frac{2}{t-3}}{1 - \frac{t}{t^2-9}}$.

Explanation. First, we check all quadratic polynomials to see if they can be factored and factor them:

$$\frac{\frac{t}{t+3} + \frac{2}{t-3}}{1 - \frac{t}{t^2-9}} = \frac{\frac{t}{t+3} + \frac{2}{t-3}}{1 - \frac{t}{(t-3)(t+3)}}$$

Next, we identify the common denominator of the three fractions, which is $(t+3)(t-3)$. We then multiply the main numerator and denominator by that expression:

$$\frac{\frac{t}{t+3} + \frac{2}{t-3}}{1 - \frac{t}{t^2-9}} = \frac{\frac{t}{t+3} + \frac{2}{t-3}}{1 - \frac{t}{(t-3)(t+3)}} \cdot \frac{(t+3)(t-3)}{(t+3)(t-3)}$$

$$= \frac{\frac{t}{t+3} \cdot (t+3)(t-3) + \frac{2}{t-3} \cdot (t+3)(t-3)}{1 \cdot (t+3)(t-3) - \frac{t}{(t-3)(t+3)} \cdot (t+3)(t-3)}$$

$$= \frac{t(t-3) + 2(t+3)}{(t+3)(t-3) - t} \quad \text{for } t \neq -3, t \neq 3$$

$$= \frac{t^2 - 3t + 2t + 6}{t^2 - 9 - t} \quad \text{for } t \neq -3, t \neq 3$$

$$= \frac{t^2 - t + 6}{t^2 - t - 9} \quad \text{for } t \neq -3, t \neq 3$$

Note that since both the numerator and denominator are prime trinomials, this expression can neither factor nor simplify any further.

Exercises

Review and Warmup   Calculate the following. Use an improper fraction in your answer.

1.
  a. $\dfrac{\frac{3}{7}}{\frac{3}{4}} = \boxed{}$

  b. $\dfrac{\frac{t}{x}}{\frac{r}{y}} = \boxed{}$

2.
  a. $\dfrac{\frac{16}{5}}{\frac{4}{7}} = \boxed{}$

  b. $\dfrac{\frac{t}{r}}{\frac{y}{x}} = \boxed{}$

3.
  a. $\dfrac{3}{\frac{4}{7}} = \boxed{}$

  b. $\dfrac{\frac{3}{4}}{7} = \boxed{}$

4.
  a. $\dfrac{2}{\frac{3}{8}} = \boxed{}$

  b. $\dfrac{\frac{2}{3}}{8} = \boxed{}$

5. $\dfrac{\frac{2}{3} + \frac{4}{5}}{\frac{1}{6}} = \boxed{}$

6. $\dfrac{\frac{5}{4} + \frac{1}{5}}{\frac{2}{3}} = \boxed{}$

7. $\dfrac{4}{\frac{1}{2} - \frac{2}{3}} = \boxed{}$

8. $\dfrac{4}{\frac{5}{4} - \frac{6}{5}} = \boxed{}$

Simplifying Complex Fractions with One Variable   Simplify this expression, and if applicable, write the restricted domain.

9. $\dfrac{\frac{6p+8}{p}}{\frac{p-5}{p}} = \boxed{}$

10. $\dfrac{\frac{3m-4}{m}}{\frac{m+6}{m}} = \boxed{}$

11. $\dfrac{\frac{k}{(k-8)^2}}{\frac{8k}{k^2-64}} = \boxed{}$

12. $\dfrac{\frac{z}{(z-4)^2}}{\frac{3z}{z^2-16}} = $ ☐

13. $\dfrac{2+\frac{1}{a}}{a+6} = $ ☐

14. $\dfrac{7+\frac{1}{a}}{a+10} = $ ☐

15. $\dfrac{2}{\frac{4}{y}+\frac{3}{y+2}} = $ ☐

16. $\dfrac{5}{\frac{1}{r}-\frac{4}{r+4}} = $ ☐

17. $\dfrac{7+\frac{1}{q-3}}{\frac{1}{q-3}-\frac{1}{6}} = $ ☐

18. $\dfrac{4+\frac{1}{q-7}}{\frac{1}{q-7}-\frac{1}{3}} = $ ☐

19. $\dfrac{\frac{1}{k+10}+\frac{1}{k-10}}{10-\frac{1}{k-10}} = $ ☐

20. $\dfrac{\frac{1}{k+6}+\frac{5}{k-6}}{8-\frac{1}{k-6}} = $ ☐

21. $\dfrac{\frac{1}{x-3}+\frac{9}{x-3}}{5-\frac{1}{x+3}} = $ ☐

22. $\dfrac{\frac{1}{a-9}+\frac{4}{a-9}}{3-\frac{1}{a+9}} = $ ☐

23. $\dfrac{\frac{6}{b-1}-8}{\frac{1}{b-1}+\frac{1}{b-9}} = $ ☐

24. $\dfrac{\frac{2}{t-1}-3}{\frac{1}{t-1}+\frac{1}{t-7}} = $ ☐

25. $\dfrac{\frac{4r}{r^2-25}-1}{\frac{5}{r+5}-\frac{6}{r-5}} = $ ☐

26. $\dfrac{\frac{6r}{r^2-4}+1}{\frac{6}{r+2}-\frac{5}{r-2}} = $ ☐

27. $\dfrac{\frac{r}{r^2-4}-\frac{1}{r^2-4}}{\frac{1}{r+4}} = $ ☐

28. $\dfrac{\frac{k}{k^2-49}-\frac{1}{k^2-49}}{\frac{1}{k+49}} = $ ☐

Simplifying Complex Fractions with More Than One Variable   Simplify this expression.

29. $\dfrac{\frac{m}{n}}{\frac{4m}{3n^2}} = $ ☐

30. $\dfrac{\frac{x}{y}}{\frac{3x}{2y^2}} = $ ☐

31. $\dfrac{\frac{ab^2}{9c}}{\frac{a}{2bc}} = $ ☐

32. $\dfrac{\frac{ab^2}{6c}}{\frac{a}{6bc}} = $ ☐

33. a. $\dfrac{\frac{y}{x}}{t} = $ ☐

   b. $\dfrac{y}{\frac{x}{t}} = $ ☐

34. a. $\dfrac{\frac{r}{y}}{t} = $ ☐

   b. $\dfrac{r}{\frac{y}{t}} = $ ☐

35. $\dfrac{\frac{4}{r}}{16+\frac{4y}{5}} = $ ☐

36. $\dfrac{\frac{2}{t}}{4+\frac{2x}{5}} = $ ☐

37. $\dfrac{\frac{4}{t}+\frac{12}{r}}{\frac{8}{t}+\frac{12}{r}} = $ ☐

38. $\dfrac{\frac{3}{t}+\frac{6}{y}}{\frac{9}{t}+\frac{6}{y}} = $ ☐

13.5 Solving Rational Equations

13.5.1 Solving Rational Equations

To start this section, we will use a scenario we have seen before in Example 13.3.2:

Julia is taking her family on a boat trip 12 miles down the river and back. The river flows at a speed of 2 miles per hour and she wants to drive the boat at a constant speed, v miles per hour downstream and back upstream. Due to the current of the river, the actual speed of travel is $v+2$ miles per hour going downstream, and $v-2$ miles per hour going upstream. If Julia plans to spend 8 hours for the whole trip, how fast should she drive the boat?

The time it takes Julia to drive the boat downstream is $\frac{12}{v+2}$ hours, and upstream is $\frac{12}{v-2}$ hours. The function to model the whole trip's time is

$$t(v) = \frac{12}{v-2} + \frac{12}{v+2}$$

where t stands for time in hours. The trip will take 8 hours, so we substitute $t(v)$ with 8, and we have:

$$\frac{12}{v-2} + \frac{12}{v+2} = 8.$$

Instead of using the function's graph, we will solve this equation algebraically. You may wish to review the technique of eliminating denominators discussed in Subsection 3.3.2. We can use the same technique with variable expressions in the denominators. To remove the fractions in this equation, we will multiply both sides of the equation by the least common denominator $(v-2)(v+2)$, and we have:

$$\frac{12}{v-2} + \frac{12}{v+2} = 8$$
$$(v+2)(v-2) \cdot \left(\frac{12}{v-2} + \frac{12}{v+2} \right) = (v+2)(v-2) \cdot 8$$
$$(v+2)(v-2) \cdot \frac{12}{v-2} + (v+2)(v-2) \cdot \frac{12}{v+2} = (v+2)(v-2) \cdot 8$$
$$12(v+2) + 12(v-2) = 8(v^2 - 4)$$
$$12v + 24 + 12v - 24 = 8v^2 - 32$$
$$24v = 8v^2 - 32$$
$$0 = 8v^2 - 24v - 32$$
$$0 = 8(v^2 - 3v - 4)$$
$$0 = 8(v-4)(v+1)$$

$$v - 4 = 0 \qquad \text{or} \qquad v + 1 = 0$$
$$v = 4 \qquad \text{or} \qquad v = -1$$

Remark 13.5.2. At this point, logically all that we know is that the only *possible* solutions are -1 and 4. Because of the step where factors were canceled, it's possible that these might not actually be solutions to the original equation. They each might be what is called an **extraneous solution**. An extraneous solution is a number that would appear to be a solution based on the solving process, but actually does not make the original equation true. Because of this, it is important that these proposed solutions be checked. Note that we're not checking to see if we made a calculation error, but are instead checking to see if the proposed solutions actually solve the original equation.

We check these values.

$$\frac{12}{-1-2} + \frac{12}{-1+2} \overset{?}{=} 8 \qquad\qquad \frac{12}{4-2} + \frac{12}{4+2} \overset{?}{=} 8$$

$$\frac{12}{-3} + \frac{12}{1} \overset{?}{=} 8 \qquad\qquad \frac{12}{2} + \frac{12}{6} \overset{?}{=} 8$$

$$-4 + 12 \overset{\checkmark}{=} 8 \qquad\qquad 6 + 2 \overset{\checkmark}{=} 8$$

Algebraically, both values do check out to be solutions. In the context of this scenario, the boat's speed can't be negative, so we only take the solution 4. If Julia drives at 4 miles per hour, the whole trip would take 8 hours. This result matches the solution in Example 13.3.2.

Let's look at another application problem.

> **Example 13.5.3** It takes Ku 3 hours to paint a room and it takes Jacob 6 hours to paint the same room. If they work together, how long would it take them to paint the room?
>
> **Explanation.** Since it takes Ku 3 hours to paint the room, he paints $\frac{1}{3}$ of the room each hour. Similarly, Jacob paints $\frac{1}{6}$ of the room each hour. If they work together, they paint $\frac{1}{3} + \frac{1}{6}$ of the room each hour.
>
> Assume it takes x hours to paint the room if Ku and Jacob work together. This implies they paint $\frac{1}{x}$ of the room together each hour. Now we can write this equation:
>
> $$\frac{1}{3} + \frac{1}{6} = \frac{1}{x}.$$
>
> To clear away denominators, we multiply both sides of the equation by the common denominator of 3, 6 and x, which is $6x$:
>
> $$\frac{1}{3} + \frac{1}{6} = \frac{1}{x}$$
> $$6x \cdot \left(\frac{1}{3} + \frac{1}{6}\right) = 6\cancel{x} \cdot \frac{1}{\cancel{x}}$$
> $$6x \cdot \frac{1}{3} + 6x \cdot \frac{1}{6} = 6$$
> $$2x + x = 6$$
> $$3x = 6$$
> $$x = 2$$
>
> Does the possible solution $x = 2$ check as an actual solution?
>
> $$\frac{1}{3} + \frac{1}{6} \overset{?}{=} \frac{1}{2}$$
> $$\frac{2}{6} + \frac{1}{6} \overset{?}{=} \frac{1}{2}$$
> $$\frac{3}{6} \overset{\checkmark}{=} \frac{1}{2}$$
>
> It does, so it is a solution. If Ku and Jacob work together, it would take them 2 hours to paint the room.

Let's look at a few more examples of solving rational equations.

Example 13.5.4 Solve for y in $\frac{2}{y+1} = \frac{3}{y}$.

Explanation. The common denominator is $y(y+1)$. We will multiply both sides of the equation by $y(y+1)$:

$$\frac{2}{y+1} = \frac{3}{y}$$

$$y(y+1) \cdot \frac{2}{y+1} = y(y+1) \cdot \frac{3}{y}$$

$$2y = 3(y+1)$$

$$2y = 3y + 3$$

$$0 = y + 3$$

$$-3 = y$$

Does the possible solution $y = -3$ check as an actual solution?

$$\frac{2}{-3+1} \overset{?}{=} \frac{3}{-3}$$

$$\frac{2}{-2} \overset{\checkmark}{=} -1$$

It checks, so -3 is a solution. We write the solution set as $\{-3\}$.

Example 13.5.5 Solve for z in $z + \frac{1}{z-4} = \frac{z-3}{z-4}$.

Explanation. The common denominator is $z - 4$. We will multiply both sides of the equation by $z - 4$:

$$z + \frac{1}{z-4} = \frac{z-3}{z-4}$$

$$(z-4) \cdot \left(z + \frac{1}{z-4} \right) = (z-4) \cdot \frac{z-3}{z-4}$$

$$(z-4) \cdot z + (z-4) \cdot \frac{1}{z-4} = z - 3$$

$$(z-4) \cdot z + 1 = z - 3$$

$$z^2 - 4z + 1 = z - 3$$

$$z^2 - 5z + 4 = 0$$

$$(z-1)(z-4) = 0$$

$$
\begin{array}{ccc}
z - 1 = 0 & \text{or} & z - 4 = 0 \\
z = 1 & \text{or} & z = 4
\end{array}
$$

Do the possible solutions $z = 1$ and $z = 4$ check as actual solutions?

$$1 + \frac{1}{1-4} \overset{?}{=} \frac{1-3}{1-4} \qquad\qquad 4 + \frac{1}{4-4} \overset{?}{=} \frac{4-3}{4-4}$$

$$1 - \frac{1}{3} \overset{\checkmark}{=} \frac{-2}{-3} \qquad\qquad 4 + \frac{1}{0} \overset{\text{no}}{=} \frac{1}{0}$$

The possible solution $z = 4$ does not actually work, since it leads to division by 0 in the equation. It is an extraneous solution. However, $z = 1$ is a valid solution. The only solution to the equation is 1, and thus we can write the solution set as $\{1\}$.

Example 13.5.6 Solve for p in $\frac{3}{p-2} + \frac{5}{p+2} = \frac{12}{p^2-4}$.

Explanation. To find the common denominator, we need to factor all denominators if possible:

$$\frac{3}{p-2} + \frac{5}{p+2} = \frac{12}{(p+2)(p-2)}$$

Now we can see the common denominator is $(p+2)(p-2)$. We will multiply both sides of the equation by $(p+2)(p-2)$:

$$\frac{3}{p-2} + \frac{5}{p+2} = \frac{12}{p^2-4}$$

$$\frac{3}{p-2} + \frac{5}{p+2} = \frac{12}{(p+2)(p-2)}$$

$$(p+2)(p-2) \cdot \left(\frac{3}{p-2} + \frac{5}{p+2} \right) = (p+2)(p-2) \cdot \frac{12}{(p+2)(p-2)}$$

$$(p+2)\cancel{(p-2)} \cdot \frac{3}{\cancel{p-2}} + \cancel{(p+2)}(p-2) \cdot \frac{5}{\cancel{p+2}} = \cancel{(p+2)}\cancel{(p-2)} \cdot \frac{12}{\cancel{(p+2)}\cancel{(p-2)}}$$

$$3(p+2) + 5(p-2) = 12$$

$$3p + 6 + 5p - 10 = 12$$

$$8p - 4 = 12$$

$$8p = 16$$

$$p = 2$$

Does the possible solution $p = 2$ check as an actual solution?

$$\frac{3}{2-2} + \frac{5}{2+2} \stackrel{?}{=} \frac{12}{2^2-4}$$

$$\frac{3}{0} + \frac{5}{4} \stackrel{no}{=} \frac{12}{0}$$

The possible solution $p = 2$ does not actually work, since it leads to division by 0 in the equation. So this is an extraneous solution, and the equation actually has no solution. We could say that its solution set is the empty set, $\emptyset$.

Example 13.5.7 Solve $C(t) = 0.35$, where $C(t) = \frac{3t}{t^2+8}$ gives a drug's concentration in milligrams per liter t hours since an injection. (This function was explored in the introduction of Section 13.1.)

Explanation. To solve $C(t) = 0.35$, we'll begin by setting up $\frac{3t}{t^2+8} = 0.35$. We'll begin by identifying that the LCD is $t^2 + 8$, and multiply each side of the equation by this:

$$\frac{3t}{t^2+8} = 0.35$$

$$\frac{3t}{t^2+8} \cdot \cancel{(t^2+8)} = 0.35 \cdot (t^2+8)$$

$$3t = 0.35\left(t^2 + 8\right)$$
$$3t = 0.35t^2 + 2.8$$

This results in a quadratic equation so we will put it in standard form and use the quadratic formula:

$$0 = 0.35t^2 - 3t + 2.8$$
$$t = \frac{-(-3) \pm \sqrt{(-3)^2 - 4(0.35)(2.8)}}{2(0.35)}$$
$$t = \frac{3 \pm \sqrt{5.08}}{0.7}$$
$$t \approx 1.066 \text{ or } t \approx 7.506$$

Each of these answers should be checked in the original equation; they both work. In context, this means that the drug concentration will reach 0.35 milligrams per liter about 1.066 hours after the injection was given, and again 7.506 hours after the injection was given.

13.5.2 Solving Rational Equations for a Specific Variable

Rational equations can contain many variables and constants and we can solve for any one of them. The process for solving still involves multiplying each side of the equation by the LCD. Instead of having a numerical answer though, our final result will contain other variables and constants.

Example 13.5.8 In physics, when two resistances, R_1 and R_2, are connected in a parallel circuit, the combined resistance, R, can be calculated by the formula

$$\frac{1}{R} = \frac{1}{R_1} + \frac{1}{R_2}.$$

Solve for R in this formula.

Explanation. The common denominator is RR_1R_2. We will multiply both sides of the equation by RR_1R_2:

$$\frac{1}{R} = \frac{1}{R_1} + \frac{1}{R_2}$$
$$\cancel{R}R_1R_2 \cdot \frac{1}{\cancel{R}} = RR_1R_2 \cdot \left(\frac{1}{R_1} + \frac{1}{R_2}\right)$$
$$R_1R_2 = R\cancel{R_1}R_2 \cdot \frac{1}{\cancel{R_1}} + RR_1\cancel{R_2} \cdot \frac{1}{\cancel{R_2}}$$
$$R_1R_2 = RR_2 + RR_1$$
$$R_1R_2 = R\left(R_2 + R_1\right)$$
$$\frac{R_1R_2}{R_2 + R_1} = R$$
$$R = \frac{R_1R_2}{R_1 + R_2}$$

Example 13.5.9 Here is the slope formula

$$m = \frac{y_2 - y_1}{x_2 - x_1}.$$

Solve for x_1 in this formula.

Explanation. The common denominator is $x_2 - x_1$. We will multiply both sides of the equation by $x_2 - x_1$:

$$m = \frac{y_2 - y_1}{x_2 - x_1}$$

$$(x_2 - x_1) \cdot m = \cancel{(x_2 - x_1)} \cdot \frac{y_2 - y_1}{\cancel{x_2 - x_1}}$$

$$mx_2 - mx_1 = y_2 - y_1$$

$$-mx_1 = y_2 - y_1 - mx_2$$

$$\frac{-mx_1}{-m} = \frac{y_2 - y_1 - mx_2}{-m}$$

$$x_1 = -\frac{y_2 - y_1 - mx_2}{m}$$

Example 13.5.10 Solve the rational equation $x = \frac{4y-1}{2y-3}$ for y.

Explanation. Our first step will be to multiply each side by the LCD, which is simply $2y - 3$. After that, we'll isolate all terms containing y, factor out y, and then finish solving for that variable.

$$x = \frac{4y - 1}{2y - 3}$$

$$x \cdot (2y - 3) = \frac{4y - 1}{\cancel{2y - 3}} \cdot \cancel{(2y - 3)}$$

$$2xy - 3x = 4y - 1$$

$$2xy = 4y - 1 + 3x$$

$$2xy - 4y = -1 + 3x$$

$$y(2x - 4) = 3x - 1$$

$$\frac{y\cancel{(2x - 4)}}{\cancel{2x - 4}} = \frac{3x - 1}{2x - 4}$$

$$y = \frac{3x - 1}{2x - 4}$$

13.5.3 Solving Rational Equations Using Technology

In some instances, it may be difficult to solve rational equations algebraically. We can instead use graphing technology to obtain approximate solutions. Let's look at one such example.

Example 13.5.11 Solve the equation $\frac{2}{x-3} = \frac{x^3}{8}$ using graphing technology.

Explanation.

We will define $f(x) = \frac{2}{x-3}$ and $g(x) = \frac{x^3}{8}$, and then look for the points of intersection.

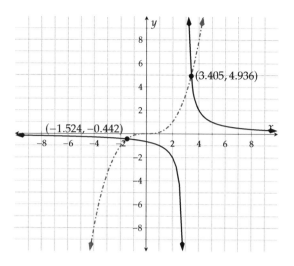

Figure 13.5.12: Graph of $f(x) = \frac{2}{x-3}$ and $g(x) = \frac{x^3}{8}$

Since the two functions intersect at approximately $(-1.524, -0.442)$ and $(3.405, 4.936)$, the solutions to $\frac{2}{x-3} = \frac{x^3}{8}$ are approximately -1.524 and 3.405. We can write the solution set as $\{-1.524\ldots, 3.405\ldots\}$ or in several other forms. It may be important to do *something* to commuincate that these solutions are approximations. Here we used $\ldots$, but you could also just say in words that the solutions are approximate.

Exercises

Review and Warmup   Solve the equation.

1. $10y + 5 = y + 50$

2. $9r + 9 = r + 33$

3. $6 = 3 - 3(a - 9)$

4. $52 = 10 - 7(b - 8)$

5. $4(A + 1) - 7(A - 7) = 35$

6. $2(C + 5) - 6(C - 2) = -2$

7. $(x + 5)^2 = 121$

8. $(x + 8)^2 = 49$

9. $x^2 + 4x - 96 = 0$

10. $x^2 - x - 90 = 0$

11. $x^2 - 6x + 1 = 8$

12. $x^2 - 14x + 64 = 19$

Solving Rational Equations   Solve the equation.

13. $\dfrac{-12}{y} = 6$

14. $\dfrac{-6}{r} = -1$

15. $\dfrac{r}{r + 3} = 4$

16. $\dfrac{t}{t - 4} = -3$

17. $\dfrac{t + 7}{5t + 7} = \dfrac{1}{19}$

18. $\dfrac{t - 2}{4t - 1} = \dfrac{1}{3}$

19. $\dfrac{-x-2}{x-1} = -\dfrac{x}{x-6}$

20. $\dfrac{-x+10}{x-6} = -\dfrac{x}{x-1}$

21. $\dfrac{6}{y} = 2 - \dfrac{4}{y}$

22. $\dfrac{6}{y} = -5 + \dfrac{16}{y}$

23. $\dfrac{4}{5A} - \dfrac{1}{4A} = -2$

24. $\dfrac{1}{5B} + \dfrac{5}{6B} = 4$

25. $\dfrac{t}{2t-10} + \dfrac{1}{t-5} = 4$

26. $\dfrac{t}{6t+24} - \dfrac{3}{t+4} = 2$

27. $\dfrac{t+8}{t^2+3} = 0$

28. $\dfrac{x+2}{x^2+6} = 0$

29. $-\dfrac{5}{x} = 0$

30. $\dfrac{7}{y} = 0$

31. $\dfrac{y+1}{y^2+8y+7} = 0$

32. $\dfrac{r-6}{r^2-10r+24} = 0$

33. $-\dfrac{9}{r} - \dfrac{6}{r+6} = 1$

34. $\dfrac{6}{r} + \dfrac{6}{r-9} = 1$

35. $\dfrac{1}{t-7} - \dfrac{7}{t^2-7t} = \dfrac{1}{6}$

36. $\dfrac{1}{t+6} + \dfrac{6}{t^2+6t} = -\dfrac{1}{9}$

37. $\dfrac{1}{x-8} - \dfrac{6}{x^2-8x} = \dfrac{1}{8}$

38. $\dfrac{1}{x-5} - \dfrac{4}{x^2-5x} = \dfrac{1}{6}$

39. $\dfrac{y+3}{y+7} - \dfrac{2}{y+5} = -1$

40. $\dfrac{y+1}{y+3} - \dfrac{8}{y+6} = -1$

Solve the equation.

41. $-\dfrac{3}{r+5} = -\left(\dfrac{6}{r-5} + \dfrac{3}{r^2-25}\right)$

42. $-\dfrac{6}{r+3} = -\left(\dfrac{4}{r-3} + \dfrac{2}{r^2-9}\right)$

43. $\dfrac{7}{r+4} - \dfrac{8}{r+1} = -\dfrac{9}{r^2+5r+4}$

44. $\dfrac{9}{t+7} - \dfrac{7}{t+3} = \dfrac{4}{t^2+10t+21}$

45. $-\dfrac{2}{t-8} + \dfrac{2t}{t-6} = -\dfrac{4}{t^2-14t+48}$

46. $\dfrac{2}{x-7} + \dfrac{2x}{x-5} = \dfrac{4}{x^2-12x+35}$

47. $\dfrac{4}{x-4} + \dfrac{8x}{x+1} = -\dfrac{8}{x^2-3x-4}$

48. $\dfrac{2}{y+5} + \dfrac{2y}{y+7} = -\dfrac{2}{y^2+12y+35}$

Solving Rational Equations for a Specific Variable

49. Solve this equation for y:

$$b = \frac{a}{y}$$

50. Solve this equation for n:

$$A = \frac{q}{n}$$

51. Solve this equation for t:

$$B = \frac{t}{r}$$

52. Solve this equation for t:

$$m = \frac{t}{C}$$

53. Solve this equation for n:

$$\frac{1}{5n} = \frac{1}{p}$$

54. Solve this equation for q:

$$\frac{1}{3q} = \frac{1}{A}$$

55. Solve this equation for r:

$$\frac{1}{y} = \frac{3}{r+5}$$

56. Solve this equation for n:

$$\frac{1}{r} = \frac{9}{n+6}$$

Solving Rational Equations Using Technology   Use technology to solve the equation

57.
$$\frac{10}{x^2+3} = \frac{x+1}{x+5}.$$

58.
$$\frac{x-9}{x^5+1} = -3x - 7.$$

59.
$$\frac{1}{x} + \frac{1}{x^2} = \frac{1}{x^3}.$$

60.
$$\frac{12x}{x-5} + \frac{3}{x+1} = \frac{x-5}{x^2}.$$

61.
$$2x - \frac{1}{x+4} = \frac{3}{x+6}.$$

62.
$$\frac{1}{x^2-1} - \frac{2}{x-4} = \frac{3}{x-2}.$$

Application Problems

63. Jessica and Derick are working together to paint a room. If Jessica paints the room alone, it would take her 9 hours to complete the job. If Derick paints the room alone, it would take him 18 hours to complete the job. Answer the following question:

If they work together, it would take them ⬚ hours to complete the job. Use a decimal in your answer if needed.

64. There are three pipes at a tank. To fill the tank, it would take Pipe A 4 hours, Pipe B 12 hours, and Pipe C 3 hours. Answer the following question:

If all three pipes are turned on, it would take ⬚ hours to fill the tank.

65. Neil and Nathan are working together to paint a room. Neil works 5.5 times as fast as Nathan does. If they work together, it took them 11 hours to complete the job. Answer the following questions:

If Neil paints the room alone, it would take him ☐ hours to complete the job.

If Nathan paints the room alone, it would take him ☐ hours to complete the job.

66. Two pipes are being used to fill a tank. Pipe A can fill the tank 4.5 times as fast as Pipe B does. When both pipes are turned on, it takes 9 hours to fill the tank. Answer the following questions:

If only Pipe A is turned on, it would take ☐ hours to fill the tank.

If only Pipe B is turned on, it would take ☐ hours to fill the tank.

67. Aleric and Carmen worked together to paint a room, and it took them 4 hours to complete the job. If they work alone, it would take Carmen 6 more hours than Aleric to complete the job. Answer the following questions:

If Aleric paints the room alone, it would take him ☐ hours to complete the job.

If Carmen paints the room alone, it would take her ☐ hours to complete the job.

68. If both Pipe A and Pipe B are turned on, it would take 2 hours to fill a tank. If each pipe is turned on alone, it takes Pipe B 3 fewer hours than Pipe A to fill the tank. Answer the following questions:

If only Pipe A is turned on, it would take ☐ hours to fill the tank.

If only Pipe B is turned on, it would take ☐ hours to fill the tank.

69. Town A and Town B are 720 miles apart. A boat traveled from Town A to Town B, and then back to Town A. Since the river flows from Town B to Town A, the boat's speed was 30 miles per hour faster when it traveled from Town B to Town A. The whole trip took 20 hours. Answer the following questions:

The boat traveled from Town A to Town B at the speed of ☐ miles per hour.

The boat traveled from Town B back to Town A at the speed of ☐ miles per hour.

70. A river flows at 13 miles per hour. A boat traveled with the current from Town A to Town B, which are 360 miles apart. Then, the boat turned around, and traveled against the current to reach Town C, which is 100 miles away from Town B. The second leg of the trip (Town B to Town C) took the same time as the first leg (Town A to Town B). During this whole trip, the boat was driving at a constant still-water speed. Answer the following question:

During this trip, the boat's speed on still water was ☐ miles.

71. A river flows at 8 miles per hour. A boat traveled with the current from Town A to Town B, which are 130 miles apart. The boat stayed overnight at Town B. The next day, the water's current stopped, and boat traveled on still water to reach Town C, which is 200 miles away from Town B. The second leg of the trip (Town B to Town C) took 5 hours longer than the first leg (Town A to Town B). During this whole trip, the boat was driving at a constant still-water speed. Find this speed.

Note that you should not consider the unreasonable answer.

During this trip, the boat's speed on still water was ⬚ miles per hour.

72. Town A and Town B are 600 miles apart. With a constant still-water speed, a boat traveled from Town A to Town B, and then back to Town A. During this whole trip, the river flew from Town A to Town B at 11 miles per hour. The whole trip took 10 hours. Answer the following question:

During this trip, the boat's speed on still water was ⬚ miles per hour.

73. Town A and Town B are 200 miles apart. With a constant still-water speed of 45 miles per hour, a boat traveled from Town A to Town B, and then back to Town A. During this whole trip, the river flew from Town B to Town A at a constant speed. The whole trip took 9 hours. Answer the following question:

During this trip, the river's speed was ⬚ miles per hour.

74. Suppose that a large pump can empty a swimming pool in 33 hr and that a small pump can empty the same pool in 49 hr. If both pumps are used at the same time, how long will it take to empty the pool?

If both pumps are used at the same time, it will take ⬚ to empty the pool.

75. The winner of a 6 mi race finishes 15.05 min ahead of the second-place runner. On average, the winner ran 0.6 $\frac{mi}{hr}$ faster than the second place runner. Find the average running speed for each runner.

The winner's average speed was ⬚ and the second-place runner's average speed was ⬚.

76. In still water a tugboat can travel 20 $\frac{mi}{hr}$. It travels 43 mi upstream and then 43 mi downstream in a total time of 4.35 hr. Find the speed of the current.

The current's speed is ⬚.

77. Without any wind an airplane flies at 246 $\frac{mi}{hr}$. The plane travels 620 mi into the wind and then returns with the wind in a total time of 5.07 hr. Find the average speed of the wind.

The wind's speed is [ ].

78. When there is a 22.4 $\frac{mi}{hr}$ wind, an airplane can fly 800 mi with the wind in the same time that it can fly 642 mi against the wind. Find the speed of the plane when there is no wind.

The plane's airspeed is [ ].

79. It takes one employee 4.5 hr longer to mow a football field than it does a more experienced employee. Together they can mow the grass in 2.8 hr. How long does it take each person to mow the football field working alone?

The more experienced worker takes [ ] to mow the field alone, and the less experienced worker takes [ ].

80. It takes one painter 19 hr longer to paint a house than it does a more experienced painter. Together they can paint the house in 25 hr. How long does it take for each painter to paint the house working alone?

The more experienced painter takes [ ] to paint the house alone, and the less experienced painter takes [ ].

13.6 Rational Functions and Equations Chapter Review

13.6.1 Introduction to Rational Functions

In Section 13.1 we learned about rational functions and explored them with tables and graphs.

Example 13.6.1 Graphs of Rational Functions. In an apocalypse, a zombie infestation begins with 1 zombie and spreads rapidly. The population of zombies can be modeled by $Z(x) = \frac{200000x+100}{5x+100}$, where x is the number of days after the apocalypse began. Use technology to graph the function and answer these questions:

   a. How many zombies are there 2 days after the apocalypse began?

   b. After how many days will the zombie population be 20,000?

   c. As time goes on, the population will level off at about how many zombies?

Explanation. We will graph the function with technology. After adjusting window settings, we have:

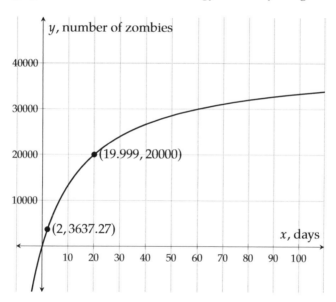

Figure 13.6.2: Graph of $y = Z(x) = \frac{200000x+100}{5x+100}$

   a. To find the number of zombies after 2 days, we locate the point $(2, 3637.27)$. Since we can only have a whole number of zombies, we round to 3,637 zombies.

   b. To find the number of days it will take for the zombie population reach 20,000, we locate the point $(19.999, 20000)$ so it will take about 20 days.

   c. When we look far to the right on the graph using technology we can see that the population will level off at about 40,000 zombies.

13.6.2 Multiplication and Division of Rational Expressions

In Section 13.2 we covered how to simplify rational expressions. It is very important to list any domain restrictions from factors that are canceled. We also multiplied and divided rational expressions.

Example 13.6.3 Simplifying Rational Expressions. Simplify the expression $\frac{8t+4t^2-12t^3}{1-t}$.

Explanation. To begin simplifying this expression, we will rewrite each polynomial in descending order. Then we'll factor out the GCF, including the constant -1 from both the numerator and denominator because their leading terms are negative.

$$\frac{8t + 4t^2 - 12t^3}{1 - t} = \frac{-12t^3 + 4t^2 + 8t}{-t + 1}$$
$$= \frac{-4t(3t^2 - t - 2)}{-(t - 1)}$$
$$= \frac{-4t(3t + 2)(t - 1)}{-(t - 1)}$$
$$= \frac{-4t(3t + 2)\cancel{(t - 1)}}{-\cancel{(t - 1)}}$$
$$= \frac{-4t(3t + 2)}{-1}, \text{ for } t \neq 1$$
$$= 4t(3t + 2), \text{ for } t \neq 1$$

Example 13.6.4 Multiplication of Rational Functions and Expressions. Multiply the rational expressions: $\frac{r^3s}{4t} \cdot \frac{2t^2}{r^2s^3}$.

Explanation. Note that we won't need to factor anything in this problem, and can simply multiply across and then simplify. With multivariable expressions, this textbook ignores domain restrictions.

$$\frac{r^3s}{4t} \cdot \frac{2t^2}{r^2s^3} = \frac{r^3s \cdot 2t^2}{4t \cdot r^2s^3}$$
$$= \frac{2r^3st^2}{4r^2s^3t}$$
$$= \frac{rt}{2s^2}$$

Example 13.6.5 Division of Rational Functions and Expressions. Divide the rational expressions: $\frac{2x^2+8xy}{x^2-4x+3} \div \frac{x^3+4x^2y}{x^2+4x-5}$.

Explanation. To divide rational expressions, we multiply by the reciprocal of the second fraction. Then we will factor and cancel any common factors. With multivariable expressions, this textbook ignores domain restrictions.

$$\frac{2x^2 + 8xy}{x^2 - 4x + 3} \div \frac{x^3 + 4x^2y}{x^2 + 4x - 5} = \frac{2x^2 + 8xy}{x^2 - 4x + 3} \cdot \frac{x^2 + 4x - 5}{x^3 + 4x^2y}$$
$$= \frac{2x\cancel{(x + 4y)}}{\cancel{(x - 1)}(x - 3)} \cdot \frac{\cancel{(x - 1)}(x + 5)}{x^2\cancel{(x + 4y)}}$$

$$= \frac{2x}{x-3} \cdot \frac{x+5}{x^2}$$
$$= \frac{2(x+5)}{x(x-3)}$$

13.6.3 Addition and Subtraction of Rational Expressions

In Section 13.3 we covered how to add and subtract rational expressions.

Example 13.6.6 Addition and Subtraction of Rational Expressions with the Same Denominator. Add the rational expressions: $\dfrac{5x}{x+5} + \dfrac{25}{x+5}$.

Explanation. These expressions already have a common denominator:

$$\frac{5x}{x+5} + \frac{25}{x+5} = \frac{5x+25}{x+5}$$
$$= \frac{5\cancel{(x+5)}}{\cancel{x+5}}$$
$$= \frac{5}{1}, \text{ for } x \neq -5$$
$$= 5, \text{ for } x \neq -5$$

Note that we didn't stop at $\frac{5x+25}{x+5}$. If possible, we must simplify the numerator and denominator.

Example 13.6.7 Addition and Subtraction of Rational Expressions with Different Denominators. Add and subtract the rational expressions: $\dfrac{6y}{y+2} + \dfrac{y}{y-2} - 7$

Explanation. The denominators can't be factored, so we'll find the least common denominator and build each expression to that denominator. Then we will be able to combine the numerators and simplify the expression.

$$\frac{6y}{y+2} + \frac{y}{y-2} - 7 = \frac{6y}{y+2} \cdot \frac{y-2}{y-2} + \frac{y}{y-2} \cdot \frac{y+2}{y+2} - 7 \cdot \frac{(y-2)(y+2)}{(y-2)(y+2)}$$
$$= \frac{6y(y-2)}{(y-2)(y+2)} + \frac{y(y+2)}{(y-2)(y+2)} - \frac{7(y-2)(y+2)}{(y-2)(y+2)}$$
$$= \frac{6y^2 - 12y + y^2 + 2y - \overset{\downarrow}{(}7(y^2 - \overset{\downarrow}{4)})}{(y-2)(y+2)}$$
$$= \frac{6y^2 - 12y + y^2 + 2y - 7y^2 + 28}{(y-2)(y+2)}$$
$$= \frac{-10y + 28}{(y-2)(y+2)}$$
$$= \frac{-2(5y - 14)}{(y-2)(y+2)}$$

13.6.4 Complex Fractions

In Section 13.4 we covered how to simplify a rational expression that has fractions in the numerator and/or denominator.

Example 13.6.8 Simplifying Complex Fractions. Simplify the complex fraction $\dfrac{\frac{2t}{t^2-9} + 3}{\frac{6}{t+3} + \frac{1}{t-3}}$.

Explanation. First, we check all quadratic polynomials to see if they can be factored and factor them:

$$\frac{\frac{2t}{t^2-9} + 3}{\frac{6}{t+3} + \frac{1}{t-3}} = \frac{\frac{2t}{(t-3)(t+3)} + 3}{\frac{6}{t+3} + \frac{1}{t-3}}$$

Next, we identify the common denominator of the three fractions, which is $(t+3)(t-3)$. We then multiply the main numerator and denominator by that expression:

$$\frac{\frac{2t}{(t-3)(t+3)} + 3}{\frac{6}{t+3} + \frac{1}{t-3}} = \frac{\frac{2t}{(t-3)(t+3)} + 3}{\frac{6}{t+3} + \frac{1}{t-3}} \cdot \frac{(t-3)(t+3)}{(t-3)(t+3)}$$

$$= \frac{\frac{2t}{(t-3)(t+3)}(t-3)(t+3) + 3(t-3)(t+3)}{\frac{6}{t+3}(t-3)(t+3) + \frac{1}{t-3}(t-3)(t+3)}$$

$$= \frac{2t + 3(t-3)(t+3)}{6(t-3) + 1(t+3)} \quad \text{for } t \neq -3, t \neq 3$$

$$= \frac{2t + 3(t^2 - 9)}{6t - 18 + t + 3} \quad \text{for } t \neq -3, t \neq 3$$

$$= \frac{2t + 3t^2 - 27}{7t - 15} \quad \text{for } t \neq -3, t \neq 3$$

$$= \frac{3t^2 + 2t - 27}{7t - 15} \quad \text{for } t \neq -3, t \neq 3$$

Both the numerator and denominator are prime polynomials so this expression can neither factor nor simplify any further.

13.6.5 Solving Rational Equations

In Section 13.5 we covered how to solve rational equations. We looked at rate problems, solved for a specified variable and used technology to solve rational equations.

Example 13.6.9 Solving Rational Equations. Two pipes are being used to fill a large tank. Pipe B can fill the tank twice as fast as Pipe A can. When both pipes are turned on, it takes 12 hours to fill the tank. Write and solve a rational equation to answer the following questions:

   a. If only Pipe A is turned on, how many hours would it take to fill the tank?

   b. If only Pipe B is turned on, how many hours would it take to fill the tank?

Explanation. Since both pipes can fill the tank in 12 hours, they fill $\frac{1}{12}$ of the tank together each hour. We will let a represent the number of hours it takes pipe A to fill the tank alone, so pipe A will fill $\frac{1}{a}$ of the tank each hour. Pipe B can fill the tank twice as fast so it fills $2 \cdot \frac{1}{a}$ of the tank each hour or $\frac{2}{a}$. When they are both turned on, they fill $\frac{1}{a} + \frac{2}{a}$ of the tank each hour.

Now we can write this equation:

$$\frac{1}{a} + \frac{2}{a} = \frac{1}{12}$$

To clear away denominators, we multiply both sides of the equation by the common denominator of 12 and a, which is $12a$:

$$\frac{1}{a} + \frac{2}{a} = \frac{1}{12}$$
$$12a \cdot \left(\frac{1}{a} + \frac{2}{a}\right) = 12a \cdot \frac{1}{12}$$
$$12a \cdot \frac{1}{a} + 12a \cdot \frac{2}{a} = 12a \cdot \frac{1}{12}$$
$$12 + 24 = a$$
$$36 = a$$
$$a = 36$$

The possible solution $a = 36$ should be checked

$$\frac{1}{36} + \frac{2}{36} \overset{?}{=} \frac{1}{12}$$
$$\frac{3}{36} \overset{\checkmark}{=} \frac{1}{12}$$

So it is a solution.

a. If only Pipe A is turned on, it would take 36 hours to fill the tank.

b. Since Pipe B can fill the tank twice as fast, it would take half the time, or 18 hours to fill the tank.

Example 13.6.10 Solving Rational Equations for a Specific Variable. Solve the rational equation $y = \frac{2x+5}{3x-1}$ for x.

Explanation. To get the x out of the denominator, our first step will be to multiply each side by the LCD, which is $3x - 1$. Then we'll isolate all terms containing x, factor out x, and then finish solving for that variable.

$$y = \frac{2x+5}{3x-1}$$
$$y \cdot (3x - 1) = \frac{2x+5}{\cancel{3x-1}} \cdot \cancel{(3x-1)}$$
$$3xy - y = 2x + 5$$
$$3xy = 2x + 5 + y$$
$$3xy - 2x = y + 5$$
$$x(3y - 2) = y + 5$$
$$\frac{x(3y-2)}{3y-2} = \frac{y+5}{3y-2}$$
$$x = \frac{y+5}{3y-2}$$

Example 13.6.11 Solving Rational Equations Using Technology. Solve the equation $\frac{1}{x+2} + 1 = \frac{10x}{x^2+5}$ using graphing technology.

Explanation.

We will define $f(x) = \frac{1}{x+2} + 1$ and $g(x) = \frac{10x}{x^2+5}$, and then find a window where we can see all of the points of intersection.

Since the two functions intersect at approximately $(-2.309, -2.235)$, $(0.76, 1.362)$ and $(8.549, 1.095)$, the solutions to $\frac{1}{x+2} + 1 = \frac{10x}{x^2+5}$ are approximately -2.309, 0.76 and 8.549. The solution set is approximately $\{-2.309\ldots, 0.76\ldots, 8.549\ldots\}$.

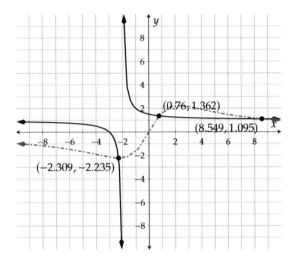

Figure 13.6.12: Graph of $f(x) = \frac{1}{x+2} + 1$ and $g(x) = \frac{10x}{x^2+5}$

Exercises

Introduction to Rational Functions

1. A function is graphed.

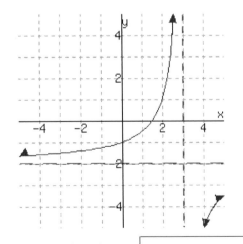

This function has domain [_____]

and range [_____].

2. A function is graphed.

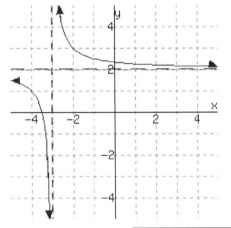

This function has domain [_____]

and range [_____].

3. The population of deer in a forest can be modeled by

$$P(x) = \frac{780x + 2660}{3x + 7}$$

where x is the number of years in the future. Answer the following questions.

   a. How many deer live in this forest this year?

   b. How many deer will live in this forest 24 years later? Round your answer to an integer.

   c. After how many years, the deer population will be 276? Round your answer to an integer.

   d. Use a calculator to answer this question: As time goes on, the population levels off at about how many deer?

4. The population of deer in a forest can be modeled by

$$P(x) = \frac{400x + 1950}{4x + 5}$$

where x is the number of years in the future. Answer the following questions.

   a. How many deer live in this forest this year?

   b. How many deer will live in this forest 27 years later? Round your answer to an integer.

   c. After how many years, the deer population will be 143? Round your answer to an integer.

   d. Use a calculator to answer this question: As time goes on, the population levels off at about how many deer?

5. In a certain store, cashiers can serve 55 customers per hour on average. If x customers arrive at the store in a given hour, then the average number of customers C waiting in line can be modeled by the function

$$C(x) = \frac{x^2}{3025 - 55x}$$

where $x < 55$.

Answer the following questions with a graphing calculator. Round your answers to integers.

   a. If 42 customers arrived in the store in the past hour, there are approximately ☐ customers waiting in line.

   b. If there are 7 customers waiting in line, approximately ☐ customers arrived in the past hour.

6. In a certain store, cashiers can serve 60 customers per hour on average. If x customers arrive at the store in a given hour, then the average number of customers C waiting in line can be modeled by the function

$$C(x) = \frac{x^2}{3600 - 60x}$$

where $x < 60$.

Answer the following questions with a graphing calculator. Round your answers to integers.

   a. If 51 customers arrived in the store in the past hour, there are approximately ☐ customers waiting in line.

   b. If there are 2 customers waiting in line, approximately ☐ customers arrived in the past hour.

7. The concentration of a drug in a patient's blood stream, in milligrams per liter, can be modeled by the function $C(t) = \frac{6t}{t^2+4}$, where t is the number of hours since the drug is injected. Answer the following question with technology. Round your answer to two decimal places if needed.

   [   ] hours since injection, the drug's concentration is at the maximum value

   of [   ] milligrams per liter.

8. The concentration of a drug in a patient's blood stream, in milligrams per liter, can be modeled by the function $C(t) = \frac{7t}{t^2+8}$, where t is the number of hours since the drug is injected. Answer the following question with technology. Round your answer to two decimal places if needed.

   [   ] hours since injection, the drug's concentration is at the maximum value

   of [   ] milligrams per liter.

Multiplication and Division of Rational Expressions

9. Simplify this expression.

   $$\frac{-t^2 + 6ty - 5y^2}{t^2 - 25y^2} = [\quad\quad\quad]$$

10. Simplify this expression.

    $$\frac{-t^2 - 8tx - 15x^2}{t^2 - 25x^2} = [\quad\quad\quad]$$

11. Simplify the function formula, and if applicable, write the restricted domain.

    $$G(x) = \frac{x^4 + 8x^3 + 16x^2}{3x^4 + 13x^3 + 4x^2}$$

    Reduced $G(x) = [\quad\quad\quad]$

12. Simplify the function formula, and if applicable, write the restricted domain.

    $$h(x) = \frac{x^4 - 4x^3 + 4x^2}{2x^4 - 3x^3 - 2x^2}$$

    Reduced $h(x) = [\quad\quad\quad]$

13. Simplify this expression, and if applicable, write the restricted domain.

    $$\frac{y^2 - 16y}{y^2 - 16} \cdot \frac{y^2 - 4y}{y^2 - 19y + 48} = [\quad\quad\quad]$$

14. Simplify this expression, and if applicable, write the restricted domain.

    $$\frac{y^2 - 4y}{y^2 - 4} \cdot \frac{y^2 - 2y}{y^2 - 3y - 4} = [\quad\quad\quad]$$

15. Simplify this expression, and if applicable, write the restricted domain.

    $$\frac{9r^2 - 25}{3r^2 + 8r + 5} \div (5 - 3r) = [\quad\quad\quad]$$

16. Simplify this expression, and if applicable, write the restricted domain.

    $$\frac{25r^2 - 9}{5r^2 + 8r + 3} \div (3 - 5r) = [\quad\quad\quad]$$

17. Simplify this expression.

    $$\frac{r^3}{r^2y + 4r} \div \frac{1}{r^2y^2 + 5ry + 4} = [\quad\quad\quad]$$

18. Simplify this expression.

    $$\frac{t^4}{t^2y + 4t} \div \frac{1}{t^2y^2 + 5ty + 4} = [\quad\quad\quad]$$

Addition and Subtraction of Rational Expressions Add or subtract the rational expressions to a single rational expression and then simplify. If applicable, state the restricted domain.

19. $\dfrac{1}{t-4} - \dfrac{8}{t^2-16} = \boxed{}$

20. $\dfrac{1}{x+1} + \dfrac{2}{x^2-1} = \boxed{}$

21. $-\dfrac{5x}{x^2-3x+2} - \dfrac{5x}{x-1} = \boxed{}$

22. $\dfrac{8y}{y^2+2y-3} + \dfrac{2y}{y+3} = \boxed{}$

23. $\dfrac{y^2-20}{y^2+5y} - \dfrac{y-4}{y} = \boxed{}$

24. $\dfrac{y^2-10}{y^2-2y} - \dfrac{y+5}{y} = \boxed{}$

Add or subtract the rational expressions to a single rational expression and then simplify.

25. $-\dfrac{4r}{3t^5} - \dfrac{6}{5rt} = \boxed{}$

26. $-\dfrac{6r}{5y^3} + \dfrac{3}{2ry} = \boxed{}$

27. $\dfrac{6tx}{t^2-4tx-5x^2} - \dfrac{t}{t-5x} = \boxed{}$

28. $-\dfrac{6ty}{t^2+14ty+40y^2} - \dfrac{t}{t+10y} = \boxed{}$

Complex Fractions

29. Calculate the following. Use an improper fraction in your answer.

   a. $\dfrac{\frac{3}{7}}{\frac{3}{8}} = \boxed{}$

   b. $\dfrac{\frac{x}{y}}{\frac{r}{t}} = \boxed{}$

30. Calculate the following. Use an improper fraction in your answer.

   a. $\dfrac{\frac{3}{8}}{\frac{3}{5}} = \boxed{}$

   b. $\dfrac{\frac{x}{t}}{\frac{y}{r}} = \boxed{}$

31. Simplify this expression, and if applicable, write the restricted domain.

   $\dfrac{\frac{8}{b-1} - 4}{\frac{1}{b-1} + \frac{1}{b-10}} = \boxed{}$

32. Simplify this expression, and if applicable, write the restricted domain.

   $\dfrac{\frac{4}{b-1} - 7}{\frac{1}{b-1} + \frac{1}{b-8}} = \boxed{}$

33. Simplify this expression, and if applicable, write the restricted domain.

   $\dfrac{\frac{2y}{y^2-9} - 1}{\frac{3}{y+3} + \frac{2}{y-3}} = \boxed{}$

34. Simplify this expression, and if applicable, write the restricted domain.

   $\dfrac{\frac{2r}{r^2-9} + 3}{\frac{3}{r+3} + \frac{1}{r-3}} = \boxed{}$

35. Simplify this expression.

$$\frac{\frac{p}{q}}{\frac{4p}{3q^2}} = \boxed{}$$

36. Simplify this expression.

$$\frac{\frac{m}{n}}{\frac{6m}{5n^2}} = \boxed{}$$

37. Simplify this expression.

$$\frac{\frac{3}{t}}{3 + \frac{3r}{2}} = \boxed{}$$

38. Simplify this expression.

$$\frac{\frac{2}{x}}{2 - \frac{2y}{3}} = \boxed{}$$

Solving Rational Equations Solve the equation.

39. $\dfrac{2}{x-1} + \dfrac{4}{x+7} = -\dfrac{2}{x^2+6x-7}$

40. $\dfrac{3}{y+7} - \dfrac{9}{y-4} = -\dfrac{9}{y^2+3y-28}$

41. $\dfrac{1}{y-5} - \dfrac{5}{y^2-5y} = \dfrac{1}{4}$

42. $\dfrac{1}{y+8} + \dfrac{8}{y^2+8y} = -\dfrac{1}{4}$

43. $-\dfrac{6}{r+9} + \dfrac{2r}{r-3} = \dfrac{8}{r^2+6r-27}$

44. $\dfrac{3}{r-1} + \dfrac{6r}{r-7} = \dfrac{9}{r^2-8r+7}$

45. $\dfrac{t-3}{t+1} - \dfrac{6}{t-4} = 6$

46. $\dfrac{t+7}{t-9} + \dfrac{6}{t+6} = -1$

47. Solve this equation for r:

$$\frac{1}{x} = \frac{3}{r+4}$$

48. Solve this equation for n:

$$\frac{1}{r} = \frac{3}{n+8}$$

49. Use technology to solve the equation

$$2x - \frac{1}{x+4} = \frac{3}{x+6}.$$

50. Use technology to solve the equation

$$\frac{1}{x^2-1} - \frac{2}{x-4} = \frac{3}{x-2}.$$

51. Two pipes are being used to fill a tank. Pipe A can fill the tank 4.5 times as fast as Pipe B does. When both pipes are turned on, it takes 18 hours to fill the tank. Answer the following questions:

If only Pipe A is turned on, it would take [ ] hours to fill the tank.

If only Pipe B is turned on, it would take [ ] hours to fill the tank.

52. Two pipes are being used to fill a tank. Pipe A can fill the tank 5.5 times as fast as Pipe B does. When both pipes are turned on, it takes 11 hours to fill the tank. Answer the following questions:

If only Pipe A is turned on, it would take [ ] hours to fill the tank.

If only Pipe B is turned on, it would take [ ] hours to fill the tank.

53. Town A and Town B are 360 miles apart. A boat traveled from Town A to Town B, and then back to Town A. Since the river flows from Town B to Town A, the boat's speed was 15 miles per hour faster when it traveled from Town B to Town A. The whole trip took 20 hours. Answer the following questions:

The boat traveled from Town A to Town B at the speed of [ ] miles per hour.

The boat traveled from Town B back to Town A at the speed of [ ] miles per hour.

54. Town A and Town B are 560 miles apart. A boat traveled from Town A to Town B, and then back to Town A. Since the river flows from Town B to Town A, the boat's speed was 30 miles per hour faster when it traveled from Town B to Town A. The whole trip took 28 hours. Answer the following questions:

The boat traveled from Town A to Town B at the speed of [ ] miles per hour.

The boat traveled from Town B back to Town A at the speed of [ ] miles per hour.

CHAPTER 14

Radical Functions and Equations

14.1 Introduction to Radical Functions

We learned the basics of square roots in Section 8.2. The study of radicals is much broader than our first attempt covered and we need to expand our investigation. To do so, we will first look at an example that makes use of a topic covered in Section 8.3.

A #10 washer has a 5.6 mm inner diameter and is 1.2 mm thick. We will let d represent the outer diameter, measured in mm.

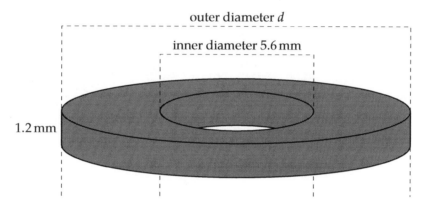

Figure 14.1.2: A Diagram of a #10 Washer

The amount of steel, M, in mg that it takes to make the washer with outer diameter d is approximated by the formula

$$M = 7.59d^2 - 238$$

Note that if you know the value of M ahead of time, this is a quadratic equation. We will now solve the equation for d using the square root method.

$$M = 7.59d^2 - 238$$
$$M + 238 = 7.59d^2$$
$$\frac{M + 238}{7.59} = d^2$$

$$d = \sqrt{\frac{M + 238}{7.59}} \qquad \text{or} \qquad d = -\sqrt{\frac{M + 238}{7.59}}$$

Since we know that the diameter cannot really be negative, our formula for d must be

$$d = \sqrt{\frac{M + 238}{7.59}}$$

This formula finds the diameter that the washer must be when you input the amount of steel used, M. Figure 14.1.3 shows a graph of this relationship.

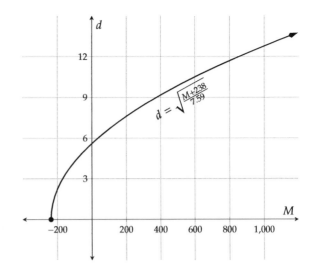

Figure 14.1.3: A Graph of $d = \sqrt{\frac{M+238}{7.59}}$

We also know that M cannot be negative, so we will cut the graph to begin at $M = 0$. This formula tells us that if we plan on using, for example, 1000 mg of steel (about as much as in a large paper clip) that we can find the the outer diameter for the washer that will be created:

$$d = \sqrt{\frac{1000 + 238}{7.59}}$$
$$\approx 12.8$$

So the washer's outer diameter must be about 12.8 mm for to have a mass of 1000 mg.

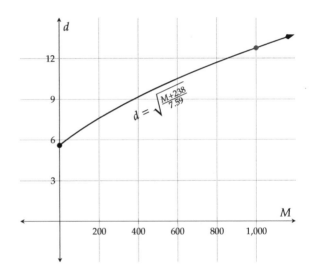

Figure 14.1.4: A Revised Graph of $d = \sqrt{\frac{M+238}{7.59}}$

Note that the vertical intercept of the graph is $(0, 5.6)$, which says that a washer that uses no steel at all (is that really a washer?) would have an outer diameter of 5.6 mm. This is the "smallest" possible washer with an inner diameter of 5.6 mm, even though it would technically be massless. Perhaps the implied domain of this function should be $(0, \infty)$ to exclude 0 mass.

Square roots often appear when we consider formulas from geometry like the washer problem, and they also show up in topics like in statistics (where $\sigma = \sqrt{\frac{1}{n} \sum_{i=1}^{n} (x_i - \mu)^2}$ finds the standard deviation), in chemistry (where $v_{rms} = \sqrt{\frac{3RT}{M_m}}$ finds the velocity of a particle), and in physics (where $m(v) = \frac{m_0}{\sqrt{1 - v^2/c^2}}$ finds the mass of an object as its velocity nears the speed of light). There are many more examples to give, but we need a firmer understanding of radicals to properly study these things, so it's time to venture into deeper waters.

14.1.1 The Square Root Function

Example 14.1.5 Gilberto is an artist who etches designs into square copper plates of different sizes. Customers can order the size they would like.

    a. Build a function that calculates the length of a plate's side given its area. Explore the function with a table and graph.

    b. One customer ordered a plate with an area of 6.25 square feet. Calculate the length of its side.

    c. Find the domain and range of the function from Part a.

Explanation.

    a. We know the formula to calculate a square's area is $A = \ell^2$, where A stands for a square's area, and ℓ is the length of the square's side. To build a function to calculate ℓ, we solve for ℓ in the formula:

$$A = \ell^2$$

$$\ell = \sqrt{A} \qquad \text{or} \qquad \ell = -\sqrt{A}$$

The formula is $\ell = \sqrt{A}$. We don't consider the negative solution in this context since negative length doesn't make sense. Since ℓ depends on A, we can use function notation and write $f(A) = \sqrt{A}$. We will make a table, plot points, and look at the graph of $\ell = \sqrt{A}$.

A	$f(A)$	Points on the Curve
0	$\sqrt{0} = 0$	$(0,0)$
1	$\sqrt{1} = 1$	$(1,1)$
4	$\sqrt{4} = 2$	$(4,2)$
6.25	$\sqrt{6.25} = 2.5$	$(6.25, 2.5)$
9	$\sqrt{9} = 3$	$(9,3)$

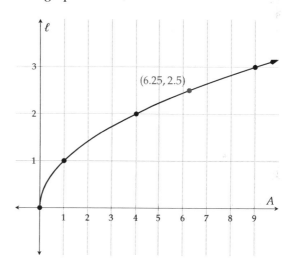

Table 14.1.6: Values of $f(A) = \sqrt{A}$ **Figure 14.1.7:** Graph of $y = \sqrt{A}$

    b. The point $(6.25, 2.5)$ implies that a square plate with an area of 6.25 square feet would have a length of 2.5 feet on each side.

    c. According to the graph, the function's domain is $[0, \infty)$ because the graph goes forever to the right from $A = 0$. This should make some sense because you cannot take the square root of a negative number. The function's range is $[0, \infty)$ because the graph seems to go up forever starting at $\ell = 0$. This should also make sense because a square root never gives you a negative number as an answer.

Fact 14.1.8 Domain of a Square Root Functions. *To algebraically find the domain of a square root function, set the radicand (the expression under the radical) greater than or equal to 0 and solve for the variable. The solution set to* *that inequality is the domain of the function.*

Example 14.1.9 Algebraically find the domain of the function g where $g(x) = \sqrt{2x - 4} + 1$ and then find the range by making a graph.

Explanation. Using Fact 14.1.8 to find the function's domain, we set the radicand greater than or equal to zero and solve:

$$2x - 4 \geq 0$$
$$2x \geq 4$$
$$x \geq 2$$

The function's domain is $[2, \infty)$ in interval notation.

To find the function's range, we use technology to look at a graph of the function. The graph shows that the function's range is $[1, \infty)$. The graph also verifies the function's domain is indeed $[2, \infty)$.

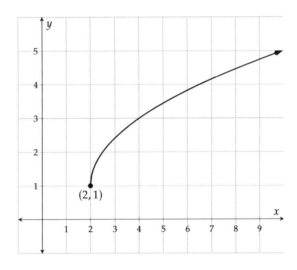

Figure 14.1.10: Graph of $g(x) = \sqrt{2x - 4} + 1$

Example 14.1.11 Algebraically find the domain and graphically find the range of the function h where $h(x) = 2 - 3\sqrt{8 - 5x}$.

Explanation. To find the function's domain, we set the radicand to be greater than or equal to zero:

$$8 - 5x \geq 0$$
$$-5x \geq -8$$
$$\frac{-5x}{-5} \leq \frac{-8}{-5}$$
$$x \leq \frac{8}{5}$$

So, the function's domain is $\left(-\infty, \frac{8}{5}\right]$. The 2 and 3 in the function do not play a role in the domain, although they do alter the range which we will find now by making a graph.

From the graph we can see that the range is all numbers below (or equal to) the y-value 2. In interval notation, this would be written $(-\infty, 2]$.

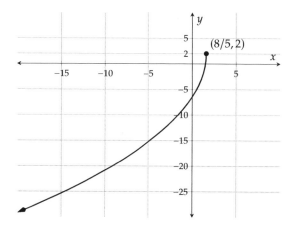

Figure 14.1.12: Graph of $h(x) = 2 - 3\sqrt{8 - 5x}$

Example 14.1.13 When an object is dropped, the time it takes to hit the ground can be modeled by

$$t = \sqrt{\frac{d}{16}}$$

where t is in seconds, and d is the initial height of the object in feet. Use graphing technology to create a graph and answer the following questions.

a. In a science experiment, Amaka's class drops a beanbag from the top of a 100-foot-tall building. How long will it take for the beanbag to hit the ground?

b. Her class then goes to a second building, drops the beanbag from the top, and uses a stopwatch to measures the time it takes to hit the ground. If it takes 3 seconds for the beanbag to hit the ground, how tall is the building?

Explanation.

With graphing technology, and after adjusting the window settings, we can see the graph of $t = \sqrt{\frac{d}{16}}$ and some important points.

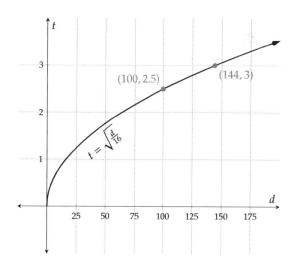

Figure 14.1.14: Graph of $t = \sqrt{\frac{d}{16}}$

a. We look for a d-value of 100 feet and find the point $(100, 2.5)$ on the graph. This means it will take the beanbag 2.5 seconds to hit the ground if it's released from the top of a 100-foot-tall building.

b. This time we look for a t-value of 3 seconds and find the point $(144, 3)$. This means the beanbag will fall approximately 144 feet in 3 seconds, so the second building is approximately 144 feet tall.

14.1.2 The Distance Formula

A square root is used in calculating the distance between two points on a coordinate plane. We learned the Pythagorean Theorem in Section 8.3.2. In a coordinate plane, we can use the Pythagorean Theorem to calculate the distance between any two points.

Example 14.1.15 Calculate the distance between $(2, 3)$ and $(5, 7)$.

Explanation.   First, we will sketch a graph of those two points.

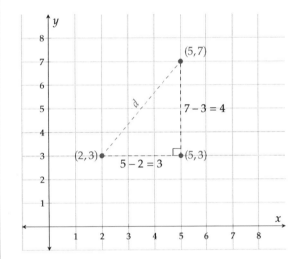

To calculate the distance between $(2, 3)$ and $(5, 7)$, we sketch a right triangle as in the figure and then use the Pythagorean Theorem:

$$d^2 = (5 - 2)^2 + (7 - 3)^2$$
$$d^2 = 3^2 + 4^2$$
$$d^2 = 9 + 16$$
$$d^2 = 25$$
$$d = \sqrt{25}$$
$$d = 5$$

Figure 14.1.16: Calculating the Distance Between $(2, 3)$ and $(5, 7)$

In conclusion, the distance between $(2, 3)$ and $(5, 7)$ is 5. Note that in our calculations, we didn't need to show $d = \pm\sqrt{25}$ because distance must have positive values.

With the same method, we can derive a formula to calculate the distance between any two points.

Example 14.1.17 With a generic first and second point, we will use subscripts to identify the first pair (x_1, y_1) and the second pair (x_2, y_2). Calculate the distance between the generic points (x_1, y_1) and (x_2, y_2).

Explanation.   First, we will sketch a graph of those two points. We will put the image from last example side by side with the new image, so it's clear that we are using the same method.

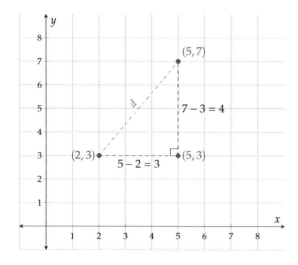

Figure 14.1.18: Calculating the Distance Between $(2,3)$ and $(5,7)$

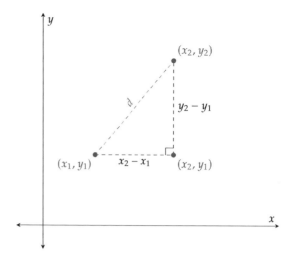

Figure 14.1.19: Calculating the Distance Between (x_1, y_1) and (x_2, y_2)

To calculate the distance between (x_1, y_1) and (x_2, y_2), we sketch a right triangle as in the figure and then use Pythagorean Theorem:

$$d^2 = (x_2 - x_1)^2 + (y_2 - y_1)^2$$
$$d = \sqrt{(x_2 - x_1)^2 + (y_2 - y_1)^2}$$

Fact 14.1.20 The Distance Formula. *The distance between two points (x_1, y_1) and (x_2, y_2), is given by the formula:*

$$d = \sqrt{(x_2 - x_1)^2 + (y_2 - y_1)^2}$$

With this formula, we can calculate the distance between two points without sketching a graph.

Example 14.1.21 Find the distance between $(-2, 4)$ and $(5, -20)$.

Explanation. To calculate the distance between $(-2, 4)$ and $(5, -20)$, we use the distance formula. It's good practice to mark each value with the corresponding variables in the formula. Again, (x_1, y_1) stands for the first point's coordinates, and (x_2, y_2) stands for the second point's coordinates:

$$(\overset{x_1}{-2}, \overset{y_1}{4}), (\overset{x_2}{5}, \overset{y_2}{-20})$$

We have:

$$d = \sqrt{(x_2 - x_1)^2 + (y_2 - y_1)^2}$$
$$d = \sqrt{(5 - (-2))^2 + ((-20) - 4)^2}$$
$$d = \sqrt{(7)^2 + (-24)^2}$$
$$d = \sqrt{49 + 576}$$
$$d = \sqrt{625}$$

$$d = 25$$

The distance between $(-2, 4)$ and $(5, -20)$ is 25 units.

Warning 14.1.22. Note that it's good practice to add parentheses around negative values when we do substitutions. For example, when we substitute x with -7 in x^2, we should write

$$x^2 = (-7)^2 = 49 \text{ correct } \checkmark$$

We should *not* write

$$x^2 = -7^2 = -49 \text{ incorrect}$$

14.1.3 Cube Root Function

The square of 2 is 4, so the square root of 4 is 2.

Similarly, the cube of 2 is 8, so the cube root of 8 is 2. We write

$$\sqrt[3]{8} = 2$$

It's helpful to memorize the first few perfect cube numbers and their cube roots:

$0^3 = 0$	$\sqrt[3]{0} = 0$
$1^3 = 1$	$\sqrt[3]{1} = 1$
$2^3 = 8$	$\sqrt[3]{8} = 2$
$3^3 = 27$	$\sqrt[3]{27} = 3$
$4^3 = 64$	$\sqrt[3]{64} = 4$
$5^3 = 125$	$\sqrt[3]{125} = 5$

One major difference between a cube root and a square root is that we can find the cube root of negative numbers. For example:

$$(-4)^3 = -64, \text{ so } \sqrt[3]{-64} = -4$$

However, $\sqrt{-64}$ is non-real and in general we cannot take the square root of a negative number.

Remark 14.1.23. Many calculators don't have a cube root button. If yours does, it might look like $\sqrt[n]{}$ and you will tell the calculator both to enter a number for the "n" as well as the radicand. Many calculators also allow you to type something like $\mathrm{root}(3,8)$ for $\sqrt[3]{8}$, for example.

Another way to calculate the cube root on a calculator is to use the exponent button (which is usually marked with the caret symbol, ^) with a *reciprocal* power. For example, to calculate $\sqrt[3]{8}$, you may type 8^(1/3). We will explain why $\sqrt[3]{8} = 8^{\frac{1}{3}}$ in Section 14.2. For now, just learn how to use a calculator to calculate the cube root of a given number.

We can also estimate the value of a cube root, like $\sqrt[3]{10}$, by knowing the perfect cubes nearby:

$$\sqrt[3]{8} = 2 \qquad\qquad \sqrt[3]{10} = ? \qquad\qquad \sqrt[3]{27} = 3$$

Since 10 is between the perfect cubes 8 and 27, $\sqrt[3]{10}$ must be between 2 and 3, and closer to 2. We can use a calculator to verify $\sqrt[3]{10} \approx 2.154$

Let's build a table and graph the cube root function.

x	$g(x) = \sqrt[3]{x}$	Points on the Curve
-8	$\sqrt[3]{-8} = -2$	$(-8, -2)$
-1	$\sqrt[3]{-1} = -1$	$(-1, -1)$
0	$\sqrt[3]{0} = 0$	$(0, 0)$
1	$\sqrt[3]{1} = 1$	$(1, 1)$
8	$\sqrt[3]{8} = 2$	$(8, 2)$

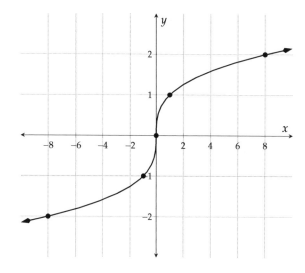

Table 14.1.24: Values of $g(x) = \sqrt[3]{x}$ **Figure 14.1.25:** Graph of $g(x) = \sqrt[3]{x}$

Both the domain and range of the *cube* root function are $(-\infty, \infty)$. Compare this with the domain and range of the *square* root function, which are each $[0, \infty)$. The reason for the difference is that we cannot take the *square* root of negative numbers, but we can take the *cube* root of negative numbers (and when we do, we get negative numbers as the output).

Remark 14.1.26. It is helpful to be able to quickly sketch the graphs of the following types of basic functions:

$$f(x) = c \qquad\qquad f(x) = mx + b \qquad\qquad f(x) = a(x - h)^2 + k \qquad\qquad f(x) = |x|$$

$$f(x) = \frac{1}{x} \qquad\qquad f(x) = \frac{1}{x^2} \qquad\qquad f(x) = \sqrt{x} \qquad\qquad f(x) = \sqrt[3]{x}$$

Now with the graph of the cube root function, you have seen all of these shapes in this book.

Example 14.1.27 Nasim makes solid copper spheres for their grounding and healing properties. A sphere's radius can be calculated by the formula $r(V) = \sqrt[3]{\frac{3V}{4\pi}}$, where $r(V)$ stands for the sphere's radius for a given volume V. If Nasim uses 2 cubic inches of copper per sphere, what diameter should he list on his website? Round your answer to two decimal places.

Explanation. First, to find the radius we will substitute 2 in for V, and we have:

$$r(V) = \sqrt[3]{\frac{3V}{4\pi}}$$

$$r(2) = \sqrt[3]{\frac{3(2)}{4\pi}}$$

$$\approx 0.78$$

The spheres will have a radius of approximately 0.78 in.

When we calculate $\sqrt[3]{\frac{3(2)}{4\pi}}$ with a calculator, we enter $(3*2/(4\square))^\wedge(1/3)$. To find the diameter, we multiply the radius by 2 to get 1.56 in. Nasim can advertise the spheres to be 1.56 inches in diameter.

14.1.4 Other Roots

Similar to the cube root, there is the fourth root, and the fifth root, and so on, as in the following examples:

$$\sqrt[4]{16} = 2 \text{ because } 2^4 = 16 \qquad \sqrt[5]{-32} = -2 \text{ because } (-2)^5 = -32 \qquad \sqrt[6]{64} = 2 \text{ because } 2^6 = 64,$$

To calculate the fifth root of -32 with a calculator, try typing $(-32)\verb|^|(1/5)$ or $\verb|root(5,-32)|$.

Definition 14.1.28. The **index** of a radical is the number "n" in $\sqrt[n]{\ }$. The symbol $\sqrt[n]{\ }$ is read "the nth root." The plural of index is **indices**, as in "we can evaluate radicals of multiple indices in a single expression."

Fact 14.1.29 Domain of Radical Functions. *To find the domain of any even indexed radical function, set the radicand greater than or equal to zero. The solution set is the domain of the function.*

The domain of any odd indexed radical of a polynomial is $(-\infty, \infty)$.

> **Example 14.1.30**
>
> a. Let $g(x) = 7 - 3\sqrt[4]{10 - 5x}$. Algebraically find g's domain and graphically find g's range.
>
> b. Let $h(x) = 4\sqrt[5]{2x - 5} + 1$. Algebraically find h's domain and graphically find h's range.
>
> **Explanation.**
>
> a. First note that the index of this function is 4. By Fact 14.1.29, to find the domain of this function we must set the radicand greater or equal to zero and solve.
>
> $$10 - 5x \geq 0$$
> $$-5x \geq -10$$
> $$x \leq \frac{-10}{-5}$$
> $$x \leq 2$$
>
> So, the domain of g must be $(-\infty, 2]$.
>
> To find the function's range, we use technology to graph the function. By the graph, the function's range is $(-\infty, 7]$.

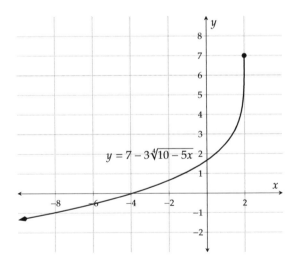

Figure 14.1.31: Graph of $y = 7 - 3\sqrt[4]{10 - 5x}$

b. First note that the index of the function h for $h(x) = 4\sqrt[5]{2x-5} + 1$ is 5. By Fact 14.1.29, the domain is $(-\infty, \infty)$.

To find the function's range, we use technology to graph the function. According to the graph, the function's range is also the set of all real numbers.

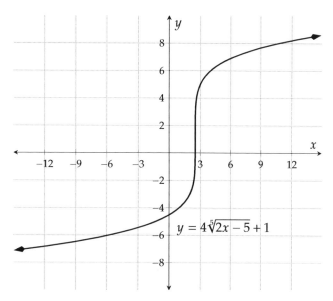

Figure 14.1.32: Graph of $y = 4\sqrt[5]{2x-5} + 1$

Exercises

Review and Warmup

1. Evaluate the following.

$-\sqrt{81} = $ [   ].

2. Evaluate the following.

$-\sqrt{121} = $ [   ].

3. Evaluate the following.

$\sqrt{-144} = $ [   ].

4. Evaluate the following.

$\sqrt{-4} = $ [   ].

5. Evaluate the following.

$\sqrt{\dfrac{4}{81}} = $ [   ].

6. Evaluate the following.

$\sqrt{\dfrac{9}{25}} = $ [   ].

7. Without using a calculator, evaluate the expression.

a. $\sqrt[3]{8} = $ [   ]

b. $\sqrt[3]{-8} = $ [   ]

c. $-\sqrt[3]{8} = $ [   ]

8. Without using a calculator, evaluate the expression.

a. $\sqrt[4]{16} = $ [   ]

b. $\sqrt[4]{-16} = $ [   ]

c. $-\sqrt[4]{16} = $ [   ]

9. Without using a calculator, estimate the value of $\sqrt{65}$:

($\square$ 8.94 $\square$ 7.94 $\square$ 7.06 $\square$ 8.06)

10. Without using a calculator, estimate the value of $\sqrt{61}$:

($\square$ 7.19 $\square$ 7.81 $\square$ 8.81 $\square$ 8.19)

11. Without using a calculator, estimate the value of $\sqrt[3]{68}$:

($\square$ 4.08 $\square$ 3.08 $\square$ 3.92 $\square$ 4.92)

12. Without using a calculator, estimate the value of $\sqrt[3]{62}$:

($\square$ 3.04 $\square$ 4.96 $\square$ 3.96 $\square$ 4.04)

Domain and Range Find the domain of the function.

13. $f(x) = \sqrt{10 - x}$

14. $g(x) = \sqrt{7 - x}$

15. $g(x) = \sqrt{3 + 20x}$

16. $h(x) = \sqrt{9 + 14x}$

17. $F(x) = \sqrt[3]{-9x - 1}$

18. $G(x) = \sqrt[3]{-7x - 7}$

19. $G(x) = \sqrt[4]{45 - 5x}$

20. $H(x) = \sqrt[4]{-9 - 3x}$

21. $K(x) = -\frac{8}{\sqrt{x+5}}$

22. $f(x) = \frac{6}{\sqrt{x-7}}$

23. $f(x) = \sqrt{x + 7}$

24. $f(x) = \sqrt{x + 5}$

25. $f(x) = \sqrt{-3 - 7x}$

26. $f(x) = \sqrt{-2 - 7x}$

27. $f(x) = \frac{1}{\sqrt{18-5x}}$

28. $f(x) = \frac{1}{\sqrt{40-56x}}$

29. $f(x) = \frac{1}{\sqrt{29x+62}}$

30. $f(x) = \frac{1}{\sqrt{2x+85}}$

31. $f(x) = \frac{x+75}{\sqrt{53x^2+4}}$

32. $f(x) = \frac{x+28}{\sqrt{26x^2+15}}$

33. $f(x) = \sqrt[11]{x + 26}$

34. $f(x) = \sqrt[9]{x + 37}$

35. $f(x) = \frac{1}{\sqrt[7]{27x+48}}$

36. $f(x) = \frac{1}{\sqrt[3]{92x+59}}$

37. $f(x) = \frac{1}{\sqrt[8]{7x+70}}$

38. $f(x) = \frac{1}{\sqrt[10]{3x+81}}$

39. Use technology to find the range of the function K defined by $K(x) = \sqrt{4 - x} - 3$.

40. Use technology to find the range of the function f defined by $f(x) = \sqrt{2 - x} - 5$.

41. Use technology to find the range of the function f defined by $f(x) = \frac{0.1}{\sqrt{x-4}} + 3$.

42. Use technology to find the range of the function g defined by $g(x) = \frac{0.5}{\sqrt{x+1}} + 1$.

Applications If an object is dropped with no initial velocity, the time since the drop, in seconds, can be calculated by the function

$$T(h) = \sqrt{\frac{2h}{g}}$$

where h is the distance the object traveled in feet. The variable g is the gravitational acceleration on earth, and we can round it to $32\frac{ft}{s^2}$ for this problem.

43. a. After [ ] seconds since the release, the object would have traveled 24 feet.

  b. After 2.9 seconds since the release, the object would have traveled [ ] feet.

44. a. After [ ] seconds since the release, the object would have traveled 29 feet.

  b. After 2.4 seconds since the release, the object would have traveled [ ] feet.

A factory manufactures toy plastic balls. For a ball with a certain volume, V in cubic centimeters, the ball's radius can be calculated by the formula

$$r(V) = \sqrt[3]{\frac{3V}{4\pi}}$$

45. a. If a ball's volume is 350 cubic centimeters, its radius must be [ ] centimeters.

  b. If a ball's radius is 3.2 centimeters, its volume would be [ ] cubic centimeters.

46. a. If a ball's volume is 400 cubic centimeters, its radius must be [ ] centimeters.

  b. If a ball's radius is 2.7 centimeters, its volume would be [ ] cubic centimeters.

The speed of a tsunami, in meters per second, can be modeled by the function $S(d) = \sqrt{9.8d}$, where d is the depth of water in meters. Answer the following question with technology.

47. A tsunami's speed at 900 meters below the sea level is [ ] meters per second.

48. A tsunami's speed at [ ] meters below the sea level is 80 meters per second.

Distance Formula

49. Find the distance between the points $(-3, 3)$ and $(-24, 23)$.

50. Find the distance between the points $(14, -9)$ and $(19, 3)$.

51. Find the distance between the points $(8, -10)$ and $(11, -4)$.

52. Find the distance between the points $(-6, 7)$ and $(-1, 17)$.

14.2 Radical Expressions and Rational Exponents

Recall that in Remark 14.1.23, we learned to calculate the cube root of a number, say $\sqrt[3]{8}$, we can type 8^(1/3) into a calculator. This suggests that $\sqrt[3]{8} = 8^{1/3}$. In this section, we will learn why this is true, and how to simplify expressions with rational exponents.

Many learners will find a review of exponent rules to be helpful before continuing with the current section. Section 2.9 covers an introduction to exponent rules, and there is more in Section 6.1. The basic rules are summarized in List 6.1.15. These rules are still true and we can use them throughout this section whenever they might help.

14.2.1 Radical Expressions and Rational Exponents

Compare the following calculations:

$$\sqrt{9} \cdot \sqrt{9} = 3 \cdot 3 \qquad\qquad 9^{1/2} \cdot 9^{1/2} = 9^{1/2+1/2}$$
$$= 9 \qquad\qquad\qquad\qquad\quad = 9^1$$
$$\qquad\qquad\qquad\qquad\qquad\qquad\qquad\qquad = 9$$

If we rewrite the above calculations with exponents, we have:

$$\left(\sqrt{9}\right)^2 = 9 \qquad\qquad\qquad \left(9^{1/2}\right)^2 = 9$$

Since $\sqrt{9}$ and $9^{1/2}$ are both positive, and squaring either of them generates the same number, we conclude that

$$\sqrt{9} = 9^{1/2}$$

We can verify this result by entering 9^(1/2) into a calculator, and we get 3. In general for any non-negative real number a, we have:

$$\sqrt{a} = a^{1/2}$$

Similarly, when a is non-negative we can prove:

$$\sqrt[3]{a} = a^{1/3} \qquad\qquad \sqrt[4]{a} = a^{1/4} \qquad\qquad \sqrt[5]{a} = a^{1/5} \qquad \cdots$$

Let's summarize this information with a new exponent rule.

Fact 14.2.2 Radicals and Rational Exponents Rule. *If n is any natural number, and a is any non-negative real number or function with non-negative outputs, then*

$$a^{1/n} = \sqrt[n]{a}.$$

Additionally, if n is an odd natural number, then even when a is negative, we still have $a^{1/n} = \sqrt[n]{a}$.

Warning 14.2.3 Exponents on Negative Bases. Some computers and calculators follow different conventions when there is an exponent on a negative base. To see an example of this, visit *WolframAlpha* and try entering cuberoot(-8), and then try (-8)^(1/3), and you will get different results. cuberoot(-8) will come out as -2, but (-8)^(1/3) will come out as a certain non-real complex number. Most likely, the graphing technology you are using *does* behave as in Fact 14.2.2, but you should confirm this.

With this relationship, we can re-write radical expressions as expressions with rational exponents.

Example 14.2.4 Evaluate $\sqrt[4]{9}$ with a calculator. Round your answer to two decimal places.

Since $\sqrt[4]{9} = 9^{1/4}$, we press the following buttons on a calculator to get the value: 9^(1/4). So, we see that $\sqrt[4]{9} \approx 1.73$.

For many examples that follow, we will not need a calculator. We will, however, need to recognize the roots in Table 14.2.5.

Square Roots	Cube Roots	4th-Roots	5th-Roots	Roots of Powers of 2
$\sqrt{1} = 1$	$\sqrt[3]{1} = 1$	$\sqrt[4]{1} = 1$	$\sqrt[5]{1} = 1$	
$\sqrt{4} = 2$	$\sqrt[3]{8} = 2$	$\sqrt[4]{16} = 2$	$\sqrt[5]{32} = 2$	$\sqrt{4} = 2$
$\sqrt{9} = 3$	$\sqrt[3]{27} = 3$	$\sqrt[4]{81} = 3$		$\sqrt[3]{8} = 2$
$\sqrt{16} = 4$	$\sqrt[3]{64} = 4$			$\sqrt[4]{16} = 2$
$\sqrt{25} = 5$	$\sqrt[3]{125} = 5$			$\sqrt[5]{32} = 2$
$\sqrt{36} = 6$				$\sqrt[6]{64} = 2$
$\sqrt{49} = 7$				$\sqrt[7]{128} = 2$
$\sqrt{64} = 8$				$\sqrt[8]{256} = 2$
$\sqrt{81} = 9$				$\sqrt[9]{512} = 2$
$\sqrt{100} = 10$				$\sqrt[10]{1024} = 2$
$\sqrt{121} = 11$				
$\sqrt{144} = 12$				

Table 14.2.5: Small Roots of Appropriate Natural Numbers

Example 14.2.6 Convert the radical expression $\sqrt[3]{5}$ into an expression with a rational exponent and simplify it if possible.

$\sqrt[3]{5} = 5^{1/3}$. No simplification is possible since the cube root of 5 is not a perfect integer appearing in Table 14.2.5.

Example 14.2.7 Write the expressions in radical form using Fact 14.2.2 and simplify the results.

a. $4^{1/2}$ c. $-16^{1/4}$ e. $(-27)^{1/3}$ g. 12^0

b. $(-9)^{1/2}$ d. $64^{-1/3}$ f. $3^{1/2} \cdot 3^{1/2}$

Explanation.

a. $4^{1/2} = \sqrt{4}$
   $= 2$

b. $(-9)^{1/2} = \sqrt{-9}$ This value is non-real.

c. Without parentheses around -16, the negative sign in this problem should be left out of the radical.

   $-16^{1/4} = -\sqrt[4]{16}$
   $= -2$

d. $64^{-1/3} = \dfrac{1}{64^{1/3}}$

$\phantom{64^{-1/3}} = \dfrac{1}{\sqrt[3]{64}}$

$\phantom{64^{-1/3}} = \dfrac{1}{4}$

e. $(-27)^{1/3} = \sqrt[3]{-27}$

$\phantom{(-27)^{1/3}} = -3$

f. $3^{1/2} \cdot 3^{1/2} = \sqrt{3} \cdot \sqrt{3}$

$\phantom{3^{1/2} \cdot 3^{1/2}} = \sqrt{3 \cdot 3}$

$\phantom{3^{1/2} \cdot 3^{1/2}} = \sqrt{9}$

$\phantom{3^{1/2} \cdot 3^{1/2}} = 3$

g. $12^0 = 1$

Fact 14.2.2 applies to variables in expressions just as much as it does to numbers.

Remark 14.2.8. In general, it is easier to do algebra with rational exponents on variables than with radicals of variables. You should use Fact 14.2.2 to convert from rational exponents to radicals on variables *only as a last step* in simplifying.

Example 14.2.9 Write the expressions as simplified as they can be using radicals.

a. $2x^{-1/2}$
b. $(5x)^{1/3}$
c. $\left(-27x^{12}\right)^{1/3}$
d. $\left(\dfrac{16x}{81y^8}\right)^{1/4}$

Explanation.

a. Note that in this example the exponent is only applied to the x. Making this type of observation should be our first step for each of these exercises.

$$2x^{-1/2} = \dfrac{2}{x^{1/2}} \qquad \text{by the Negative Exponent Rule}$$

$$\phantom{2x^{-1/2}} = \dfrac{2}{\sqrt{x}} \qquad \text{by the Radicals and Rational Exponents Rule}$$

b. In this exercise, the exponent applies to both the 5 and x.

$$(5x)^{1/3} = \sqrt[3]{5x} \qquad \text{by the Radicals and Rational Exponents Rule}$$

We could choose to simplify our answer in a different way. Note that neither one is technically preferred over the other except that perhaps the first way is simpler.

$$(5x)^{1/3} = 5^{1/3}x^{1/3} \qquad \text{by the Product to a Power Rule}$$

$$\phantom{(5x)^{1/3}} = \sqrt[3]{5}\sqrt[3]{x} \qquad \text{by the Radicals and Rational Exponents Rule}$$

c. As in the previous exercise, we have a choice as to how to simplify this expression. Here we should note that we *do* know what the cube root of -27 is, so we will take the path to splitting up the expression, using the Product to a Power Rule, before applying the root.

$$
\begin{aligned}
\left(-27x^{12}\right)^{1/3} &= (-27)^{1/3} \cdot \left(x^{12}\right)^{1/3} && \text{by the Product to a Power Rule} \\
&= (-27)^{1/3} \cdot \left(x^{12 \cdot 1/3}\right) && \text{by the Power to a Power Rule} \\
&= \sqrt[3]{-27} \cdot x^4 && \text{by the Radicals and Rational Exponents Rule} \\
&= -3x^4
\end{aligned}
$$

d. We'll use the exponent rule for a fraction raised to a power.

$$
\begin{aligned}
\left(\frac{16x}{81y^8}\right)^{1/4} &= \frac{(16x)^{1/4}}{\left(81y^8\right)^{1/4}} && \text{by the Quotient to a Power Rule} \\[2mm]
&= \frac{16^{1/4} \cdot x^{1/4}}{81^{1/4} \cdot \left(y^8\right)^{1/4}} && \text{by the Product to a Power Rule} \\[2mm]
&= \frac{16^{1/4} \cdot x^{1/4}}{81^{1/4} \cdot y^2} \\[2mm]
&= \frac{\sqrt[4]{16} \cdot \sqrt[4]{x}}{\sqrt[4]{81} \cdot y^2} && \text{by the Radicals and Rational Exponents Rule} \\[2mm]
&= \frac{2\sqrt[4]{x}}{3y^2}
\end{aligned}
$$

Fact 14.2.2 describes what can be done when there is a fractional exponent and the numerator is a 1. The numerator doesn't have to be a 1 though and we need guidance for that situation.

Fact 14.2.10 Radicals and Rational Exponents Rule. *If m and n are natural numbers such that $\frac{m}{n}$ is a reduced fraction, and a is any non-negative real number or function that takes non-negative values, then*

$$
a^{m/n} = \sqrt[n]{a^m} = \left(\sqrt[n]{a}\right)^m.
$$

Additionally, if n is an odd natural number, then even when a is negative, we still have $a^{m/n} = \sqrt[n]{a^m} = \left(\sqrt[n]{a}\right)^m$.

Remark 14.2.11. By Fact 14.2.10, there are two ways to express $a^{m/n}$ as a radical, both

$$
a^{m/n} = \sqrt[n]{a^m} \qquad\qquad \text{and} \qquad\qquad a^{m/n} = \left(\sqrt[n]{a}\right)^m
$$

There are different times to use each formula. In general, use $a^{m/n} = \sqrt[n]{a^m}$ for variables and $a^{m/n} = \left(\sqrt[n]{a}\right)^m$ for numbers.

Example 14.2.12

a. Consider the expression $27^{4/3}$. Use both versions of Fact 14.2.10 to explain part of Remark 14.2.11.

b. Consider the expression $x^{4/3}$. Use both versions of Fact 14.2.10 to explain the other part of Remark 14.2.11.

Explanation.

a. The expression $27^{4/3}$ can be evaluated in the following two ways by Fact 14.2.10.

$$27^{4/3} = \sqrt[3]{27^4} \qquad \text{by the first part of the Radicals and Rational Exponents Rule}$$
$$= \sqrt[3]{531441}$$
$$= 81$$

or

$$27^{4/3} = \left(\sqrt[3]{27}\right)^4 \qquad \text{by the second part of the Radicals and Rational Exponents Rule}$$
$$= 3^4$$
$$= 81$$

The calculations using $a^{m/n} = \left(\sqrt[n]{a}\right)^m$ worked with smaller numbers and can be done without a calculator. This is why we made the general recommendation in Remark 14.2.11.

b. The expression $x^{4/3}$ can be evaluated in the following two ways by Fact 14.2.10.

$$x^{4/3} = \sqrt[3]{x^4} \qquad \text{by the first part of Radicals and Rational Exponents Rule}$$

or

$$x^{4/3} = \left(\sqrt[3]{x}\right)^4 \qquad \text{by the second part of the Radicals and Rational Exponents Rule}$$

In this case, the simplification using $a^{m/n} = \sqrt[n]{a^m}$ is just shorter looking and easier to write. This is why we made the general recommendation in Remark 14.2.11.

Example 14.2.13 Simplify the expressions using Fact 14.2.10.

a. $8^{2/3}$

b. $16^{-3/2}$

c. $-16^{3/4}$

d. $\left(-\frac{27}{64}\right)^{2/3}$

Explanation.

a.

$$8^{2/3} = \left(\sqrt[3]{8}\right)^2 \qquad \text{by the second part of the Radicals and Rational Exponents Rule}$$
$$= 2^2$$
$$= 4$$

b.

$$16^{-3/2} = \frac{1}{16^{3/2}} \qquad \text{by the Negative Exponent Rule}$$
$$= \frac{1}{\left(\sqrt{16}\right)^3} \qquad \text{by the second part of the Radicals and Rational Exponents Rule}$$
$$= \frac{1}{4^3}$$
$$= \frac{1}{64}$$

c.

$$-16^{3/4} = -\left(\sqrt[4]{16}\right)^3 \qquad \text{by the second part of the Radicals and Rational Exponents Rule}$$
$$= -2^3$$
$$= -8$$

d. In this problem the negative can be associated with either the numerator or the denominator, but not both. We choose the numerator.

$$\left(-\frac{27}{64}\right)^{2/3} = \left(\sqrt[3]{-\frac{27}{64}}\right)^2 \qquad \text{by the second part of the Radicals and Rational Exponents Rule}$$
$$= \left(\frac{\sqrt[3]{-27}}{\sqrt[3]{64}}\right)^2$$
$$= \left(\frac{-3}{4}\right)^2$$
$$= \frac{(-3)^2}{(4)^2}$$
$$= \frac{9}{16}$$

While we are looking at the algebra of $x^{m/n}$, we should briefly examine a graph to see what this type of function can look like. Fractional powers can make some fairly interesting graphs. We invite you to play with these graphs on your favorite graphing program.

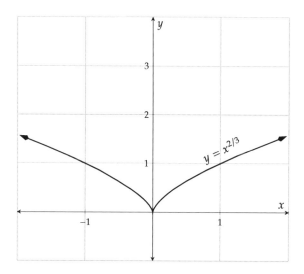

Figure 14.2.14: A Graph of $y = x^{2/3}$

14.2.2 More Expressions with Rational Exponents

To recap, here is a "complete" list of exponent rules.

Product Rule $a^n \cdot a^m = a^{n+m}$

Power to a Power Rule $(a^n)^m = a^{n \cdot m}$

Product to a Power Rule $(ab)^n = a^n \cdot b^n$

Quotient Rule $\dfrac{a^n}{a^m} = a^{n-m}$, as long as $a \neq 0$

Quotient to a Power Rule $\left(\dfrac{a}{b}\right)^n = \dfrac{a^n}{b^n}$, as long as $b \neq 0$

Zero Exponent Rule $a^0 = 1$ for $a \neq 0$

Negative Exponent Rule $a^{-n} = \dfrac{1}{a^n}$

Negative Exponent Reciprocal Rule $\dfrac{1}{a^{-n}} = a^n$

Negative Exponent on Fraction Rule $\left(\dfrac{x}{y}\right)^{-n} = \left(\dfrac{y}{x}\right)^n$

Radical and Rational Exponent Rule $x^{1/n} = \sqrt[n]{x}$

Radical and Rational Exponent Rule $x^{m/n} = \left(\sqrt[n]{x}\right)^m$, usually for numbers

Radical and Rational Exponent Rule $x^{m/n} = \sqrt[n]{x^m}$, usually for variables

List 14.2.15: Complete List of Exponent Rules

Example 14.2.16 Convert the following radical expressions into expressions with rational exponents, and simplify them if possible.

a. $\dfrac{1}{\sqrt{x}}$

b. $\dfrac{1}{\sqrt[3]{25}}$

Explanation.

a.

$$\frac{1}{\sqrt{x}} = \frac{1}{x^{1/2}}$$ by the Radicals and Rational Exponents Rule

$$= x^{-1/2}$$ by the Negative Exponent Rule

b.

$$\frac{1}{\sqrt[3]{25}} = \frac{1}{25^{1/3}}$$ by the Radicals and Rational Exponents Rule

$$= \frac{1}{\left(5^2\right)^{1/3}}$$

$$= \frac{1}{5^{2 \cdot 1/3}}$$ by the Power to a Power Rule

$$= \frac{1}{5^{2/3}}$$

$$= 5^{-2/3}$$ by the Negative Exponent Rule

Learners of these simplifications often find it challenging, so we now include a plethora of examples of varying difficulty.

Example 14.2.17 Use exponent properties in List 14.2.15 to simplify the expressions, and write all final versions using radicals.

a. $2w^{7/8}$ d. $\left(-8p^6\right)^{5/3}$ g. $\frac{\sqrt{z}}{\sqrt[3]{z}}$ i. $3\left(c^{1/2} + d^{1/2}\right)^2$

b. $\frac{1}{2}y^{-1/2}$ e. $\sqrt{x^3} \cdot \sqrt[4]{x}$ h. $\sqrt{\sqrt[4]{q}}$ j. $3\left(4k^{2/3}\right)^{-1/2}$

c. $(27b)^{2/3}$ f. $h^{1/3} + h^{1/3} + h^{1/3}$

Explanation.

a.

$$2w^{7/8} = 2\sqrt[8]{w^7}$$ by the Radicals and Rational Exponents Rule

b.

$$\frac{1}{2}y^{-1/2} = \frac{1}{2}\frac{1}{y^{1/2}}$$ by the Negative Exponent Rule

$$= \frac{1}{2}\frac{1}{\sqrt{y}}$$ by the Radicals and Rational Exponents Rule

$$= \frac{1}{2\sqrt{y}}$$

c.

$$(27b)^{2/3} = (27)^{2/3} \cdot (b)^{2/3} \qquad \text{by the Product to a Power Rule}$$

$$= \left(\sqrt[3]{27}\right)^2 \cdot \sqrt[3]{b^2} \qquad \text{by the Radicals and Rational Exponents Rule}$$

$$= 3^2 \cdot \sqrt[3]{b^2}$$

$$= 9\sqrt[3]{b^2}$$

d.

$$\left(-8p^6\right)^{5/3} = (-8)^{5/3} \cdot \left(p^6\right)^{5/3} \qquad \text{by the Product to a Power Rule}$$

$$= (-8)^{5/3} \cdot p^{6 \cdot 5/3} \qquad \text{by the Power to a Power Rule}$$

$$= \left(\sqrt[3]{-8}\right)^5 \cdot p^{10} \qquad \text{by the Radicals and Rational Exponents Rule}$$

$$= (-2)^5 \cdot p^{10}$$

$$= -32p^{10}$$

e.

$$\sqrt{x^3} \cdot \sqrt[4]{x} = x^{3/2} \cdot x^{1/4} \qquad \text{by the Radicals and Rational Exponents Rule}$$

$$= x^{3/2+1/4} \qquad \text{by the Product Rule}$$

$$= x^{6/4+1/4}$$

$$= x^{7/4}$$

$$= \sqrt[4]{x^7} \qquad \text{by the Radicals and Rational Exponents Rule}$$

f.

$$h^{1/3} + h^{1/3} + h^{1/3} = 3h^{1/3}$$

$$= 3\sqrt[3]{h} \qquad \text{by the Radicals and Rational Exponents Rule}$$

g.

$$\frac{\sqrt{z}}{\sqrt[3]{z}} = \frac{z^{1/2}}{z^{1/3}} \qquad \text{by the Radicals and Rational Exponents Rule}$$

$$= z^{1/2-1/3} \qquad \text{by the Quotient Rule}$$

$$= z^{3/6-2/6}$$

$$= z^{1/6}$$

$$= \sqrt[6]{z} \qquad \text{by the Radicals and Rational Exponents Rule}$$

h.

$$\sqrt{\sqrt[4]{q}} = \sqrt{q^{1/4}} \qquad \text{by the Radicals and Rational Exponents Rule}$$

$$= \left(q^{1/4}\right)^{1/2} \qquad \text{by the Radicals and Rational Exponents Rule}$$

$$= q^{1/4 \cdot 1/2} \qquad \text{by the Power to a Power Rule}$$
$$= q^{1/8}$$
$$= \sqrt[8]{q} \qquad \text{by the Radicals and Rational Exponents Rule}$$

i.

$$3\left(c^{1/2} + d^{1/2}\right)^2 = 3\left(c^{1/2} + d^{1/2}\right)\left(c^{1/2} + d^{1/2}\right)$$

$$= 3\left(\left(c^{1/2}\right)^2 + 2c^{1/2} \cdot d^{1/2} + \left(d^{1/2}\right)^2\right)$$

$$= 3\left(c^{1/2 \cdot 2} + 2c^{1/2} \cdot d^{1/2} + d^{1/2 \cdot 2}\right)$$

$$= 3\left(c + 2c^{1/2} \cdot d^{1/2} + d\right)$$

$$= 3\left(c + 2(cd)^{1/2} + d\right) \qquad \text{by the Product to a Power Rule}$$

$$= 3\left(c + 2\sqrt{cd} + d\right) \qquad \text{by the Radicals and Rational Exponents Rule}$$

$$= 3c + 6\sqrt{cd} + 3d$$

j.

$$3\left(4k^{2/3}\right)^{-1/2} = \frac{3}{\left(4k^{2/3}\right)^{1/2}} \qquad \text{by the Negative Exponent Rule}$$

$$= \frac{3}{4^{1/2}\left(k^{2/3}\right)^{1/2}} \qquad \text{by the Product to a Power Rule}$$

$$= \frac{3}{4^{1/2}k^{2/3 \cdot 1/2}} \qquad \text{by the Power to a Power Rule}$$

$$= \frac{3}{4^{1/2}k^{1/3}}$$

$$= \frac{3}{\sqrt{4}\sqrt[3]{k}} \qquad \text{by the Radicals and Rational Exponents Rule}$$

$$= \frac{3}{2\sqrt[3]{k}}$$

We will end a with a short application on rational exponents. Kepler's Laws of Orbital Motion[1] describe how planets orbit stars and how satellites orbit planets. In particular, his third law has a rational exponent, which we will now explore.

Example 14.2.18 Kepler and the Satellite. Kepler's third law of motion says that for objects with a roughly circular orbit that the time (in hours) that it takes to make one full revolution around the planet, T, is proportional to three-halves power of the distance (in kilometers) from the center of the planet to the satellite, r. For the Earth, it looks like this:

$$T = \frac{2\pi}{\sqrt{G \cdot M_E}}r^{3/2}$$

---

[1]en.wikipedia.org/wiki/Kepler%27s_laws_of_planetary_motion

In this case, both G and M_E are constants. G stands for the universal gravitational constant[a] where G is about 8.65×10^{-13} $\frac{\text{km}^3}{\text{kg}\,\text{h}^2}$ and M_E stands for the mass of the Earth[b] where M_E is about 5.972×10^{24} kg. Inputting these values into this formula yields a simplified version that looks like this:

$$T \approx 2.76 \times 10^{-6} r^{3/2}$$

Most satellites orbit in what is called low Earth orbit[c], including the international space station which orbits at about 340 km above from Earth's surface. The Earth's average radius is about 6380 km. Find the period of the international space station.

Explanation. The formula has already been identified, but the input takes just a little thought. The formula uses r as the distance from the center of the Earth to the satellite, so to find r we need to combine the radius of the Earth and the distance to the satellite above the surface of the Earth.

$$r = 340 + 6380$$
$$= 6720$$

Now we can input this value into the formula and evaluate.

$$T \approx 2.76 \cdot 10^{-6} r^{3/2}$$
$$\approx 2.76 \cdot 10^{-6} (6720)^{3/2}$$
$$\approx 2.76 \cdot 10^{-6} \left(\sqrt{6720} \right)^3$$
$$\approx 1.52$$

The formula tells us that it takes a little more than an hour and a half for the ISS to orbit the Earth! That works out to 15 or 16 sunrises per day.

---

[a]en.wikipedia.org/wiki/Gravitational_constant
[b]en.wikipedia.org/wiki/Earth_mass
[c]en.wikipedia.org/wiki/Low_Earth_orbit

Exercises

Review and Warmup

1. Evaluate the following.

　　a. $\sqrt{36} =$ [　　　　]

　　b. $\sqrt{1} =$ [　　　　]

　　c. $\sqrt{4} =$ [　　　　]

2. Evaluate the following.

　　a. $\sqrt{49} =$ [　　　　]

　　b. $\sqrt{81} =$ [　　　　]

　　c. $\sqrt{36} =$ [　　　　]

3. Evaluate the following.

　　a. $\sqrt{\dfrac{81}{16}} =$ [　　　　]

　　b. $\sqrt{-\dfrac{144}{121}} =$ [　　　　]

4. Evaluate the following.

a. $\sqrt{\dfrac{100}{49}} = \boxed{}$

b. $\sqrt{-\dfrac{16}{9}} = \boxed{}$

5. Evaluate the following.

Do not use a calculator.

a. $\sqrt{144} = \boxed{}$

b. $\sqrt{1.44} = \boxed{}$

c. $\sqrt{14400} = \boxed{}$

6. Evaluate the following.

Do not use a calculator.

a. $\sqrt{4} = \boxed{}$

b. $\sqrt{0.04} = \boxed{}$

c. $\sqrt{400} = \boxed{}$

7. Use the properties of exponents to simplify the expression.

$r^4 \cdot r^{19}$

8. Use the properties of exponents to simplify the expression.

$x^6 \cdot x^{12}$

9. Use the properties of exponents to simplify the expression.

$\left(y^5\right)^4$

10. Use the properties of exponents to simplify the expression.

$\left(t^7\right)^{11}$

11. Use the properties of exponents to simplify the expression.

$\left(\dfrac{5x^7}{8}\right)^3 = \boxed{}$

12. Use the properties of exponents to simplify the expression.

$\left(\dfrac{7x^8}{6}\right)^2 = \boxed{}$

13. Use the properties of exponents to simplify the expression.

$\left(-10x^{10}\right)^3$

14. Use the properties of exponents to simplify the expression.

$\left(-6t^{11}\right)^2$

15. Use the properties of exponents to simplify the expression.

$\dfrac{t^3}{t} = \boxed{}$

16. Use the properties of exponents to simplify the expression.

$\dfrac{y^5}{y^4} = \boxed{}$

17. Rewrite the expression simplified and using only positive exponents.

$x^{-10} \cdot x^8 = \boxed{}$

18. Rewrite the expression simplified and using only positive exponents.

$y^{-4} \cdot y^2 = \boxed{}$

19. Rewrite the expression simplified and using only positive exponents.

$\left(-6y^{-16}\right) \cdot \left(2y^{10}\right) = \boxed{}$

20. Rewrite the expression simplified and using only positive exponents.

$\left(-3r^{-10}\right) \cdot \left(6r^2\right) = \boxed{}$

Calculations Without using a calculator, evaluate the expression.

21.
   a. $64^{\frac{1}{2}} = \boxed{}$

   b. $(-64)^{\frac{1}{2}} = \boxed{}$

   c. $-64^{\frac{1}{2}} = \boxed{}$

22.
   a. $81^{\frac{1}{2}} = \boxed{}$

   b. $(-81)^{\frac{1}{2}} = \boxed{}$

   c. $-81^{\frac{1}{2}} = \boxed{}$

23.
   a. $125^{\frac{1}{3}} = \boxed{}$

   b. $(-125)^{\frac{1}{3}} = \boxed{}$

   c. $-125^{\frac{1}{3}} = \boxed{}$

24.
   a. $8^{\frac{1}{3}} = \boxed{}$

   b. $(-8)^{\frac{1}{3}} = \boxed{}$

   c. $-8^{\frac{1}{3}} = \boxed{}$

25. $8^{-\frac{2}{3}} = \boxed{}$

26. $32^{-\frac{2}{5}} = \boxed{}$

27. $\left(\dfrac{1}{27}\right)^{-\frac{2}{3}} = \boxed{}$

28. $\left(\dfrac{1}{27}\right)^{-\frac{2}{3}} = \boxed{}$

29. $\sqrt[4]{16^3} = \boxed{}$

30. $\sqrt[4]{81^3} = \boxed{}$

31. $\sqrt[5]{1024} = \boxed{}$

32. $\sqrt[3]{64} = \boxed{}$

33.
   a. $\sqrt[3]{1} = \boxed{}$

   b. $\sqrt[3]{-1} = \boxed{}$

   c. $-\sqrt[3]{1} = \boxed{}$

34.
   a. $\sqrt[3]{8} = \boxed{}$

   b. $\sqrt[3]{-8} = \boxed{}$

   c. $-\sqrt[3]{8} = \boxed{}$

35.
   a. $\sqrt[4]{1} = \boxed{}$

   b. $\sqrt[4]{-1} = \boxed{}$

   c. $-\sqrt[4]{1} = \boxed{}$

36.
   a. $\sqrt[4]{16} = \boxed{}$

   b. $\sqrt[4]{-16} = \boxed{}$

   c. $-\sqrt[4]{16} = \boxed{}$

37. $\sqrt[4]{10^3} = \boxed{}$

38. $\sqrt[3]{8^2} = \boxed{}$

39. $\sqrt[3]{-\dfrac{27}{64}} = \boxed{}.$

40. $\sqrt[3]{-\dfrac{27}{125}} = \boxed{}.$

Convert Radicals to Fractional Exponents Use rational exponents to write the expression.

41. $\sqrt{b} = \boxed{}$

42. $\sqrt[8]{x} = \boxed{}$

43. $\sqrt[5]{5y + 7} = \boxed{}$

44. $\sqrt[4]{2z + 1} = \boxed{}$

45. $\sqrt[7]{t} = \boxed{}$

46. $\sqrt[4]{r} = \boxed{}$

47. $\dfrac{1}{\sqrt{m^3}} = $ [_____]   **48.** $\dfrac{1}{\sqrt[7]{n^4}} = $ [_____]

Convert Fractional Exponents to Radicals   Convert the expression to radical notation.

49. $a^{\frac{3}{4}} = $ [_____]   **50.** $b^{\frac{5}{6}} = $ [_____]   **51.** $x^{\frac{5}{6}} = $ [_____]

52. $r^{\frac{2}{3}} = $ [_____]   **53.** $16^{\frac{1}{6}}z^{\frac{5}{6}} = $ [_____]   **54.** $5^{\frac{1}{5}}t^{\frac{4}{5}} = $ [_____]

55. Convert $r^{\frac{3}{7}}$ to a radical.

  ⊙ $\sqrt[3]{r^7}$

  ⊙ $\sqrt[7]{r^3}$

56. Convert $m^{\frac{5}{6}}$ to a radical.

  ⊙ $\sqrt[6]{m^5}$

  ⊙ $\sqrt[5]{m^6}$

57. Convert $n^{-\frac{5}{8}}$ to a radical.

  ⊙ $\dfrac{1}{\sqrt[5]{n^8}}$

  ⊙ $-\sqrt[8]{n^5}$

  ⊙ $-\sqrt[5]{n^8}$

  ⊙ $\dfrac{1}{\sqrt[8]{n^5}}$

58. Convert $a^{-\frac{4}{7}}$ to a radical.

  ⊙ $-\sqrt[4]{a^7}$

  ⊙ $-\sqrt[7]{a^4}$

  ⊙ $\dfrac{1}{\sqrt[4]{a^7}}$

  ⊙ $\dfrac{1}{\sqrt[7]{a^4}}$

59. Convert $5^{\frac{3}{7}}b^{\frac{5}{7}}$ to a radical.

  ⊙ $5^3 \cdot \sqrt[7]{b^5}$

  ⊙ $\sqrt[3]{5^7} \cdot \sqrt[5]{b^7}$

  ⊙ $\sqrt[7]{5^3 b^5}$

  ⊙ $5^7 \cdot \sqrt[5]{b^7}$

60. Convert $2^{\frac{4}{5}}x^{\frac{2}{5}}$ to a radical.

  ⊙ $\sqrt[5]{2^4 x^2}$

  ⊙ $\sqrt[4]{2^5} \cdot \sqrt{x^5}$

  ⊙ $2^4 \cdot \sqrt[5]{x^2}$

  ⊙ $2^5 \cdot \sqrt{x^5}$

Simplifying Expressions with Rational Exponents   Simplify the expression, answering with rational exponents and not radicals.

61. $\sqrt[3]{y}\,\sqrt[3]{y} = $ [_____]   **62.** $\sqrt[9]{z}\,\sqrt[9]{z} = $ [_____]   **63.** $\sqrt[5]{32t^3} = $ [_____]

64. $\sqrt[3]{125r} = $ [_____]   **65.** $\dfrac{\sqrt[3]{8m}}{\sqrt[6]{m^5}} = $ [_____]   **66.** $\dfrac{\sqrt{16n}}{\sqrt[6]{n^5}} = $ [_____]

67. $\dfrac{\sqrt[5]{32a^4}}{\sqrt[10]{a^3}} = $ [_____]   **68.** $\dfrac{\sqrt[3]{8b^2}}{\sqrt[6]{b}} = $ [_____]   **69.** $\sqrt{c} \cdot \sqrt[6]{c^5} = $ [_____]

70. $\sqrt[5]{y} \cdot \sqrt[10]{y^3} =$ [   ]   **71.** $\sqrt[3]{\sqrt[4]{z}} =$ [   ]   **72.** $\sqrt[4]{\sqrt[3]{t}} =$ [   ]

73. $\sqrt{c}\,\sqrt[6]{c} =$ [   ]   **74.** $\sqrt{n}\,\sqrt[7]{n} =$ [   ]

14.3 More on Rationalizing the Denominator

In Section 8.2, we learned how to rationalize the denominator in simple expressions like $\frac{1}{\sqrt{2}}$. We will briefly review this topic and then extend the concept to the next level.

14.3.1 A Review of Rationalizing the Denominator

To remove radicals from the denominator of $\frac{1}{\sqrt{2}}$, we multiply the numerator and denominator by $\sqrt{2}$:

$$\frac{1}{\sqrt{2}} = \frac{1}{\sqrt{2}} \cdot \frac{\sqrt{2}}{\sqrt{2}}$$
$$= \frac{\sqrt{2}}{2}$$

We used the property:
$$\sqrt{x} \cdot \sqrt{x} = x, \text{ where } x \text{ is positive}$$

Example 14.3.2 Rationalize the denominator of the expressions.

a. $\frac{3}{\sqrt{6}}$

b. $\frac{\sqrt{5}}{\sqrt{72}}$

Explanation.

a. To rationalize the denominator of $\frac{3}{\sqrt{6}}$, we take the expression and multiply by a special version of 1 to make the radical in the denominator cancel.

$$\frac{3}{\sqrt{6}} = \frac{3}{\sqrt{6}} \cdot \frac{\sqrt{6}}{\sqrt{6}}$$
$$= \frac{3\sqrt{6}}{6}$$
$$= \frac{\sqrt{6}}{2}$$

b. Rationalizing the denominator of $\frac{\sqrt{5}}{\sqrt{72}}$ is slightly trickier. We could go the brute force method and multiply both the numerator and denominator by $\sqrt{72}$, and it would be effective; however, we should note that the $\sqrt{72}$ in the denominator can be *reduced* first. This will simplify future algebra.

$$\frac{\sqrt{5}}{\sqrt{72}} = \frac{\sqrt{5}}{\sqrt{36 \cdot 2}}$$
$$= \frac{\sqrt{5}}{\sqrt{36} \cdot \sqrt{2}}$$
$$= \frac{\sqrt{5}}{6 \cdot \sqrt{2}}$$

Now all that remains is to multiply the numerator and denominator by $\sqrt{2}$.

$$= \frac{\sqrt{5}}{6 \cdot \sqrt{2}} \cdot \frac{\sqrt{2}}{\sqrt{2}}$$

$$= \frac{\sqrt{10}}{6 \cdot 2}$$

$$= \frac{\sqrt{10}}{12}$$

14.3.2 Rationalize Denominator with Difference of Squares Formula

How can be remove the radical from the denominator of $\frac{1}{\sqrt{2}+1}$? Let's try multiplying the numerator and denominator by $\sqrt{2}$:

$$\frac{1}{\sqrt{2}+1} = \frac{1}{\left(\sqrt{2}+1\right)} \cdot \frac{\sqrt{2}}{\sqrt{2}}$$

$$= \frac{\sqrt{2}}{\sqrt{2} \cdot \sqrt{2} + 1 \cdot \sqrt{2}}$$

$$= \frac{\sqrt{2}}{2 + \sqrt{2}}$$

We removed one radical from the denominator, but created another. We need to find another method. The difference of squares formula will help:

$$(a + b)(a - b) = a^2 - b^2$$

Those two squares in $a^2 - b^2$ can remove square roots. To remove the radical from the denominator of $\frac{1}{\sqrt{2}+1}$, we multiply the numerator and denominator by $\sqrt{2} - 1$:

$$\frac{1}{\sqrt{2}+1} = \frac{1}{\left(\sqrt{2}+1\right)} \cdot \frac{\left(\sqrt{2}-1\right)}{\left(\sqrt{2}-1\right)}$$

$$= \frac{\sqrt{2}-1}{\left(\sqrt{2}\right)^2 - (1)^2}$$

$$= \frac{\sqrt{2}-1}{2-1}$$

$$= \frac{\sqrt{2}-1}{1}$$

$$= \sqrt{2}-1$$

Let's look at a few more examples.

Example 14.3.3 Rationalize the denominator in $\frac{\sqrt{7}-\sqrt{2}}{\sqrt{5}+\sqrt{3}}$.

Explanation. To remove radicals in $\sqrt{5}+\sqrt{3}$ with the difference of squares formula, we multiply it with

$\sqrt{5} - \sqrt{3}$.

$$\frac{\sqrt{7} - \sqrt{2}}{\sqrt{5} + \sqrt{3}} = \frac{\sqrt{7} - \sqrt{2}}{\sqrt{5} + \sqrt{3}} \cdot \frac{\left(\sqrt{5} - \sqrt{3}\right)}{\left(\sqrt{5} - \sqrt{3}\right)}$$

$$= \frac{\sqrt{7} \cdot \sqrt{5} - \sqrt{7} \cdot \sqrt{3} - \sqrt{2} \cdot \sqrt{5} - \sqrt{2} \cdot -\sqrt{3}}{\left(\sqrt{5}\right)^2 - \left(\sqrt{3}\right)^2}$$

$$= \frac{\sqrt{35} - \sqrt{21} - \sqrt{10} + \sqrt{6}}{5 - 3}$$

$$= \frac{\sqrt{35} - \sqrt{21} - \sqrt{10} + \sqrt{6}}{2}$$

Example 14.3.4 Rationalize the denominator in $\frac{\sqrt{3}}{3 - 2\sqrt{3}}$.

Explanation. To remove the radical in $3 - 2\sqrt{3}$ with the difference of squares formula, we multiply it with $3 + 2\sqrt{3}$.

$$\frac{\sqrt{3}}{3 - 2\sqrt{3}} = \frac{\sqrt{3}}{(3 - 2\sqrt{3})} \cdot \frac{(3 + 2\sqrt{3})}{(3 + 2\sqrt{3})}$$

$$= \frac{3 \cdot \sqrt{3} + 2\sqrt{3} \cdot \sqrt{3}}{(3)^2 - \left(2\sqrt{3}\right)^2}$$

$$= \frac{3\sqrt{3} + 2 \cdot 3}{9 - 2^2 \left(\sqrt{3}\right)^2}$$

$$= \frac{3\sqrt{3} + 6}{9 - 4(3)}$$

$$= \frac{3\left(\sqrt{3} + 2\right)}{9 - 12}$$

$$= \frac{3\left(\sqrt{3} + 2\right)}{-3}$$

$$= \frac{\sqrt{3} + 2}{-1}$$

$$= -\sqrt{3} - 2$$

Exercises

Review and Warmup Rationalize the denominator and simplify the expression.

1. $\dfrac{1}{\sqrt{6}} = \boxed{}$ **2.** $\dfrac{1}{\sqrt{7}} = \boxed{}$ **3.** $\dfrac{30}{\sqrt{10}} = \boxed{}$ **4.** $\dfrac{20}{\sqrt{10}} = \boxed{}$

5. $\dfrac{1}{\sqrt{28}} =$ ⬚   **6.** $\dfrac{1}{\sqrt{45}} =$ ⬚   **7.** $\dfrac{8}{\sqrt{180}} =$ ⬚   **8.** $\dfrac{9}{\sqrt{72}} =$ ⬚

Further Rationalizing a Denominator   Rationalize the denominator and simplify the expression.

9. $\dfrac{3}{\sqrt{m}} =$ ⬚   **10.** $\dfrac{1}{\sqrt{n}} =$ ⬚   **11.** $\sqrt{\dfrac{13}{14}} =$ ⬚

12. $\sqrt{\dfrac{14}{15}} =$ ⬚   **13.** $\sqrt{\dfrac{11}{72}} =$ ⬚   **14.** $\sqrt{\dfrac{5}{24}} =$ ⬚

Rationalizing the Denominator Using the Difference of Squares Formula   Rationalize the denominator and simplify the expression.

15. $\dfrac{3}{\sqrt{11}+8} =$ ⬚   **16.** $\dfrac{4}{\sqrt{2}+3} =$ ⬚   **17.** $\dfrac{5}{\sqrt{17}+6} =$ ⬚   **18.** $\dfrac{3}{\sqrt{6}+5} =$ ⬚

19. $\dfrac{\sqrt{2}-12}{\sqrt{11}+10} =$ ⬚   **20.** $\dfrac{\sqrt{5}-13}{\sqrt{13}+8} =$ ⬚   **21.** $\dfrac{\sqrt{2}-14}{\sqrt{7}+5} =$ ⬚   **22.** $\dfrac{\sqrt{5}-15}{\sqrt{13}+3} =$ ⬚

14.4 Solving Radical Equations

In this section, we will learn how to solve equations involving radicals.

14.4.1 Solving Radical Equations

One common application of radicals is the Pythagorean Theorem. We already saw some examples in earlier sections. We will look at some other applications of radicals in this section.

The formula $T = 2\pi\sqrt{\frac{L}{g}}$ is used to calculate the period of a pendulum and is attributed to the scientist Christiaan Huygens[1]. In the formula, T stands for the pendulum's period (how long one back-and-forth oscillation takes) in seconds, L stands for the pendulum's length in meters, and g is approximately $9.8\,\frac{m}{s^2}$ which is the gravitational acceleration constant on Earth.

An engineer is designing a pendulum. Its period must be 10 seconds. How long should the pendulum's length be?

We will substitute 10 into the formula for T and also the value of g, and then solve for L:

$$10 = 2\pi\sqrt{\frac{L}{9.8}}$$

$$\frac{1}{2\pi} \cdot 10 = \frac{1}{2\pi} \cdot 2\pi\sqrt{\frac{L}{9.8}}$$

$$\frac{5}{\pi} = \sqrt{\frac{L}{9.8}}$$

$$\left(\frac{5}{\pi}\right)^2 = \left(\sqrt{\frac{L}{9.8}}\right)^2 \qquad \text{canceling square root by squaring both sides}$$

$$\frac{25}{\pi^2} = \frac{L}{9.8}$$

$$9.8 \cdot \frac{25}{\pi^2} = 9.8 \cdot \frac{L}{9.8}$$

$$24.82 \approx L$$

To build a pendulum with a period of 10 seconds, its length should be approximately 24.82 meters.

Remark 14.4.2. The basic strategy to solve radical equations is to isolate the radical on one side of the equation and then square both sides to cancel the radical.

Remark 14.4.3. Squaring both sides of an equation is "dangerous," as it could create **extraneous solutions**, which will not make the equation true. For example, if we square both sides of $1 = -1$, we have:

$$1 = -1 \qquad \text{false}$$

$$(1)^2 = (-1)^2 \qquad \text{square both sides} \ldots$$

$$1 = 1 \qquad \text{true}$$

By squaring both sides of an equation, we turned a false equation into a true one. This is why we *must check solutions* when we square both sides of an equation.

---

[1]en.wikipedia.org/wiki/Christiaan_Huygens#Pendulums

Example 14.4.4 Solve the equation $1 + \sqrt{y - 1} = 4$ for y.

Explanation. We will isolate the radical first, and then square both sides.

$$1 + \sqrt{y - 1} = 4$$
$$\sqrt{y - 1} = 3$$
$$\left(\sqrt{y - 1}\right)^2 = 3^2$$
$$y - 1 = 9$$
$$y = 10$$

Because we squared both sides of an equation, we must check the solution.

$$1 + \sqrt{10 - 1} \overset{?}{=} 4$$
$$1 + \sqrt{9} \overset{?}{=} 4$$
$$1 + 3 \overset{\checkmark}{=} 4$$

So, 10 is the solution to the equation $1 + \sqrt{y - 1} = 4$.

Example 14.4.5 Solve the equation $5 + \sqrt{q} = 3$ for q.

Explanation. First, isolate the radical and square both sides.

$$5 + \sqrt{q} = 3$$
$$\sqrt{q} = -2$$
$$\left(\sqrt{q}\right)^2 = (-2)^2$$
$$q = 4$$

Because we squared both sides of an equation, we must check the solution.

$$5 + \sqrt{4} \overset{?}{=} 3$$
$$5 + 2 \overset{?}{=} 3$$
$$7 \overset{no}{=} 3$$

Thus, the potential solution -2 is actually extraneous and we have no real solutions to the equation $5 + \sqrt{q} = 3$. The solution set is the empty set, $\emptyset$.

Remark 14.4.6. In the previous example, it would be legitimate to observe that there are no solutions at earlier stages. From the very beginning, how could 5 plus a positive quantity result in 3? Or at the second step, since square roots are non-negative, how could a square root equal -2?

You do not have to be able to make these observations. If you follow the general steps for solving radical equations *and* you remember to check the possible solutions you find, then that will be enough.

Example 14.4.7 Solve for z in $\sqrt{z} + 2 = z$.

Explanation. We will isolate the radical first, and then square both sides.

$$\sqrt{z} + 2 = z$$
$$\sqrt{z} = z - 2$$
$$\left(\sqrt{z}\right)^2 = (z-2)^2$$
$$z = z^2 - 4z + 4$$
$$0 = z^2 - 5z + 4$$
$$0 = (z-1)(z-4)$$

$$z - 1 = 0 \qquad \text{or} \qquad z - 4 = 0$$
$$z = 1 \qquad \text{or} \qquad z = 4$$

Because we squared both sides of an equation, we must check both solutions.

$$\sqrt{1} + 2 \overset{?}{=} 1 \qquad\qquad \sqrt{4} + 2 \overset{?}{=} 4$$
$$1 + 2 \overset{\text{no}}{=} 1 \qquad\qquad 2 + 2 \overset{\checkmark}{=} 4$$

It turned out that 1 is an extraneous solution, but 4 is a valid solution. So the equation has one solution: 4. The solution set is $\{4\}$.

Sometimes, we need to square both sides of an equation *twice* before finding the solutions, like in the next example.

Example 14.4.8 Solve the equation $\sqrt{p-5} = 5 - \sqrt{p}$ for p.

Explanation. We cannot isolate two radicals, so we will simply square both sides, and later try to isolate the remaining radical.

$$\sqrt{p-5} = 5 - \sqrt{p}$$
$$\left(\sqrt{p-5}\right)^2 = \left(5 - \sqrt{p}\right)^2$$
$$p - 5 = 25 - 10\sqrt{p} + p \qquad \text{after expanding the binomial squared}$$
$$-5 = 25 - 10\sqrt{p}$$
$$-30 = -10\sqrt{p}$$
$$3 = \sqrt{p}$$
$$3^2 = \left(\sqrt{p}\right)^2$$
$$9 = p$$

Because we squared both sides of an equation, we must check the solution.

$$\sqrt{9-5} \overset{?}{=} 5 - \sqrt{9}$$
$$\sqrt{4} \overset{?}{=} 5 - 3$$
$$2 \overset{\checkmark}{=} 2$$

So 9 is the solution. The solution set is {9}.

Example 14.4.9 Solve the equation $\sqrt{2n-6} = 1 + \sqrt{n-2}$ for n.

Explanation.   We cannot isolate two radicals, so we will simply square both sides, and later try to isolate the remaining radical.

$$\sqrt{2n-6} = 1 + \sqrt{n-2}$$
$$\left(\sqrt{2n-6}\right)^2 = \left(1 + \sqrt{n-2}\right)^2$$
$$2n - 6 = 1^2 + 2\sqrt{n-2} + \left(\sqrt{n-2}\right)^2$$
$$2n - 6 = 1 + 2\sqrt{n-2} + n - 2$$
$$2n - 6 = 2\sqrt{n-2} + n - 1$$
$$n - 5 = 2\sqrt{n-2}$$

Note here that we can leave the factor of 2 next to the radical. We will square the 2 also.

$$(n-5)^2 = \left(2\sqrt{n-2}\right)^2$$
$$n^2 - 10n + 25 = 4(n-2)$$
$$n^2 - 10n + 25 = 4n - 8$$
$$n^2 - 14n + 33 = 0$$
$$(n-11)(n-3) = 0$$

$$n - 11 = 0 \qquad \text{or} \qquad n - 3 = 0$$
$$n = 11 \qquad \text{or} \qquad n = 3$$

So our two potential solutions are 11 and 3. We should now verify that they truly are solutions.

$$\sqrt{2(11)-6} \overset{?}{=} 1 + \sqrt{11-2} \qquad\qquad \sqrt{2(3)-6} \overset{?}{=} 1 + \sqrt{3-2}$$
$$\sqrt{22-6} \overset{?}{=} 1 + \sqrt{9} \qquad\qquad \sqrt{6-6} \overset{?}{=} 1 + \sqrt{1}$$
$$\sqrt{16} \overset{?}{=} 1 + 3 \qquad\qquad \sqrt{0} \overset{?}{=} 1 + 1$$
$$4 \overset{\checkmark}{=} 4 \qquad\qquad 0 \overset{\text{no}}{=} 2$$

So, 11 is the only solution. The solution set is {11}.

Let's look at an example of solving an equation with a cube root. There is very little difference between solving a cube-root equation and solving a square-root equation. Instead of *squaring* both sides, you *cube* both sides.

Example 14.4.10 Solve for q in $\sqrt[3]{2-q} + 2 = 5$.

Explanation.

$$\sqrt[3]{2-q} + 2 = 5$$
$$\sqrt[3]{2-q} = 3$$
$$\left(\sqrt[3]{2-q}\right)^3 = 3^3$$
$$2 - q = 27$$
$$-q = 25$$
$$q = -25$$

Unlike squaring both sides of an equation, raising both sides of an equation to the 3rd power will not create extraneous solutions. It's still good practice to check solution, though. This part is left as exercise.

14.4.2 Solving a Radical Equation with a Variable

We also need to be able to solve radical equations with other variables, like in the next example. The strategy is the same: isolate the radical, and then raise both sides to a certain power to cancel the radical.

Example 14.4.11 The study of black holes has resulted in some interesting science. One fundamental concept about black holes is that there is a distance close enough to the black hole that not even light can escape, called the Schwarzschild radius[a] or the event horizon radius. To find the Schwarzschild radius, R_s, we set the formula for the escape velocity equal to the speed of light, c, and we get $c = \sqrt{\frac{2GM}{R_s}}$ which we need to solve for R_s. Note that G is a constant, and M is the mass of the black hole.

Explanation. We will start by taking the equation $c = \sqrt{\frac{2GM}{R_s}}$ and applying our standard radical-equation-solving techniques. Isolate the radical and square both sides:

$$c = \sqrt{\frac{2GM}{R_s}}$$
$$c^2 = \left(\sqrt{\frac{2GM}{R_s}}\right)^2$$
$$c^2 = \frac{2GM}{R_s}$$
$$R_s \cdot c^2 = R_s \cdot \frac{2GM}{R_s}$$
$$R_s c^2 = 2GM$$
$$\frac{R_s c^2}{c^2} = \frac{2GM}{c^2}$$
$$R_s = \frac{2GM}{c^2}$$

So, the Schwarzschild radius can be found using the formula $R_s = \frac{2GM}{c^2}$.

---

[a]en.wikipedia.org/wiki/Schwarzschild_radius

Example 14.4.12 The term redshift[a] refers to the Doppler effect[b] for light. When an object (like a star) is moving away from Earth at very fast speeds, the wavelength of the light emitted by the star is increased due to the distance between the planets increasing (and the constant speed of light). Increased wavelength makes light "redder." The opposite phenomenon is called blueshift[c]. It turns out that the formula to calculate the redshift for a star moving away from the Earth uses square roots:

$$f_r = f_s \cdot \sqrt{\frac{c - v}{c + v}}$$

where c stands for the constant speed of light in a vacuum, f_r represents the frequency of the light that the receiver on Earth sees, f_s represents the frequency of light that the source star emits, and v is the velocity that the star moving away from Earth. Solve this equation for v.

Explanation. We will take the original equation $f_r = f_s \cdot \sqrt{\frac{c-v}{c+v}}$ and follow the steps to solving a radical equation. We could isolate the radical and then square both sides, but in this case isolating the radical is not necessary. If we begin by squaring both sides, that too will eliminate the radical.

$$f_r = f_s \cdot \sqrt{\frac{c - v}{c + v}}$$

$$(f_r)^2 = \left(f_s \cdot \sqrt{\frac{c - v}{c + v}}\right)^2$$

$$f_r^2 = f_s^2 \cdot \frac{c - v}{c + v}$$

$$f_r^2 \cdot (c + v) = f_s^2 \cdot \frac{c - v}{c + v} \cdot (c + v)$$

$$f_r^2(c + v) = f_s^2(c - v)$$

$$f_r^2 c + f_r^2 v = f_s^2 c - f_s^2 v$$

$$f_r^2 v + f_s^2 v = f_s^2 c - f_r^2 c$$

$$\left(f_s^2 + f_r^2\right) v = \left(f_s^2 - f_r^2\right) c$$

$$v = \frac{f_s^2 - f_r^2}{f_s^2 + f_r^2} c$$

This formula will tell us the velocity of the star away from Earth if we can know the respective frequencies of the starlight. This formula is used to demonstrate that the universe is expanding[d].

---

[a]en.wikipedia.org/wiki/Redshift
[b]en.wikipedia.org/wiki/Doppler_effect
[c]https://en.wikipedia.org/wiki/Blueshift
[d]en.wikipedia.org/wiki/Metric_expansion_of_space

14.4.3 Graphing Technology

We can use technology to solve equations by finding where two graphs intersect.

Example 14.4.13 Solve the equation $1 - x = \sqrt{x + 5}$ with technology.

Explanation.

We define $f(x) = 1 - x$ and $g(x) = \sqrt{x + 5}$, and then look for the intersection(s) of the graphs. Since the two functions intersect at $(-1, 2)$, the solution to $1 - x = \sqrt{x + 5}$ is -1. The solution set is $\{-1\}$.

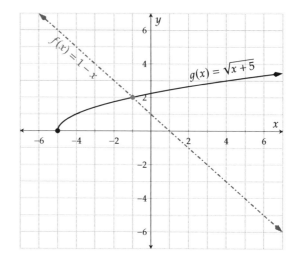

Figure 14.4.14: Graph of $f(x) = 1 - x$ and $g(x) = \sqrt{x + 5}$

Before we finish with this example, we would like to illustrate why there are sometimes extraneous solutions to radical equations. It has to do with the squaring-both-sides step of the solving process.

A graph of a radical, for example $y = \sqrt{x + 5}$, actually graphs as *half* of a sideways parabola, as you can see in Figure 14.4.14. When we square both sides of that equation, we get $y^2 = x + 5$ which actually graphs as a *complete* sideways parabola.

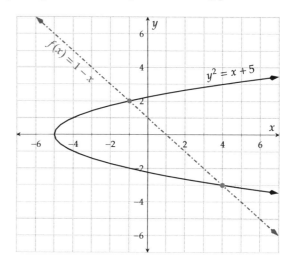

Figure 14.4.15: Graph of $f(x) = 1 - x$ and $y^2 = x + 5$

The curve and the line now intersect *twice*! This second solution (which is 4, by the way) is the extraneous solution that we would have found had we solved $1 - x = \sqrt{x + 5}$ algebraically.

Example 14.4.16 Solve the equation $3 + \sqrt{x + 4} = x - \sqrt{x - 4}$ graphically using technology.

Explanation.

To solve the equation graphically, first we will assign the left side of the equation the label $m(x) = 3 + \sqrt{x+4}$ and the right side $n(x) = x - \sqrt{x-4}$. Next we will make graphs of both m and n on the same grid and look for their intersection point(s). Since the two functions intersect at about $(8.75, 6.571)$, the solution to $3 + \sqrt{x+4} = x - \sqrt{x-4}$ is 8.75. The solution set is $\{8.75\}$.

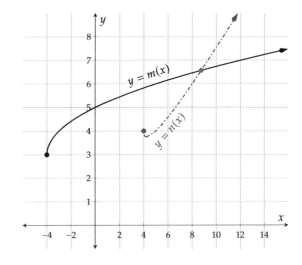

Figure 14.4.17: Graph of $m(x) = 3 + \sqrt{x+4}$ and $n(x) = x - \sqrt{x-4}$

Exercises

Review and Warmup   Solve the equation.

1. $-7y + 3 = -y - 51$ **2.** $-4r + 6 = -r - 12$ **3.** $-155 = -5(10 - 3a)$ **4.** $-72 = -4(4 - 2c)$

5. $-24 = 3 - 9(A - 2)$ **6.** $18 = 9 - 3(C - 2)$ **7.** $(x + 5)^2 = 49$ **8.** $(x + 8)^2 = 16$

9. $x^2 + 21x + 108 = 0$ **10.** $x^2 - x - 90 = 0$ **11.** $x^2 - 16x + 54 = -9$ **12.** $x^2 + 3x - 32 = 8$

Solving Radical Equations   Solve the equation.

13. $\sqrt{y} = 7$ **14.** $\sqrt{r} = 3$ **15.** $\sqrt{5r} = 25$ **16.** $\sqrt{3t} = 9$

17. $2\sqrt{t} = 10$ **18.** $4\sqrt{t} = 8$ **19.** $-3\sqrt{x} = 12$ **20.** $-2\sqrt{x} = 4$

21. $-2\sqrt{-1-y} + 9 = -5$ **22.** $-5\sqrt{8-y} + 9 = -11$ **23.** $\sqrt{2r + 80} = r$ **24.** $\sqrt{4r + 21} = r$

25. $\sqrt{t + 6} = t$ **26.** $\sqrt{t + 72} = t$ **27.** $t = \sqrt{t + 1} + 5$ **28.** $x = \sqrt{x - 3} + 9$

29. $\sqrt{x + 8} = \sqrt{x} + 2$ **30.** $\sqrt{y + 3} = \sqrt{y} + 1$ **31.** $\sqrt{y + 3} = -1 - \sqrt{y}$ **32.** $\sqrt{r - 7} = 1 - \sqrt{r}$

33. $\sqrt{6r} = 6$ **34.** $\sqrt{3t} = 9$ **35.** $\sqrt[3]{t-9} = 4$ **36.** $\sqrt[3]{t-6} = 8$

37. $\sqrt{3x+3} + 9 = 17$ **38.** $\sqrt{8x+6} + 7 = 15$ **39.** $\sqrt{y} + 42 = y$ **40.** $\sqrt{y} + 12 = y$

41. $\sqrt[3]{r-9} = 6$ **42.** $\sqrt[3]{r-5} = -6$ **43.** $r = \sqrt{r+3} + 9$ **44.** $t = \sqrt{t+1} + 89$

45. $\sqrt{38-t} = t+4$ **46.** $\sqrt{148-x} = x+8$

Solving Radical Equations Using Technology Use technology to solve the equation

47.

$$\sqrt{x-2.1} = \sqrt{x} - 4.$$

48.

$$\sqrt{x-2} = \sqrt{x} - 2.1.$$

Solving Radical Equations with Variables

49. Solve the equation for R. Assume that R is positive.

$$Z = \sqrt{L^2 + R^2}$$

$R = \boxed{}$.

50. According to the Pythagorean Theorem, the length c of the hypothenuse of a rectangular triangle can be found through the following equation:

$$c = \sqrt{a^2 + b^2}$$

Solve the equation for the length a of one of the triangle's legs.

$a = \boxed{}$.

51. In an electric circuit, resonance occurs when the frequency f, inductance L, and capacitance C fulfill the following equation:

$$f = \frac{1}{2\pi\sqrt{LC}}$$

Solve the equation for the inductance L.

The frequency is measured in Hertz, the inductance in Henry, and the capacitance in Farad.

$L = \boxed{}$.

52. A pendulum has the length L. The time period T that it takes to once swing back and forth can be found with the following formula:

$$T = 2\pi\sqrt{\frac{L}{32}}$$

Solve the equation for the length L.

The length is measured in feet and the time period in seconds.

$L = \boxed{}$.

Radical Equation Applications   According to the Pythagorean Theorem, the length c of the hypothenuse of a rectangular triangle can be found through the following equation.

$$c = \sqrt{a^2 + b^2}$$

53. If a rectangular triangle has a hypothenuse of 41 ft and one leg is 40 ft long, how long is the third side of the triangle?

    The third side of the triangle is [____] long.

54. If a rectangular triangle has a hypothenuse of 41 ft and one leg is 40 ft long, how long is the third side of the triangle?

    The third side of the triangle is [____] long.

In a coordinate system, the distance r of a point (x, y) from the origin $(0, 0)$ is given by the following equation.

$$r = \sqrt{x^2 + y^2}$$

55. If a point in a coordinate system is 5 cm away from the origin and its x coordinate is 4 cm, what is its y coordinate? Assume that y is positive.

    $y =$ [____].

56. If a point in a coordinate system is 5 cm away from the origin and its x coordinate is 4 cm, what is its y coordinate? Assume that y is positive.

    $y =$ [____].

57. A pendulum has the length L ft. The time period T that it takes to once swing back and forth is 4 s. Use the following formula to find its length.

    $$T = 2\pi\sqrt{\frac{L}{32}}$$

    The pendulum is [____] long.

58. A pendulum has the length L ft. The time period T that it takes to once swing back and forth is 6 s. Use the following formula to find its length.

    $$T = 2\pi\sqrt{\frac{L}{32}}$$

    The pendulum is [____] long.

Challenge   Solve for x.

59.
$$\sqrt{1 + \sqrt{6}} = \sqrt{2 + \sqrt{\frac{1}{\sqrt{x}}} - 1}$$

60.
$$\sqrt{1 + \sqrt{7}} = \sqrt{2 + \sqrt{\frac{1}{\sqrt{x}}} - 1}$$

14.5 Radical Functions and Equations Chapter Review

14.5.1 Introduction to Radical Functions

In Section 14.1 we covered the square root and other root functions. We learned how to find the domain of radical functions algebraically and the range graphically. We also saw the distance formula which is an application of square roots.

> **Example 14.5.1 The Square Root Function.** Algebraically find the domain of the function f where $f(x) = \sqrt{3x - 1} + 2$ and then find the range by making a graph.
>
> **Explanation.** Using Fact 14.1.29 to find the function's domain, we set the radicand greater than or equal to zero and solve:
>
> $$3x - 1 \geq 0$$
> $$3x \geq 1$$
> $$x \geq \frac{1}{3}$$
>
> The function's domain is $\left[\frac{1}{3}, \infty\right)$ in interval notation.
>
> To find the function's range, we use technology to look at a graph of the function. The graph shows that the function's range is $[2, \infty)$. The graph also verifies the function's domain is indeed $\left[\frac{1}{3}, \infty\right)$.
>
>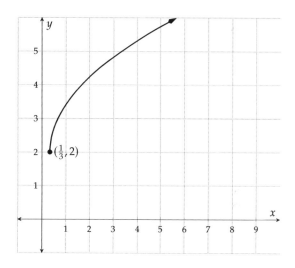
>
> **Figure 14.5.2:** Graph of $f(x) = \sqrt{3x - 1} + 2$

> **Example 14.5.3 The Distance Formula.** Find the distance between $(6, -13)$ and $(-4, 17)$.
>
> **Explanation.** To calculate the distance between $(6, -13)$ and $(-4, 17)$, we use the distance formula. It's good practice to mark each value with the corresponding variables in the formula:
>
> $$\overset{x_1}{(6}, \overset{y_1}{-13)}, \overset{x_2}{(-4}, \overset{y_2}{17)}$$
>
> We have:
>
> $$d = \sqrt{(x_2 - x_1)^2 + (y_2 - y_1)^2}$$

$$d = \sqrt{(-4-6)^2 + (17-(-13))^2}$$
$$d = \sqrt{(-10)^2 + (30)^2}$$
$$d = \sqrt{100 + 900}$$
$$d = \sqrt{1000}$$
$$d = \sqrt{100} \cdot \sqrt{10}$$
$$d = 10\sqrt{10}$$

The distance between $(6,-13)$ and $(-4,17)$ is $10\sqrt{10}$ or approximately 31.62 units.

Example 14.5.4 The Cube Root Function. Algebraically find the domain and graphically find the range of the function g where $g(x) = -2\sqrt[3]{x+6} - 1$.

Explanation. First note that the index of the function g for $g(x) = -2\sqrt[3]{x+6} - 1$ is odd. By Fact 14.1.29, the domain is $(-\infty, \infty)$.

To find the function's range, we use technology to graph the function. According to the graph, the function's range is also $(-\infty, \infty)$.

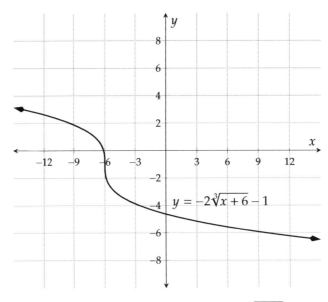

Figure 14.5.5: Graph of $y = g(x) = -2\sqrt[3]{x+6} - 1$

Example 14.5.6 Other Roots. Algebraically find the domain and graphically find the range of the function h where $h(x) = 8 - \frac{5}{2}\sqrt[4]{6-2x}$.

Explanation. First note that the index of this function is 4, which is even. By Fact 14.1.29, to find the domain of this function we must set the radicand greater or equal to zero and solve.

$$6 - 2x \geq 0$$
$$-2x \geq -6$$

$$x \leq \frac{-6}{-2}$$
$$x \leq 3$$

So, the domain of h is $(-\infty, 3]$.

To find the function's range, we use technology to graph the function. By the graph, the function's range is $(-\infty, 8]$.

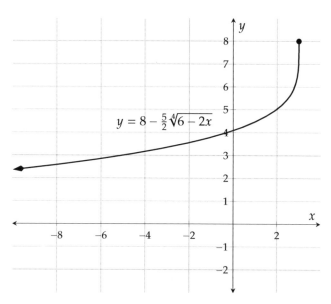

Figure 14.5.7: Graph of $y = h(x) = 8 - \frac{5}{2}\sqrt[4]{6 - 2x}$

14.5.2 Radical Expressions and Rational Exponents

In Section 14.2 we learned the rational exponent rule and added it to our list of exponent rules.

Example 14.5.8 Radical Expressions and Rational Exponents. Simplify the expressions using Fact 14.2.2 or Fact 14.2.10.

a. $100^{1/2}$　　　　b. $(-64)^{-1/3}$　　　　c. $-81^{3/4}$　　　　d. $\left(-\frac{1}{27}\right)^{2/3}$

Explanation.

a. $100^{1/2} = \left(\sqrt{100}\right)$
$\phantom{100^{1/2}} = 10$

b. $(-64)^{-1/3} = \dfrac{1}{(-64)^{1/3}}$

$= \dfrac{1}{\left(\sqrt[3]{(-64)}\right)}$

$= \dfrac{1}{-4}$

c. $-81^{3/4} = -\left(\sqrt[4]{81}\right)^3$

$= -3^3$

$= -27$

d. In this problem the negative can be associated with either the numerator or the denominator, but not both. We choose the numerator.

$$\left(-\dfrac{1}{27}\right)^{2/3} = \left(\sqrt[3]{-\dfrac{1}{27}}\right)^2$$

$$= \left(\dfrac{\sqrt[3]{-1}}{\sqrt[3]{27}}\right)^2$$

$$= \left(\dfrac{-1}{3}\right)^2$$

$$= \dfrac{(-1)^2}{(3)^2}$$

$$= \dfrac{1}{9}$$

Example 14.5.9 More Expressions with Rational Exponents. Use exponent properties in List 14.2.15 to simplify the expressions, and write all final versions using radicals.

a. $7z^{5/9}$

b. $\frac{5}{4}x^{-2/3}$

c. $\left(-9q^5\right)^{4/5}$

d. $\sqrt{y^5} \cdot \sqrt[4]{y^2}$

e. $\dfrac{\sqrt{t^3}}{\sqrt[3]{t^2}}$

f. $\sqrt{\sqrt[3]{x}}$

g. $5\left(4 + a^{1/2}\right)^2$

h. $-6\left(2p^{-5/2}\right)^{3/5}$

Explanation.

a. $7z^{5/9} = 7\sqrt[9]{z^5}$

b. $\dfrac{5}{4}x^{-2/3} = \dfrac{5}{4} \cdot \dfrac{1}{x^{2/3}}$

$\qquad = \dfrac{5}{4} \cdot \dfrac{1}{\sqrt[3]{x^2}}$

$\qquad = \dfrac{5}{4\sqrt[3]{x^2}}$

c. $\left(-9q^5\right)^{4/5} = (-9)^{4/5} \cdot \left(q^5\right)^{4/5}$

$\qquad = (-9)^{4/5} \cdot q^{5 \cdot 4/5}$

$\qquad = \left(\sqrt[5]{-9}\right)^4 \cdot q^4$

$\qquad = \left(q\sqrt[5]{-9}\right)^4$

d. $\sqrt{y^5} \cdot \sqrt[4]{y^2} = y^{5/2} \cdot y^{2/4}$

$\qquad = y^{5/2 + 2/4}$

$\qquad = y^{10/4 + 1/4}$

$\qquad = x^{11/4}$

$\qquad = \sqrt[4]{x^{11}}$

e. $\dfrac{\sqrt{t^3}}{\sqrt[3]{t^2}} = \dfrac{t^{3/2}}{t^{2/3}}$

$\qquad = t^{3/2 - 2/3}$

$\qquad = t^{9/6 - 4/6}$

$\qquad = t^{5/6}$

$\qquad = \sqrt[6]{t^5}$

f. $\sqrt{\sqrt[3]{x}} = \sqrt{x^{1/3}}$

$\qquad = \left(x^{1/3}\right)^{1/2}$

$\qquad = x^{1/3 \cdot 1/2}$

$\qquad = x^{1/6}$

$\qquad = \sqrt[6]{x}$

g. $5\left(4 + a^{1/2}\right)^2 = 5\left(4 + a^{1/2}\right)\left(4 + a^{1/2}\right)$

$\qquad = 5\left(4^2 + 2 \cdot 4 \cdot a^{1/2} + \left(a^{1/2}\right)^2\right)$

$\qquad = 5\left(16 + 8a^{1/2} + a^{1/2 \cdot 2}\right)$

$\qquad = 5\left(16 + 8a^{1/2} + a\right)$

$\qquad = 5\left(16 + 8\sqrt{a} + a\right)$

$\qquad = 80 + 40\sqrt{a} + 5a$

h. $-6\left(2p^{-5/2}\right)^{3/5} = -6 \cdot 2^{3/5} \cdot p^{-5/2 \cdot 3/5}$

$\qquad = -6 \cdot 2^{3/5} \cdot p^{-3/2}$

$\qquad = -\dfrac{6 \cdot 2^{3/5}}{p^{3/2}}$

$\qquad = -\dfrac{6\sqrt[5]{2^3}}{\sqrt{p^3}}$

$\qquad = -\dfrac{6\sqrt[5]{8}}{\sqrt{p^3}}$

14.5.3 More on Rationalizing the Denominator

In Section 14.3 we covered how to rationalize the denominator when it contains a single square root or a binomial.

Example 14.5.10 A Review of Rationalizing the Denominator. Rationalize the denominator of the expressions.

a. $\dfrac{12}{\sqrt{3}}$

b. $\dfrac{\sqrt{5}}{\sqrt{75}}$

Explanation.

a.

$$\frac{12}{\sqrt{3}} = \frac{12}{\sqrt{3}} \cdot \frac{\sqrt{3}}{\sqrt{3}}$$

$$= \frac{12\sqrt{3}}{3}$$
$$= 4\sqrt{3}$$

b. First we will simplify $\sqrt{75}$.

$$\frac{\sqrt{5}}{\sqrt{75}} = \frac{\sqrt{5}}{\sqrt{25 \cdot 3}}$$
$$= \frac{\sqrt{5}}{\sqrt{25} \cdot \sqrt{3}}$$
$$= \frac{\sqrt{5}}{5\sqrt{3}}$$

Now we can rationalize the denominator by multiplying the numerator and denominator by $\sqrt{3}$.

$$= \frac{\sqrt{5}}{5\sqrt{3}} \cdot \frac{\sqrt{3}}{\sqrt{3}}$$
$$= \frac{\sqrt{15}}{5 \cdot 3}$$
$$= \frac{\sqrt{15}}{15}$$

Example 14.5.11 Rationalize Denominator with Difference of Squares Formula. Rationalize the denominator in $\frac{\sqrt{6}-\sqrt{5}}{\sqrt{3}+\sqrt{2}}$.

Explanation. To remove radicals in $\sqrt{3}+\sqrt{2}$ with the difference of squares formula, we multiply it with $\sqrt{3}-\sqrt{2}$.

$$\frac{\sqrt{6}-\sqrt{5}}{\sqrt{3}+\sqrt{2}} = \frac{\sqrt{6}-\sqrt{5}}{\sqrt{3}+\sqrt{2}} \cdot \frac{\left(\sqrt{3}-\sqrt{2}\right)}{\left(\sqrt{3}-\sqrt{2}\right)}$$
$$= \frac{\sqrt{6}\cdot\sqrt{3} - \sqrt{6}\cdot\sqrt{2} - \sqrt{5}\cdot\sqrt{3} - \sqrt{5}\cdot-\sqrt{2}}{\left(\sqrt{3}\right)^2 - \left(\sqrt{2}\right)^2}$$
$$= \frac{\sqrt{18} - \sqrt{12} - \sqrt{15} + \sqrt{10}}{9-4}$$
$$= \frac{3\sqrt{2} - 2\sqrt{3} - \sqrt{15} + \sqrt{10}}{5}$$

14.5.4 Solving Radical Equations

In Section 14.4 we covered solving equations that contain a radical. We learned about extraneous solutions and the need to check our solutions.

Example 14.5.12 Solving Radical Equations. Solve for r in $r = 9 + \sqrt{r+3}$.

Explanation. We will isolate the radical first, and then square both sides.

$$r = 9 + \sqrt{r+3}$$
$$r - 9 = \sqrt{r+3}$$
$$(r-9)^2 = \left(\sqrt{r+3}\right)^2$$
$$r^2 - 18r + 81 = r + 3$$
$$r^2 - 19r + 78 = 0$$
$$(r-6)(r-13) = 0$$

$$r - 6 = 0 \qquad \text{or } r - 13 \qquad = 0$$
$$r = 6 \qquad \text{or } r \qquad = 13$$

Because we squared both sides of an equation, we must check both solutions.

$$6 \stackrel{?}{=} 9 + \sqrt{6+3} \qquad\qquad 13 \stackrel{?}{=} 9 + \sqrt{13+3}$$
$$6 \stackrel{?}{=} 9 + \sqrt{9} \qquad\qquad 13 \stackrel{?}{=} 9 + \sqrt{16}$$
$$6 \stackrel{no}{=} 9 + 3 \qquad\qquad 13 \stackrel{\checkmark}{=} 9 + 4$$

It turns out 6 is an extraneous solution and 13 is a valid solution. So the equation has one solution: 13. The solution set is $\{13\}$.

Example 14.5.13 Solving Radical Equations that Require Squaring Twice. Solve the equation $\sqrt{t+9} = -1 - \sqrt{t}$ for t.

Explanation. We cannot isolate two radicals, so we will simply square both sides, and later try to isolate the remaining radical.

$$\sqrt{t+9} = -1 - \sqrt{t}$$
$$\left(\sqrt{t+9}\right)^2 = \left(-1 - \sqrt{t}\right)^2$$
$$t + 9 = 1 + 2\sqrt{t} + t \qquad \text{after expanding the binomial squared}$$
$$9 = 1 + 2\sqrt{t}$$
$$8 = 2\sqrt{t}$$
$$4 = \sqrt{t}$$
$$(4)^2 = \left(\sqrt{t}\right)^2$$
$$16 = t$$

Because we squared both sides of an equation, we must check the solution by substituting 16 into

$\sqrt{t+9} = -1 - \sqrt{t}$, and we have:

$$\sqrt{t+9} = -1 - \sqrt{t}$$

$$\sqrt{16+9} \overset{?}{=} -1 - \sqrt{16}$$

$$\sqrt{25} \overset{?}{=} -1 - 4$$

$$5 \overset{no}{=} -5$$

Our solution did not check so there is no solution to this equation. The solution set is the empty set, which can be denotes $\{\ \}$ or $\emptyset$.

Exercises

Introduction to Radical Functions Find the domain of the function.

1. $H(x) = \sqrt{8-x}$

2. $H(x) = \sqrt{5-x}$

3. $K(x) = \sqrt[3]{-9x+10}$

4. $f(x) = \sqrt[3]{5x-2}$

5. $g(x) = \sqrt[4]{18-3x}$

6. $h(x) = \sqrt[4]{-12-2x}$

7. Use technology to find the range of the function h defined by $h(x) = \sqrt{2-x} + 1$.

8. Use technology to find the range of the function F defined by $F(x) = \sqrt{-2-x} - 5$.

If an object is dropped with no initial velocity, the time since the drop, in seconds, can be calculated by the function

$$T(h) = \sqrt{\frac{2h}{g}}$$

where h is the distance the object traveled in feet. The variable g is the gravitational acceleration on earth, and we can round it to $32\frac{ft}{s^2}$ for this problem.

9. a. After ⬚ seconds since the release, the object would have traveled 35 feet.

b. After 5 seconds since the release, the object would have traveled ⬚ feet.

10. a. After ⬚ seconds since the release, the object would have traveled 40 feet.

b. After 3.6 seconds since the release, the object would have traveled ⬚ feet.

11. Find the distance between the points $(-8, -2)$ and $(57, 70)$.

12. Find the distance between the points $(-10, -15)$ and $(45, 33)$.

Radical Expressions and Rational Exponents Without using a calculator, evaluate the expression.

13. $\sqrt[4]{3^3} = \boxed{}$ **14.** $\sqrt[3]{13^2} = \boxed{}$ **15.** $\sqrt[3]{-\dfrac{8}{125}} = \boxed{}$. **16.** $\sqrt[3]{-\dfrac{1}{8}} = \boxed{}$.

17. Use rational exponents to write the expression.

$\sqrt[5]{3m + 9} = \boxed{}$

18. Use rational exponents to write the expression.

$\sqrt[4]{9n + 3} = \boxed{}$

19. Convert the expression to radical notation.

$14^{\frac{1}{5}} a^{\frac{4}{5}} = \boxed{}$

20. Convert the expression to radical notation.

$3^{\frac{1}{3}} b^{\frac{2}{3}} = \boxed{}$

21. Convert $c^{-\frac{5}{7}}$ to a radical.

⊙ $-\sqrt[7]{c^5}$ ⊙ $\dfrac{1}{\sqrt[7]{c^5}}$

⊙ $\dfrac{1}{\sqrt[5]{c^7}}$ ⊙ $-\sqrt[5]{c^7}$

22. Convert $x^{-\frac{3}{8}}$ to a radical.

⊙ $\dfrac{1}{\sqrt[3]{x^8}}$ ⊙ $\dfrac{1}{\sqrt[8]{x^3}}$

⊙ $-\sqrt[3]{x^3}$ ⊙ $-\sqrt[3]{x^8}$

Simplify the expression, answering with rational exponents and not radicals.

23. $\dfrac{\sqrt{25y}}{\sqrt[10]{y^3}} = \boxed{}$ **24.** $\dfrac{\sqrt{9z}}{\sqrt[6]{z^5}} = \boxed{}$ **25.** $\sqrt[5]{t} \cdot \sqrt[10]{t^3} = \boxed{}$

26. $\sqrt[5]{m} \cdot \sqrt[10]{m^3} = \boxed{}$ **27.** $\sqrt{a}\sqrt[7]{a} = \boxed{}$ **28.** $\sqrt{t}\sqrt[8]{t} = \boxed{}$

More on Rationalizing the Denominator Rationalize the denominator and simplify the expression.

29. $\dfrac{9}{\sqrt{72}} = \boxed{}$ **30.** $\dfrac{4}{\sqrt{216}} = \boxed{}$ **31.** $\sqrt{\dfrac{7}{12}} = \boxed{}$ **32.** $\sqrt{\dfrac{11}{28}} = \boxed{}$

33. $\dfrac{4}{\sqrt{23} + 6} = \boxed{}$ **34.** $\dfrac{5}{\sqrt{7} + 9} = \boxed{}$ **35.** $\dfrac{\sqrt{5} - 10}{\sqrt{7} + 9} = \boxed{}$ **36.** $\dfrac{\sqrt{3} - 11}{\sqrt{13} + 6} = \boxed{}$

Solving Radical Equations Solve the equation.

37. $r = \sqrt{r + 2} + 4$ **38.** $t = \sqrt{t + 3} + 3$ **39.** $\sqrt{t + 8} = \sqrt{t} - 4$ **40.** $\sqrt{x + 8} = \sqrt{x} + 2$

41. $\sqrt{x + 110} = x$ **42.** $\sqrt{y + 56} = y$ **43.** $y = \sqrt{y + 3} + 17$ **44.** $r = \sqrt{r + 1} + 109$

45. $\sqrt{66 - r} = r + 6$ **46.** $\sqrt{29 - r} = r + 1$

47. Use technology to solve the equation $\sqrt{x - 1.9} = \sqrt{x} - 0.2$.

48. Use technology to solve the equation $\sqrt{x + 2.3} = \sqrt{x} - 3$.

49. According to the Pythagorean Theorem, the length c of the hypothenuse of a rectangular triangle can be found through the following equation:

$$c = \sqrt{a^2 + b^2}$$

Solve the equation for the length a of one of the triangle's legs.

$a = $ ⬚.

50. In a coordinate system, the distance r from a point (x, y) to the origin $(0, 0)$ is given by the following equation:

$$r = \sqrt{x^2 + y^2}$$

Solve the equation for the coordinate y. Assume that y is positive.

$y = $ ⬚.

According to the Pythagorean Theorem, the length c of the hypothenuse of a rectangular triangle can be found through the following equation.

$$c = \sqrt{a^2 + b^2}$$

51. If a rectangular triangle has a hypothenuse of 13 ft and one leg is 12 ft long, how long is the third side of the triangle?

The third side of the triangle is ⬚ long.

52. If a rectangular triangle has a hypothenuse of 17 ft and one leg is 15 ft long, how long is the third side of the triangle?

The third side of the triangle is ⬚ long.

53. A pendulum has the length L ft. The time period T that it takes to once swing back and forth is 6 s. Use the following formula to find its length.

$$T = 2\pi\sqrt{\frac{L}{32}}$$

The pendulum is ⬚ long.

54. A pendulum has the length L ft. The time period T that it takes to once swing back and forth is 8 s. Use the following formula to find its length.

$$T = 2\pi\sqrt{\frac{L}{32}}$$

The pendulum is ⬚ long.

Index

30708711R00188

Made in the USA
San Bernardino, CA
28 March 2019